Topics in
Current Physics

44

Topics in Current Physics
Founded by Helmut K. V. Lotsch

Persistent Spectral Hole-Burning: Science and Applications

Edited by W. E. Moerner

With Contributions by
G. C. Bjorklund   D. Haarer   J. M. Hayes
R. Jankowiak   W. Lenth   R. M. Macfarlane
W. E. Moerner   K. K. Rebane   L. A. Rebane
R. M. Shelby   A. J. Sievers   G. J. Small

With 146 Figures

Springer-Verlag Berlin Heidelberg New York
London Paris Tokyo

Dr. W. E. Moerner

IBM Research Division, Almaden Research Center, 650 Harry Road,
San Jose, CA 95120-6099, USA

ISBN-13:978-3-642-83292-5   e-ISBN-13:978-3-642-83290-1
DOI: 10.1007/978-3-642-83290-1

Library of Congress Cataloging-in-Publication Data. Persistent spectral hole-burning. (Topics in current physics ; 44) Includes bibliographies and index. 1. Optical hole burning—Congresses. 2. Energy-band theory of solids—Congresses. I. Moerner, W. E. (William Esco), 1953– . II. Bjorklund, G. C. (Gary C.) III. Series. QC176.8.O58P47 1988   530.4′1   87-37675

© Springer-Verlag Berlin Heidelberg 1988
Softcover reprint of the hardcover 1st edition 1988

2153/3150-543210

Preface

Almost fifteen years have now elapsed since the first observations of persistent spectral hole-burning in inhomogeneously broadened absorption lines in solids. The fact that the spectral shape of an inhomogeneously broadened line can be locally modified for long periods of time has led to a large number of investigations of low-temperature photophysics and photochemistry that would not have been possible otherwise. Using hole-burning, important information has been obtained about a variety of interactions, including excited-state dephasing processes, host-guest dynamics, proton tunnelling, low-frequency excitation in amorphous hosts, relaxation mechanisms for vibrational modes, photochemical mechanisms at liquid helium temperatures, and external field perturbations. At the same time, the possibility that persistent spectral holes might be used to store digital information has led to the study of materials and configurations for frequency-domain optical storage and related possible applications.

This is the first full-length book on persistent spectral hole-burning. The goal is to provide a broadly based survey of the scientific principles and applications of persistent spectral hole-burning. Since the topic is quite interdisciplinary, the book is intended for researchers, graduate students, and advanced undergraduates in the fields of chemical physics, solid-state physics, laser spectroscopy, solid-state photochemistry, and high-performance optical storage and optical processing. It should leave the reader with a detailed appreciation of the generality of the phenomenon of persistent spectral hole-burning, the power of the technique in studying microscopic dynamics and mechanisms of phototransformation in low-temperature solids, and the central materials and engineering issues involved in applications. Since persistent spectral hole-burning is a currently expanding area of research, ample references are provided in order to facilitate entry into the current literature.

The presentation is organized roughly by class of mechanism giving rise to the formation of persistent spectral holes. A brief introduction to the effect and the various mechanisms (Chap.1) is followed by a thorough description of the basic principles and methods of hole-burning (Chap.2) with examples from photochemical processes in crystalline and amorphous hosts. The principal effects occurring during photochemical hole-burning in electronic transitions, with emphasis on the proton tunnelling phenomena active in hydrogen-bonded polymers and glasses, are then described (Chap.3). Persistent spectral hole-burning in inorganic materials due either to photoionization or to photophysical effects is then summarized (Chap.4), followed by a detailed discussion of nonphotochemical hole-burning mechanisms for electronic transitions in amorphous solids (Chap.5) with emphasis on dephasing mechanisms and the two-level-system model. At infrared wavelengths, excitation of molecular vibrational modes can lead to the formation of persistent spectral holes either through conformational changes or through molecular reorientation in crystalline materials (Chap.6). The book concludes with a description of potential applications to data storage and optical processing using frequency-domain, time-domain, holographic, and electric-field techniques (Chap.7).

It is with great pleasure that the editor expresses his gratitude to Dr. H. Lotsch of Springer-Verlag for his patience and assistance throughout this project, to IBM Corporation for its support, to the impressive international group of contributors with whom it has been an honor and privilege to collaborate for their patience and hard work, and to his wife, Sharon Moerner, for her unfaltering assistance and encouragement.

San Jose, California   W.E. Moerner
December 1987

Contents

List of Contributors

Bjorklund, Gary C.
IBM Research Division, Almaden Research Center, 650 Harry Road,
San Jose, CA 95120-6099, USA

Haarer, Dietrich
Universität Bayreuth, Postfach 101251,
D-8580 Bayreuth, Fed. Rep. of Germany

Hayes, John M.
Department of Chemistry, Iowa State University, Ames, IA 50011, USA

Jankowiak, Ryszard
Department of Chemistry, Iowa State University, Ames, IA 50011, USA

Lenth, Wilfried
IBM Research Division, Almaden Research Center, 650 Harry Road,
San Jose, CA 95120-6099, USA

Macfarlane, Roger M.
IBM Research Division, Almaden Research Center, 650 Harry Road,
San Jose, CA 95120-6099, USA

Moerner, W. E.
IBM Research Division, Almaden Research Center, 650 Harry Road,
San Jose, CA 95120-6099, USA

Rebane, Karl K.
Academy of Sciences of the Estonian SSR,
SU-200106 Tallinn, Kohtu Str.6, USSR

Rebane, Ljubov A.
Institute of Chemical Physics and Biophysics,
SU-200105 Tallinn, Lenini boulv.10, USSR

Shelby, Robert M.
IBM Research Division, Almaden Research Center, 650 Harry Road,
San Jose, CA 95120-6099, USA

Sievers, Albert J.
Laboratory for Atomic and Solid State Physics, Clark Hall,
Cornell University, Ithaca, NY 14853-2501, USA

Small, Gerald J.
Department of Chemistry, Iowa State University, Ames, IA 50011, USA

1. Introduction

W. E. Moerner

With 2 Figures

This chapter introduces the general concepts of persistent spectral hole-burning and summarizes the significance of this phenomenon for science and possible applications. The history of hole-burning is reviewed, and general classes of mechanisms leading to persistent spectral hole-burning are outlined in tabular form. A synopsis of the other chapters in the book is presented.

1.1 Fundamental Requirements for Persistent Spectral Hole-Burning

Persistent spectral hole-burning (PSHB) is a process whereby normally smooth inhomogeneously broadened absorption lines in solids at low temperatures can be spectrally modified for time periods longer than the lifetime of any excited state. Given a transparent solid containing absorbing guest centers, the basic requirements for persistent spectral hole-burning are:

1. the optical absorption of the guest under consideration must be inhomogeneously broadened,
2. there must exist more than one ground state configuration of the total system (host + guest),
3. the optical absorption frequencies from the various ground states must differ by more than the linewidth of the tunable light source in use (which is typically a laser),
4. there must exist an optical pumping pathway that connects the ground state configurations, and
5. the relaxation among the ground states must be slower than the decay rate of any excited state.

This set of requirements may seem restrictive, but in fact the many diverse examples of PSHB described in this book show that PSHB is an ubiquitous feature of laser spectroscopy of solids at liquid helium temperatures. PSHB experiments can, on the one hand, yield important information about host-guest interactions, dephasing, local microscopic environments, and photochemistry, and on the other hand, they offer a variety of possible applications to optical data storage and optical signal processing.

The first and most essential requirement for PSHB, inhomogeneous broadening, is summarized in Fig. 1.1. If an absolutely perfect crystal could be fabricated in which the absorbing centers all expe-

1

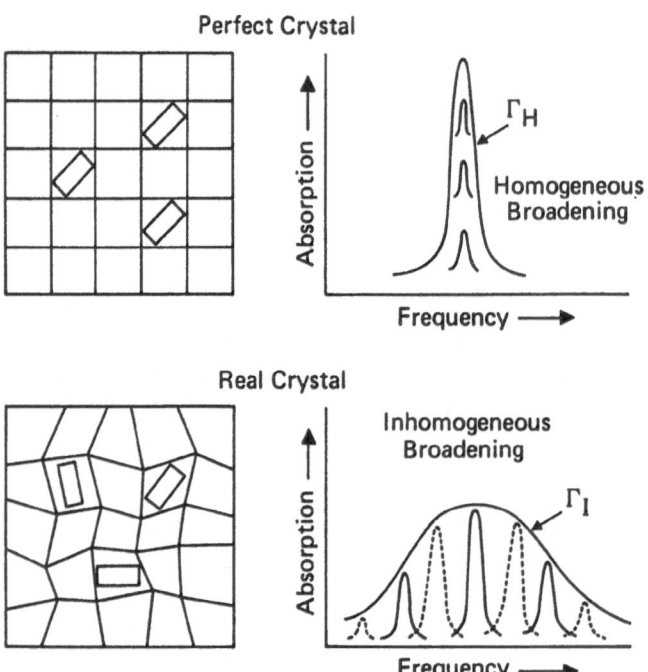

Perfect Crystal

Homogeneous
Broadening

Real Crystal

Inhomogeneous
Broadening

Fig. 1.1. (Upper half) Schematic of absorbers dispersed in a perfect crystal. At low temperatures, the absorption line is homogeneously broadened with width Γ_H. (Lower half) Illustration of inhomogeneously broadened line profile in real solid matrices. The distribution of local environments leads to a distribution of center frequencies of absorption. The resulting lineshape has width Γ_I

rience identical local environments, then the absorption line due to the entire assembly would appear as in the upper right half of the figure. The width of this line is just equal to the width of the optical absorption line for a single center, Γ_H, which is called the homogeneous width. The word "homogeneous" is appropriate since this width is determined by excited-state dephasing interactions or lifetime effects that are identical for all centers. At liquid helium temperatures where the broadening due to phonons and other excitations is minimized, homogeneous widths can become extremely small compared to the frequency of the transition itself. The exact value of Γ_H depends heavily on the specific transition under consideration, and is usually in the range 1 kHz – 100 MHz for zero-phonon transitions in centers with weak electron-phonon coupling. This is to be compared to the 10 – 1000 THz frequencies typical of optical and infrared transitions.

   Of course, absolutely perfect host materials do not exist, and real crystals have a distribution of local environments due to strains, other impurities, dislocations, and other imperfections. A distribution of local environments is a property of glassy or polymeric hosts by definition. In these physically realizable situations, the lower half of Fig. 1.1 applies. Even though all the absorbing centers are identical, the local environments around the centers are not. The unavoidable distribution of local environments leads to a distribution of center frequencies for the various absorbing centers. The resulting absorption profile is said to be inhomogeneously broadened with width Γ_I, because the width is determined by interactions that vary from absorbing center to absorbing center. Depending upon the host material, the inhomogeneous width may vary from a few hundred MHz to a few hundred cm^{-1} (1 cm^{-1} = 30,000 MHz). The critical property of inhomogeneous lines that leads to PSHB is the fact that narrowband excitation with a tunable light source at a particular frequency within the inhomogeneous line excites only those centers that are in resonance with the exciting light. In most real materials, the inhomogeneous linewidth is so large that it obscures many physical effects such as external field perturbations and certainly the value of the homogeneous width. Therefore, perhaps the most useful feature of PSHB is that it allows measurements to be performed inside the inhomogeneously broadened line.

The strength of the inhomogeneous broadening may be measured by the ratio of the inhomogeneous width to the homogeneous width, f_ω, where $f_\omega = \Gamma_I/\Gamma_H$. This factor can range from 1 to 10^4 or more, depending upon the host material. This book is chiefly concerned with optical transitions that show strong inhomogeneous broadening, i.e., $f_\omega >> 1$. Therefore, either large inhomogeneous widths or small homogeneous widths are generally required. The former requirement is easy to satisfy in amorphous hosts, although f_ω values greater than 1000 at low temperatures can also occur in crystalline hosts with the proper host-guest combination. The latter situation of small homogeneous widths usually occurs for zero-phonon optical transitions of ions, aggregate color centers, or molecules with small electron-phonon coupling at liquid helium temperatures. In fact, the need for small homogeneous widths is the principal reason why PSHB experiments are performed at low temperatures, since at elevated temperatures, the homogeneous width grows dramatically due to phonon interactions and usually becomes larger than the (temperature-independent) inhomogeneous width. The detailed structure of homogeneous lines and their temperature dependence is described in more detail in Chaps. 2 − 6 of this book.

Requirements 2 − 5 for PSHB relate more directly to the actual mechanism for hole formation, and can be satisfied in a variety of ways. For example, the multiple ground states of requirement 2 can occur by rotation of the molecule in the host, relaxation of the two-level systems characteristic of amorphous hosts, or by direct photochemical production of new products from the original absorbing center. Requirement 3 (absorption energies differing by more than the laser linewidth) is quite easy to achieve due to the extremely high resolution available from modern laser sources. For example, assuming a laser linewidth of $1-100$ MHz typical of semiconductor diode and dye lasers, extremely weak differences between the ground states suffice to produce the required absorption shift. Requirement 4 (existence of photoinduced transformations between the various ground states) is satisfied by definition for systems that undergo photochemistry. PSHB by nonphotochemical processes can also occur when photostable guest centers have local configurations of the nearby host that can be altered by excitation of the guest. Examples of such nonphotochemical processes will be briefly described below and in detail in Chaps. 5 and 6. Finally, requirement 5 (slow relaxation) basically amounts to a definition of "persistent" for the purposes of this book. Slow relaxation is often easy to achieve at liquid helium temperatures, because many reverse reactions are thermally activated. Transient spectral holes lasting only as long as some excited state lifetime also occur, but will not be of central interest here.

When requirements $1-5$ are met, persistent spectral holes form in a fashion illustrated in Fig. 1.2. Part (a) of the figure shows the

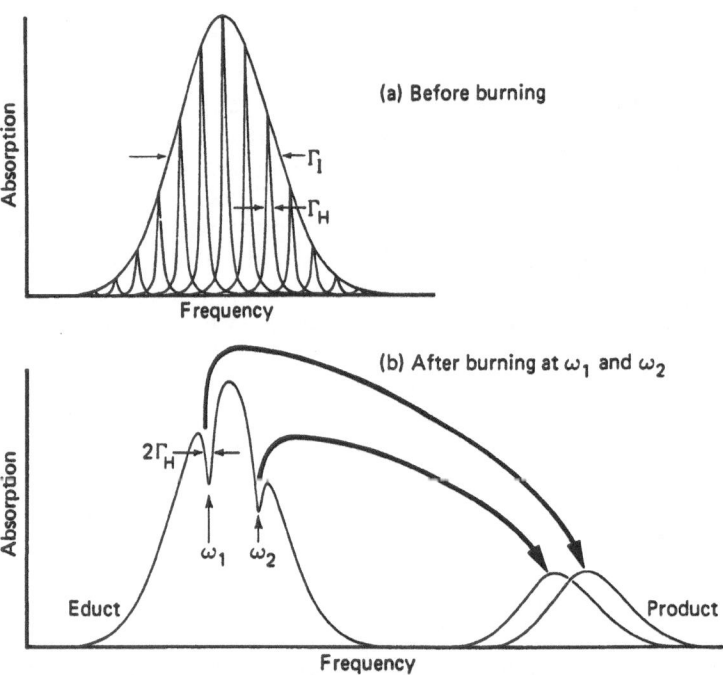

Fig. 1.2. Illustration of the persistent spectral hole formation process. The upper part of the figure shows an inhomogeneously broadened line as described in Fig. 1.1, before any hole-burning has occurred. The lower part of the figure shows how laser irradiation at ω_1 and ω_2 in general produces a decrease in absorption at these frequencies, and a corresponding appearance of product absorption in another region of frequency space

inhomogeneously broadened absorption line before PSHB. If a narrowband light source (such as a laser) is held fixed at ω_1 for a sufficient time period, the optical absorption at ω_1 decreases over a narrow range (part (b) of the figure). This decrease in absorption due to photoinduced transformations between the various ground states is called a "spectral hole". The terminology dates from the early days of magnetic resonance research when transient spectral holes were formed in inhomogeneously broadened magnetic resonance absorptions [1.1]. It is important to realize that detection of the drop in absorption due to a spectral hole amounts to probing the ground state distribution to see which centers have *not* been removed from the inhomogeneous profile. In the ideal case, the hole width is equal to twice the homogeneous width, because the light interacts with a spectral range equal to one homogeneous width during initial excitation, transforming those centers in resonance into a "product" absorbing in a different range of frequency space. Each of the centers removed from the "educt" absorption line carries with it an absorption one homogeneous width wide, so that detection of the remaining educt absorption yields a spectral hole with width $2\Gamma_H$. In many physical systems, the actual observed hole width may be even wider due to a variety of effects such as spectral diffusion between the hole formation and detection steps (see Chaps. 3 and 4).

1.2 Significance for Science and Applications

PSHB can be used to great advantage for scientific studies of the local environments of absorbing centers in solids. The advantages arise because hole-burning allows precise, high-resolution spectral measurements to be performed on essentially homogeneous groups of centers without interference or resolution limitations due to the broad inhomogeneous line profile. Because spectral holes are usually narrow, the positions and widths of the holes are naturally extremely sensitive to weak perturbations caused by internal or external interactions. Consequently, PSHB allows detailed measurement of phenomena that often cannot be observed in any other fashion. Several excellent reviews of various aspects of the subject have appeared in the literature [1.2 – 6].

Table 1.1 lists some of the microscopic interactions that can be probed by PSHB. The left column lists some of the hole properties that can be measured, and the middle column presents some of the physical phenomena that determine each hole property. The right column indicates which chapters discuss examples of these phenomena. Clearly, a wealth of structural and dynamical information about the host-defect system can be obtained from PSHB studies.

5

Table 1.1: Phenomena probed by persistent spectral hole-burning

Spectral hole property	Phenomena probed	Chapter
Temperature dependence of hole width	Dephasing mechanism - phonon, libron, TLS, local mode couplings	2-6
Time dependence of hole width	Spectral diffusion, barrier heights	3
Temperature dependence of hole position	Dephasing mechanism	-
Hole growth rate	Kinetics and order of photoinduced process, quantum efficiency[a], bottleneck lifetimes	2,3,6, 7
Hole lifetime (spontaneous hole filling)	Barrier heights to reverse reaction	3-6
Laser-induced hole filling, erasing effects	Amorphous host dynamics, photochemical mechanism	4-6
Shifts and splittings in external fields	E-field: Stark effect, site symmetry H-field: Zeeman effect, degeneracy of transition Strain field: stress coupling coefficients, symmetry	3,4,6
Satellite hole positions	Vibronic and other excited state splittings	5,6
Antihole positions	Nature and number of product states, ground state splittings	2,6

[a] The quantum efficiency, η, is the probability of photoreaction per photon absorbed by the center. See [1.7,8]

Soon after the first observation of PSHB in 1974 (see next section), it was quickly realized that in addition to the scientific promise, spectral holes could also be used to provide an additional dimension beyond spatial dimensions for the optical storage of digital data. The idea of using transient spectral holes for data storage was patented by *Szabo* [1.9], and *Castro* et al. extended the idea to long-term data storage using persistent spectral holes [1.10]. The presence (or absence) of a hole at a given frequency within the inhomogeneous line could be used to encode a digital "1" (or "0"). Due to the size of f_ω, many thousands of bits could perhaps be stored in one diffraction-limited laser spot. This storage scheme has been called "frequency-domain optical storage" (FDOS), and it would offer extremely high areal densities as well as fast random access by beam deflection and laser frequency tuning. In fact, FDOS is the potential application of PSHB that has received the most attention in recent research. However, other applications can also be envisioned such as pulse shaping and optical filtering; FDOS together with these other applications will be discussed in detail in Chap. 7. Therefore, for scientific as well as application-oriented reasons, PSHB research is being actively pursued at a variety of laboratories all over the world.

1.3 Historical Overview and Survey of Mechanisms

As was mentioned above, transient spectral holes were observed as early as 1948 for inhomogeneously broadened magnetic resonance absorption lines in solids [1.1]. PSHB in optical transitions had to wait at least for the invention of the laser in order to have a tunable, coherent, high-brightness source for narrowband excitation. As early as 1961, Nobel laureate *A. L. Schawlow* remarked at a laser spectroscopy conference to the effect that now that lasers had been invented, (transient) spectral hole-burning should occur for laser-excited transitions which are saturated by the exciting light [1.11]. It was to be many years, however, before such effects were actually observed. The first hint that PSHB might be possible came from observations of narrowing of fluorescence excitation profiles when laser sources were used to excite the fluorescence of an inhomogeneously broadened system. Such "fluorescence line-narrowing" (FLN) effects first reported by *Szabo* for ruby established that narrow packets of centers within inhomogeneously broadened lines could be selectively excited by laser radiation [1.12]. Later, *Personov* et al. demonstrated FLN for organic molecules in both polycrystalline and glassy hosts [1.13]. An important survey of laser spectroscopy of solids with emphasis on FLN was published in 1981 [1.14]. These FLN studies can thus be be regarded as forerunners of the direct observation of persistent spectral changes due to PSHB.

Transient spectral hole-burning was first observed by *Szabo* in 1974 for the case of the R lines in ruby [1.15]. Almost at the same time, persistent spectral hole formation (PSHB) was observed by two Russian groups: by *Gorokhovskii* et al. for free base phthalocyanine in a n-octane Shpol'skii matrix [1.16], and by *Kharlamov* et al. for perylene and 9-aminoacridine molecules in a glassy ethanol matrix [1.17]. These early observations of PSHB in several organic systems suggested that the process might be fairly widespread in guests with photochemical reactions or in hosts with an amorphous character.

After these pioneering observations, many researchers sought to establish the broadest class of mechanisms and materials that undergo PSHB. To date, upwards of 50 materials have shown the effect, in organic and inorganic crystals, polymers, and glasses. Here the various classes of mechanisms leading to PSHB will be briefly surveyed. On a general level, microscopic mechanisms leading to PSHB can be roughly classified into two categories: photochemical and nonphotochemical (or photophysical). Photochemical mechanisms usually involve some internal change in the guest center itself, such as bond-breaking, ionization, isomerization, tautomerization, and so forth. Nonphotochemical mechanisms arise from a change in the environment around the center, or perhaps a reorientation of the center itself with respect to the local environment. Although this classification scheme is not completely rigorous, it does provide a good starting point for a discussion of the mechanisms. In the remainder of this section, a selection of mechanisms for PSHB will be presented using Table 1.2, and the reader is urged to consult the listed chapters in this book as well as the references for further detail and additional examples. The references state one or two early results and several recent studies for each case. While the reference list presented here is by no means exhaustive, it should provide a good starting point for study of the literature.

Starting at the top of Table 1.2, the reversible proton tautomerization process in free-base phthalocyanines and porphyrins had been studied as early as 1971 using broadband spectroscopy [1.18,19]. The mechanism is light-induced tautomerization of the pair of central imino protons in the molecule. It was later recognized that this photochemical mechanism is a route to PSHB that can operate in

Table 1.2: Survey of mechanisms for persistent spectral hole-burning

Mechanism	Examples	Reference
Photochemical		
Proton tautomerization	Free-base phthalocyanines and porphyrins in n-alkanes, polymers	Chaps. 2,3, [1.16,20-25]
Photoionization and trapping	Color centers, rare earth and transition metal ions, trapped electrons in organic glasses, donor-acceptor systems	Chaps. 4,7, [1.26-33]
Hydrogen bond rearrangements	Quinizarin in glasses	Chap. 3, [1.3,34-36]
Photodecomposition	s-tetrazine, dimethyl-s-tetrazine in crystals and polymers	Chaps. 3,7, [1.37-40]
Conformer interconversion	1,2-difluoroethane in rare gas matrices	Chap. 6, [1.41]
Nonphotochemical (photophysical)		
Host TLS transitions	Many photostable molecules and ions in glasses and polymers	Chap. 5, [1.17,42-55]
Host H-bond rearrangements	Pentacene, thioindigo in hydrogen-bonded organic crystals	Chap. 5, [1.56-58]
Photoinduced reorientation of guest	ReO_4^-, CN^- in alkali halides	Chap. 6, [1.59-61]

virtually any kind of matrix [1.16,20]. In a sense, free-base por-phyrins and phthalocyanines can be viewed as bistable devices, in that either of the two tautomers can be stable at low temperatures and light (or heat) can induce transformations between the two forms. This mechanism has provided many materials for the study of dephasing and temperature dependence of hole widths [1.21,22] and a variety of other effects, such as spectral diffusion [1.23], hole-burning bottlenecks [1.24], and Stark effects [1.25].

A second fairly general mechanism for PSHB is photoionization of the defect and subsequent trapping of the ejected electron. The basic process is thought to involve photoinduced tunneling of the optically excited electron from the center to a nearby trap, because usually the optical excitation energy is less than the actual ionization threshold. The ionized center often has quite different energy levels, which generally yields a drop in absorption at the excitation wave-length. Aggregate color centers in inorganic crystals provide a number of examples of this mechanism [1.26,27]. Photoionization has also been observed for rare earth and transition metal ions in crystals [1.28-30], trapped electrons in organic glasses [1.31], J-band aggregates [1.32], and recently for donor-acceptor systems in po-lymers [1.33].

Hydrogen bond rearrangements form another photochemical mechanism for PSHB. (Here rearrangements within the molecule or between the molecule and the host are of interest; spectral holes caused by host hydrogen bond rearrangements alone are mentioned below as nonphotochemical mechanisms.) This process usually re-quires a molecule with an adjacent pair of carbonyl−hydroxide substituents in which a hydrogen atom can form a hydrogen bond to one or the other of two oxygen atoms. 1,4-dihydroxy-anthra-quinone (quinizarin) serves as the principal example of this class of mechanisms. This molecule and its derivatives have been extensively studied in alcohol and boric acid glasses [1.3,34], polymers [1.35], and even in amorphous silica [1.36].

Photochemical hole-burning can also occur when the absorbing center simply decomposes upon photoexcitation. The first example of PSHB by this process was provided by s-tetrazine and dimethyl-s-tetrazine in crystals and polymers [1.37−39]. Recently, two-color photodecomposition has been reported for a tetracene-anthracene adduct [1.40].

The preceeding mechanisms for PSHB have all involved excita-tion of electronic transitions. In fact, spectral holes can also form in some systems if vibrational modes of the electronic ground state are excited. For example, infrared-induced conformer interconversion has been shown to lead to spectral hole formation for 1,2-difluoroethane in rare gas matrices [1.41]. Using CO_2 or lead salt diode lasers, molecules were transformed between *trans* and *gauche* conformations at low temperatures, yielding spectral holes.

Because the molecular changes in this case are fairly subtle compared to photodissociation or photoionization, this mechanism is one in which it is not so easy to clearly classify as photochemical or non-photochemical.

Those mechanisms for PSHB listed in the second half of Table 1.2 that do not result in an intrinsic chemical change of the absorbing center itself are called nonphotochemical or photophysical mechanisms. Nonphotochemical hole-burning (NPHB) [1.42] usually requires a change in the local environment around the guest or perhaps a reorientation of the guest with respect to the host. In most cases, the product absorption for nonphotochemical processes is not very far removed in frequency space from the original educt absorption. In fact, in some materials the product absorption appears as regions of increased absorption immediately adjacent to the hole [1.43]. In addition, the energy barriers preventing reverse reactions may not be very high, so that nonphotochemical holes often irreversibly disappear upon warming above liquid helium temperatures, whereas the changes caused by photochemical hole-burning either require even higher temperatures for reversal, or are irreversible.

A quite general mechanism for NPHB occurs when photostable molecules are incorporated into amorphous host materials and excitation of the guest molecules induces transitions among the two-level systems (TLS) of the host. In fact, this mechanism for perylene and 9-aminoacridine in ethanol glasses provided one of the very first examples of PSHB [1.17], although its generality and theoretical interpretation were not fully recognized until the work of *Small* et al. later [1.44,45]. The effect appears to be quite widespread, and has been studied for perylene in glasses and polymers [1.17,46], tetracene in glasses [1.47,48], resorufin in polymers [1.49], and dye molecules in hydrogen-bonded polymers [1.50,51], to give a partial list. For certain hosts, the rich vibronic structure hidden within the inhomogeneous line can be conveniently probed using NPHB techniques [1.50].

A similar mechanism exists for photophysical hole-burning of ions in glasses and polymers. Here again, the absorbing center itself undergoes no intrinsic change, and transitions among the TLS's of the amorphous matrix or rearrangements of the nearby host structure are responsible for the hole formation. The effect has been observed for Eu^{3+}, Pr^{3+}, and Nd^{3+} in silicate glasses [1.52−54] and even for Pr^{3+} and Nd^{3+} in polymer films [1.55]. These holes can persist for times much longer than the spin-lattice relaxation times because the relevant back reaction involves TLS tunnelling times of the nearby host.

NPHB can also occur for photostable guest molecules in crystalline host matrices, if the host crystal has hydrogen bonds that can be altered during excitation of the guest. This novel form of NPHB has been studied for pentacene in benzoic acid crystals [1.56,57] as well as for thioindigo in benzoic acid [1.58].

An interesting question is whether or not NPHB can occur for stable absorbing centers in rigid crystalline hosts. For the case of tetrahedral ReO_4^- ions in alkali halides, hole-burning does occur due to light-induced reorientations of the molecule in the host [1.43,59]. This mechanism also provides an example of NPHB in which electronic transitions are not involved – the excitation leading to hole-burning is an internal vibrational mode of the ReO_4^- molecule. For certain choices of secondary dopant, antiholes are also observed for this interesting system [1.60]. In recent measurements, infrared hole-burning has been observed for CN^- in alkali halide hosts, again due to molecular reorientation [1.61].

To conclude this brief summary of mechanisms for PSHB, it is clear that hole formation is a fairly general process for zero-phonon excitations in solids at low temperatures, and that it gives information about multiple stable ground states of the coupled host-guest system that may not be observable at elevated temperatures. Even though a wide variety of mechanisms and host-guest combinations have been considered in recent research, much remains to be done. Through further study, additional information can be obtained about spectral properties and dynamics within the inhomogeneous line that cannot be obtained using other techniques. The fact that small, persistent, localized changes can be made at will in inhomogeneously broadened lines using PSHB, is an intriguing process that deserves continued attention in the research community.

1.4 Synopsis of the Book

Throughout this volume, the abbreviation "PSHB" will be used to refer to persistent spectral hole-burning in general, when no distinction need be made about the type of photoreaction involved. "PHB" refers to photochemical hole-burning, and "NPHB" to non-photochemical (or photophysical) hole-burning. The novel case of PSHB in the infrared will be referred to as "PIRSH" formation, for persistent infrared spectral hole formation. Other abbreviations will be defined where appropriate.

Chapter 2, "Basic Principles and Methods of Spectral Hole-Burning," begins with a detailed description of the homogeneous spectrum of an electron-vibrational transition including the purely electronic and as well as the vibronic zero-phonon components. Using these fundamental elements, inhomogeneous lines are described and the process of persistent spectral hole-burning is summarized. After outlining some of the kinetic aspects of spectral hole formation, several spectroscopic applications of PSHB spectroscopy are described, including line broadening and dephasing in crystals and glassy matrices, off-resonant hole-burning, and the particular properties of PSHB in chlorophyll-like molecules. Several relatively new methods of hole detection are described, including Doppler scanning and holographic techniques. The chapter concludes with a brief review of how persistent spectral hole-burning can be used to create holograms in the time and spatial domains.

Chapter 3, "Photochemical Hole-Burning in Electronic Transitions: Proton Tunneling Mechanisms and Phenomena in Polymers and Glasses," concentrates on the general class of mechanisms leading to photochemical hole-burning in glasses and polymers. The analogy of PSHB with the hole-burning effects observed in magnetic resonance experiments provides a useful framework for discussing the difference between chemical depletion and transient saturation processes. The bulk of the chapter concentrates on the spectroscopic analysis of photochemical hole-burning experiments in amorphous host media. Particular emphasis is placed on the physical and chemical phenomena that can be studied by such an analysis: fast relaxation, excited state dephasing, and spectral diffusion. The chapter concludes with a stimulating discussion of external field effects in hole-burning spectroscopy, including electric field (Stark) effects as well as the effects of strain fields.

Chapter 4, "Persistent Spectral Hole-Burning in Inorganic Materials," considers the varied mechanisms at visible and near infrared wavelengths that lead to photochemical hole-burning in inorganic materials. Color centers in crystalline alkali halide hosts provide a large number of zero-phonon lines in the visible and near infrared, and these materials appear to undergo PSHB via a photoionization mechanism. A variety of rare earth and transition metal ions in glasses and crystals also show the formation of persistent spectral holes with characteristic linewidths and temperature dependences, and in some cases, the mechanisms are photon-gated. In general, PSHB in these materials reveals information about the local structure and dynamics of the defect centers, as well as the nature of traps and ionization thresholds (position of deep levels) in the material.

Chapter 5, "Two-Level-System Relaxation in Amorphous Solids as Probed by Nonphotochemical Hole-Burning in Electronic Transitions," provides a comprehensive summary of a different class of mechanisms, nonphotochemical hole-burning (NPHB) processes in electronic transitions. NPHB can occur in hydrogen-bonded crystals, in amorphous polyacene films, and for photostable molecules and ions in organic glasses and polymers. The central scientific phenomena that have been studied in detail by NPHB are optical linewidths and their temperature dependence. The novel power laws observed clearly indicate that two-level systems of the amorphous host are crucial in the dephasing process. Diagonal and off-diagonal modulation can both lead to dephasing, and a new density of states function for the two-level systems is proposed. Finally, the novel phenomenon of laser-induced hole-filling is described and ascribed to a division of the host dynamics into intrinsic and extrinsic two-level-system effects.

Chapter 6, "Persistent Infrared Spectral Hole-Burning for Impurity Vibrational Modes in Solids," further illustrates the generality of PSHB by describing the unusual mechanisms that can operate when nonphotochemical hole-burning is performed on impurity vibrational modes in solids with infrared lasers. The first example of persistent infrared spectral hole-burning is provided by the 1,2-difluoroethane molecule in rare gas matrices where the mechanism has been shown to involve conformational and reorientational effects. The system composed of tetrahedral perrhenate (ReO_4^-) molecules in alkali halide crystals provided the first example of NPHB for a vibrational mode in a crystalline host. Again, molecular reorientation is responsible for the novel hole-burning in this material. Finally, molecules with extremely low symmetry such as CN^- also show PSHB under certain conditions, and this result has opened up a large class of molecules in crystals as candidates for hole-burning studies of linewidths, dephasing, and mechanisms.

The concluding chapter, Chapter 7, "Frequency Domain Optical Storage and Other Applications of Persistent Spectral Hole-Burning", describes several possible applications of PSHB with particular emphasis on the application that has received the most attention to date, namely, frequency-domain optical storage (FDOS). Engineering and systems issues dictate the form a FDOS system might take, and in this chapter optimal properties of single-photon materials are derived. The limitations of monophotonic mechanisms underscore the need for photon-gated processes, and the examples of photon gating known to date are summarized. Other methods of organizing an optical storage system based on PSHB are also described, including holographic, time domain, and electric field schemes. Applications of PSHB to spectral filtering and optical waveform processing are briefly mentioned.

The diversity of mechanisms and physical phenomena that occur in PSHB experiments as well as the possible applications of PSHB suggest that this field will continue to grow in the future.

References

1.1 N. Bloembergen, E. M. Purcell, R. V. Pound: Phys. Rev. 71, 679 (1948)
1.2 L. A. Rebane, A. A. Gorokhovskii, J. V. Kikas: Appl. Phys. B29, 235-250 (1982), and references therein
1.3 J. Friedrich, D. Haarer: Angew. Chemie 23, 113 (1984), and references therein
1.4 W. E. Moerner: J. Molec. Elec., 1, 55 (1985)
1.5 M. Maier, Appl. Phys. B 41, 73 (1986)

1.6 A special issue of the Journal of Luminescence has been devoted to the topic of optical linewidths in glasses, see J. Lum. **36**, 179-329 (1987)

1.7 W. E. Moerner, M. Gehrtz, A. L. Huston: J. Phys. Chem. **88**, 6459 (1984)

1.8 L. Kador, G. Schulte, D. Haarer: J. Phys. Chem. **90**, 1264 (1986)

1.9 A. Szabo: "Frequency selective optical memory," U. S. Patent No. 3,896,420 (1975)

1.10 G. Castro, D. Haarer, R. M. Macfarlane, H. P. Trommsdorff: "Frequency selective optical data storage system," U. S. Patent No. 4,101,976, (1978)

1.11 A. L. Schawlow: in *Advances in Quantum Electronics*, (Proc. Second Int. Conf. on Quant. Electr.), ed. by J. R. Singer, (Columbia Univ., New York 1961), pp. 50-64

1.12 A. Szabo: Phys. Rev. Lett. **25**, 924 (1970)

1.13 R. I. Personov, E. I. Al'Shits, L. A. Bykovskaya: Opt. Commun. **6**, 169 (1972)

1.14 W. M. Yen, P. M. Selzer (eds.): *Laser Spectroscopy of Solids*, Springer Topics in Applied Physics, Vol. 49 (Springer, Berlin, Heidelberg 1981)

1.15 A. Szabo: Phys. Rev. B **11**, 4512 (1975)

1.16 A. A. Gorokhovskii, R. K. Kaarli, L. A. Rebane: JETP Lett. **20**, 216 (1974)

1.17 B. M. Kharlamov, R. I. Personov, L. A. Bykovskaya: Opt. Commun. **12**, 191 (1974)

1.18 O. N. Korotaev, R. I. Personov: Opt. Spektr. **32**, 900 (1971)

1.19 K. N. Solov'ev, I. E. Zalesskii, V. N. Kotlo, S. F. Shkirman: Zh. ETF Pis. Red. **17**, 463 (1973)

1.20 S. Völker, J. H. van der Waals: Molec. Phys. **32**, 1703 (1976)

1.21 K. K. Rebane, A. A. Gorokhovskii: J. Lum. **36**, 237 (1987)

1.22 S. Völker: J. Lum. **36**, 251 (1987)

1.23 Th. Sesselmann, W. Richter, D. Haarer: J. Lum. **36**, 263 (1987)

1.24 M. Romagnoli, W. E. Moerner, F. M. Schellenberg, M. D. Levenson, G. C. Bjorklund: J. Opt. Soc. Am. B: Optical Physics 1, 341 (1984)

1.25 A. J. Meixner, A. Renn, S. E. Bucher, U. P. Wild: J. Phys. Chem. **90**, 6777 (1986)

1.26 R. M. Macfarlane, R. M. Shelby: Phys. Rev. Lett. **42**, 788 (1979)

1.27 R. M. Macfarlane, R. T. Harley, R. M. Shelby: Rad. Effects **72**, 1 (1983), and references therein

1.28 R. M. Macfarlane, R. M. Shelby: Opt. Lett. **9**, 533 (1984)

1.29 A. Winnacker, R. M. Shelby, R. M. Macfarlane: Opt. Lett. **10**, 350 (1985)

1.30 R. M. Macfarlane, J. C. Vial: Phys. Rev. B **34**, 1 (1986).

1.31 S. L. Hager, J. E. Willard: J. Chem. Phys. **61**, 3244 (1974)

1.32 S. De Boer, K. J. Vink, D. A. Wiersma: Chem. Phys. Lett. **137**, 99 (1987)

1.33 T. P. Carter, C. Bräuchle, V. Y. Lee, M. Manavi, W. E. Moerner: Opt. Lett. **12**, 370 (1987)

1.34 W. Breinl, J. Friedrich, D. Haarer: Phys. Rev. B **34**, 7271 (1986)

1.35 K. Horie, K. Hirao, K. Kuroki, T. Naito, I. Mita: J. Fac. Engr., Univ. Tokyo (B) **39**, 51 (1987)

1.36 T. Tani, H. Namikawa, K. Arai, A. Makishima: J. Appl. Phys. **58**, 3559 (1985)

1.37 H. de Vries, D. A. Wiersma: Phys. Rev. Lett. **36**, 91 (1976)

1.38 D. M. Burland, D. Haarer: IBM J. Res. Devel. **23**, 534 (1979)

1.39 R. M. Hochstrasser, D. S. King: J. Am. Chem. Soc. **97**, 4760 (1975)

1.40 M. Iannone, G. W. Scott, D. Brinza, D. R. Coulter: J. Chem. Phys. **85**, 4863 (1986)

1.41 P. Felder, H. H. Günthard: Chem. Phys. **85**, 1 (1984)

1.42 G. J. Small: in *Spectroscopy and Excitation Dynamics of Condensed Molecular Systems*, V. M. Agranovitch and R. M. Hochstrasser, editors, (North-Holland, Amsterdam 1983), pp. 515-554

1.43 W. E. Moerner, A. R. Chraplyvy, A. J. Sievers, R. H. Silsbee: Phys. Rev. B **28**, 7244 (1983), and references therein

1.44   J. M. Hayes, R. P. Stout, G. J. Small: J. Chem. Phys. **73**, 4129 (1980)
1.45   R. Jankowiak, L. Shu, M. J. Kenney, G. J. Small: J. Lum. **36**, 293 (1987)
1.46   U. Bogner, P. Schätz, R. Seel, M. Maier: Chem. Phys. Lett. **102**, 267 (1983)
1.47   A. A. Gorokhovskii, Y. V. Kikas, V. V. Pal'm, L. A. Rebane: Sov. Phys. Sol. St. **23**, 602 (1981)
1.48   R. Janowiak, H. Bässler: Chem. Phys. Lett. **101**, 274 (1983)
1.49   A. P. Marchetti, M. Scozzafava, R. H. Young: Chem. Phys. Lett. **51**, 424 (1977)
1.50   T. P. Carter, B. L. Fearey, J. M. Hayes, G. J. Small: Chem. Phys. Lett. **102**, 272 (1983)
1.51   T. P. Carter, G. J. Small: J. Phys. Chem. **90**, 1997 (1986)
1.52   R. M. Macfarlane, R. M. Shelby: Opt. Commun. **45**, 46 (1983)
1.53   R. M. Macfarlane, R. M. Shelby: in *Proceedings of the NATO Advanced Research Workshop on Coherence and Energy Transfer in Glasses*, (Plenum, New York 1984)
1.54   R. M. Shelby: Opt. Lett. **8**, 88 (1983)
1.55   B. L. Fearey, T. P. Carter, G. J. Small: Chem. Phys. **101**, 279 (1986)
1.56   R. W. Olson, H. W. H. Lee, F. G. Patterson, M. D. Fayer, R. M. Shelby, D. P. Burum, R. M. Macfarlane: J. Chem. Phys. **77**, 2283 (1982)
1.57   H. W. H. Lee, C. A. Walsh, M. D. Fayer: J. Chem. Phys. **82**, 3948 (1985)
1.58   J. M. Clemens, R. M. Hochstrasser, H. P. Tromssdorff: J. Chem. Phys. **80**, 1744 (1984)
1.59   W. E. Moerner, A. J. Sievers, R. H. Silsbee, A. R. Chraplyvy, D. K. Lambert: Phys. Rev. Lett. **49**, 398 (1982)
1.60   T. R. Gosnell, A. J. Sievers, R. H. Silsbee: Phys. Rev. Lett. **52**, 303 (1984)
1.61   R. C. Spitzer, W. P. Ambrose, A. J. Sievers: Opt. Lett. **11**, 428 (1986)

2. Basic Principles and Methods of Persistent Spectral Hole-Burning

K. K. Rebane and L. A. Rebane

With 28 Figures

In this chapter characteristic features of the purely electronic and vibronic zero-phonon lines in the homogeneous and inhomogeneous spectra of impurity centers are presented. Persistent spectral hole-burning (PSHB) is considered as a method to control optical properties of matter (absoprtion coefficient and index of refraction) and, via illumination at high spectral selectivity, to control the inhomogeneous distribution of impurities in matrices. The specific methods of PSHB and the resulting high-resolution low-temperature matrix spectroscopy of molecules are reviewed mainly for large organic molecules including chlorophyll and its analogues as impurties in glassy and polymeric solid matrices. The holographic detection of spectral holes and hole-burning time- and space-domain holography of ultrafast processes of nano- and picosecond duration are reported.

2.1 Background

The cornerstone of persistent spectral hole-burning is the zero-phonon line (ZPL). Among ZPLs the purely electronic zero-phonon line (PEL) is of special interest. The majority of the persistent spectral hole-burning (PSHB) investigations and all of them performed at the highest resolution have been done on PELs. PELs attain very narrow homogeneous linewidths and high peak intensities at liquid helium temperature. This holds not only for impurity ions or atoms in single crystals (e.g., rare-earth impurity ions in single crystals of flourite), where the presence of narrow lines (linewidths about 1 cm^{-1}) had been well known earlier, but also for a large body of small and complex molecules in various solid matrices - polycrystals, glasses, and polymer films.

The characteristics of ZPLs that are important for PSHB are well understood in all these systems; they follow the same theory as that for vibrational structure of impurity optical spectra [2.1-3]. An amazing conclusion of the theory is that there is a rather deep analogy between the characteristic features of optical spectra and Mössbauer γ-resonance spectra, well known for their very high resolution. This is why after the theoretical paper by *Trifonov* [2.4] the PEL acquired a second name - "the optical analog of the Mössbauer line" [2.5]. In [2.5], these features, i.e., sharp ZPLs accompanied by wide phonon sidebands with a characteristic temperature dependence (Fig.2.1), were demonstrated experimentally for

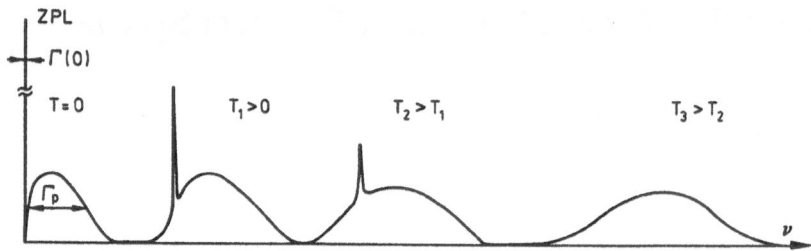

Fig.2.1. Purely electronic zero-phonon line and phonon sideband in the impurity homogeneous absorption spectrum and its dependence on temperature T

localized excitations in CdS crystals. In a theoretical paper [2.6] the analogy between the vibrational structure of optical spectra and Mössbauer spectra was utilized to treat models where localized vibrations are present, and the results were applied to the interpretation of the spectra of the Shpol'skii systems (i.e., polycyclic molecules as impurities in frozen n-alkane matrices [2.7]). Among early work on the influence of the vibrations of a solid matrix on the spectra of impurity molecules and the presence of very sharp ZPLs the paper by *Krivoglaz* and *Pekar* must be mentioned [2.8].

It was believed that the theory was correct; however, in the case of large molecules two important points had no experimental confirmation for a number of years: (i) No one had ever detected any phonon sidebands in the Shpol'skii spectra; (ii) the observed lines were at least 10^3-10^4 times broader than the very narrow widths determined by the lifetime of the excited electronic state. The latter discrepancy holds also for the narrowest lines in the spectra of rare-earth ions in single crystals.

Vibronic ZPLs accompanied by structured phonon sidebands were found in luminescence spectra of small molecular ions such as O_2^-, S_2^-, and NO_2^-, introduced in alkali halide single crystals and interpreted in accordance with the theory of impurity centers having intramolecular high-frequency local modes [2.9-11]. The presence of the phonon wings in a Shpol'skii spectrum was first demonstrated for perylene in n-hexane [2.12].

The intrinsically narrow linewidth of the PEL, down to the 10^{-2}-10^{-3} cm^{-1} values predicted by theory was first observed for ruby in site-selection studies [2.13] and for large molecules in hole-burning experiments [2.14]. The latter method is nowadays most favoured and in many cases the only method for high-resolution frequency-domain study of PELs. The complementary method, time-domain photon-echoes [2.15], also demonstrates that in accordance with theory the widths of PELs are governed by the lifetime of the excited electronic state. The time-domain experiments become more favourable for longer excited state lifetimes and narrower homogeneous linewidths.

Today there is no doubt of the presence of PELs, vibronic ZPLs and associated phonon sidebands in the low temperature spectra of a large number of various impurity atoms, ions and molecules in different matrices. The conditions of their existence and principal properties can be reasonably well understood in the framework of a fairly simple theory [2.1,6] (for further details, see [2.2,3]). The main task of the theory in understanding PSHB is to describe the spectrum of one single impurity in the solid matrix - the *homogeneous* spectrum. The presence of PELs in spectra opens new avenues for high-resolution spectroscopy and frequency-selective photochemistry.

Because of their small linewidths, ZPLs and especially PELs are extremely sensitive to inhomogeneities in the matrix structure. The inhomogeneous broadening (IB) is tremendous in comparison with the low-temperature homogeneous linewidth (HLW). In fact, IB is the reason why the PELs and vibronic ZPLs often do not show up in spectra at all. Even in luminescence spectra of small molecules in single crystal matrices with pronounced vibronic structure measured under conventional broad-band excitation, the linewidths of PELs are, because of IB, about 1 cm^{-1}, i.e., 10^3-10^4 times broader than the corresponding HLW. The enormous IB of PELs is a serious experimental difficulty in comparison to the situation with Mössbauer spectroscopy, and this is the reason that for a long time the extremely small linewidths and high peak intensities of homogeneous PELs were considered to be of only pure theoretical interest.

The development of laser spectroscopy has provided two effective and relatively simple methods to overcome the problem of IB: the site-selective excitation of fluorescence or fluorescence line narrowing (FLN), and the photo-induced burning of spectral holes or dips (PSHB). These methods increase the importance of PELs in molecular and solid state spectroscopy and eliminate IB-induced disadvantages of optical PELs in comparison with the Mössbauer lines.

PSHB has some essential advantages over the site-selection method, FLN. One is the convenience of studying the homogeneous PEL itself, which is quite complicated with the method of site-selective excitation of flourescence. (In the latter case, the scattered laser light masks PEL luminescence). Further, PSHB is not only a powerful method for performing spectroscopy without IB, but it also provides the possibility of inducing spectrally selective changes in solid matter. Actually, PSHB turns the inhomogeneous broadening of the impurity absoption band into a useful feature: owing to PSHB one can selectively excite single homogeneous PELs within a broad band of frequencies. In other words, the phenomenon of PSHB implies that an inhomogeneous broad band represents an ensemble of a large number of very sharp and intense resonances (HLWs $\simeq 10^{-3}$-10^{-4} cm^{-1}) covering a broad band of frequencies (IB $\simeq 100$-1000 cm^{-1}). It is clear that such unique systems of narrow optical resonances built-in by nature provide new possibilities for novel physical phenomena. Among the possible applications are narrow-line optical filters and spectral devices for optical data processing and information storage (high-density optical memories, time-domain holography), which are described in more detail in Chap.7.

Actually there are two complementary methods of hole-burning. The first involves illumination of the sample with a narrow laser line ω_L. This radiation provides selective excitation of those impurities whose PELs are in resonance with ω_L. After a number of excitation cycles something generally happens to the impurity molecule or its environment. The PEL frequency changes to some extent, and the molecules go out of resonance with ω_L. Because of the decrease of the number of molecules absorbing resonantly a sharp hole or dip is created in the absoption spectrum at ω_L. This may be called frequency-domain hole-burning. It allows extremely high-resolution spectroscopy (Sect.2.6 and [2.16]) and is the basis for several applications, such as narrow-band filters and optical storage [2.17], mentioned above.

The second method, time domain hole-burning, is produced by picosecond pulses having a spectral width of about 3-5 cm^{-1}, i.e. many thousands of times larger than the homogeneous width of the PEL. The hole shape stores the intensity profile of the Fourier components of the pulse. This approach leads to photochemically accumulated photon echoes, and further to time-domain holography (Sects.2.8 and 7.4.1).

2.2 Homogeneous Spectrum of an Electron-Vibrational Transition

The *homogeneous* spectrum is the spectrum of one single impurity atom or molecule. An ensemble of impurities even at low concentrations will show the same spectrum only in an ideal case when all the impurity centers are absolutely identical, and experience absolutely identical interactions with the surrounding matrix. This is never the case with the conventional solid matrices, and actual impurity spectra observed are always *inhomogeneous*. The basic properties of the inhomogeneous spectra, which we shall consider in the next section, are also determined by the homogeneous spectrum.

2.2.1 Integrated Intensities of Purely Electronic Lines and Phonon Sidebands, Electron-Phonon Interactions and Temperature Dependence

Let us consider an impurity center where no local vibrations are present and focus our attention on the electronic transition accompanied by creation-annihilation of phonons of the (matrix) lattice modes. Due to the acoustic phonons whose frequencies begin from zero, the energy spectrum of the transition covers a more or less broad-band interval; it is always continuous. Nevertheless, theory [2.1] tells us that the homogeneous spectrum of the transition is as shown in Fig.2.1. At T=0 there is a sharp PEL of high intensity, the ZPL, which corresponds to the electronic transition without creating any phonon (there is nothing to annihilate at T=0). The PEL linewidth $\Gamma(0)$ is determined by the excited electronic decay time

$T_1(0)$

$$\Gamma(0) = 1/2\pi c T_1(0) \text{ cm}^{-1} . \tag{2.1}$$

There is no Doppler broadening, because the impurity center is attached to the huge mass of the matrix. The Mössbauer γ-resonance line is free from Doppler broadening for the same reason.

For allowed optical transitions in the absence of non-radiative decay processes we have $T_1 \simeq 10^{-7}$-10^{-8} s, and, correspondingly, $\Gamma(0) \simeq 10^{-4}$-10^{-3} cm^{-1}. The PELs are even narrower for forbidden transitions, but in these cases more subtle details of the theory have to be taken into account when estimating the actual values of Γ [2.1].

At $T \neq 0$ the PEL broadens due to dephasing processes caused by phonons (or other elementary excitations) excited by the thermal motion. Dephasing by phonons is due to electron-phonon interactions and may be considered as quasi-elastic scattering of a phonon at the impurity, resulting in a change of the phonon's direction of propagation and in a very small, if any, change in its energy. On the other hand, the phonon scattering event results in an abrupt change of the phase of the wave function of the excited electronic state. The electron is left in the excited electronic energy level ℓ, but the time-dependent part of its wave function, $\exp(iE_\ell t/\hbar)$, is changed to $\exp[i(E_\ell t/\hbar)+i\delta]$, δ being a random phase shift. This means that the lifetime of the quantum state, i.e., the state with a fixed wave function is shortened, and the corresponding homogeneous linewidth has to broaden in comparison with the value determined by the excited-state energy decay time. On the other hand, for an ensemble of impurities this dephasing means that the coherence time of the ensemble of excited impurities is shortened, too. Zero-point phonons cause no dephasing. The density of thermal phonons increases with temperature, and that is the main source of the temperature broadening of PELs.

It is reasonable to introduce two characteristic periods, the energy relaxation time T_1 and the phase relaxation (pure dephasing) time T_2^* [1]. Relaxation in energy results in phase memory loss as well, and the overall coherence time T_2 of the excited electronic state, which governs the homogeneous linewidth, may be written as follows

$$\Gamma(T) = \frac{1}{\pi c T_2(T)} = \frac{1}{\pi c}\left[\frac{1}{2T_1(T)} + \frac{1}{T_2^*(T)}\right] . \tag{2.2}$$

When non-radiative decay processes are present, $1/T_1 = 1/T_{1opt} + 1/T_{1q}$, where the first term stands for purely radiative decay, and the second one for non-radiative decay. Usually the temperature dependence

---

[1] T_1 and T_2^* are also called the longitudinal and transverse relaxation times, respectively. The names arise from their relation to the decay of diagonal and off-diagonal elements of the density matrix [2.18].

of T_{1opt} may be neglected, while T_{1q} and T_2^* depend strongly on temperature. Further, at liquid-helium temperatures the non-radiative processes are usually absent ($T_{1q} = \infty$) or have already reached their low-temperature quantum-tunneling-limited value ($T_{1q} = $ const), and thus dephasing remains as the main source of the PEL's temperature dependence. The contribution of dephasing to the PEL's linewidth increases considerably with temperature. For the same molecules in glassy matrices or polymeric films the contribution of dephasing exceeds, by several times, the decay broadening even at 1.8 K (Sect.2.6.1. and Chaps.3 and 5).

The PEL is accompanied by a broad continuous band, the phonon sideband (sometimes called the phonon wing), corresponding to the transitions in which phonons are created (T = 0) or both created and annihilated (T > 0). The detailed structure and the width of phonon sidebands vary depending on the local lattice dynamics at the impurity center, i.e. on the phonon density of states in the matrix and the disturbance of the phonons in the vicinity of the impurity. In addition, the phonon sideband depends on the strength of the coupling of the electronic transition to the vibrations, i.e. on the changes in local dynamics caused by the change of the electronic state. Multiphonon transitions are quite often not only present, but even predominant, and so the width Γ_p of the phonon sideband is not necessarily limited by the width of the phonon spectrum. On the other hand, if the interaction takes place only with the long-wavelength phonons (which may be the case for large-radius impurity centers), the phonon sideband may be considerably narrower than the acoustic phonon spectrum. Nevertheless, the latter can serve as a starting point, and we shall use the estimate $\Gamma_p \simeq 10\text{-}1000$ cm^{-1}. Thus the purely electronic ZPL is very sharp: it is $10^4\text{-}10^7$ times narrower than the phonon sideband attached to it.

For high-resolution spectroscopy, several parameters of the PEL are of interest: the maximum intensity, the linewidth, and the integrated line strength. The relative integrated intensity of a PEL at T = 0 is given by the Debye-Waller factor (DWF) $\alpha(0)$

$$\alpha(0) = S_0/(S_0+S_p) = \exp(-n_{st}) , \qquad \text{with} \qquad (2.3)$$

$$n_{st} = \sum_{s=1}^{N} P_s/(\hbar\omega_s) ,$$

where S_0 and S_p are the integrated intensities of the PEL and the phonon sideband, respectively; n_{st} is the average dimensionless Stokes' shift for the vibrational mode s in the electronic transition under consideration. For the basic model [2.1] we have $2P_s = m_s\omega_s^2 q_{s0}^2$, m_s being the mass, ω_s the frequency, and q_{s0} the change in the equilibrum position of the oscillator.

With increasing temperature the Debye-Waller factor $\alpha(T)$ decreases, in most cases monotonically and quite rapidly. For the basic model the result is

22

$$\alpha(T) = \frac{S_0(T)}{S_0(T)+S_p(T)} = \exp\left[-\sum_{s=1}^{N}(2\bar{i}_s+1)n_{st}\right], \qquad (2.4)$$

where the average number of phonons in the lattice mode s at temperature T is given by $(\bar{i}_s+1/2) = (1/2)\coth[\hbar\omega_s/(2k_BT)]$, k_B being Boltzmann's constant. At high temperatures $k_BT \gg \hbar\omega_s$, $\bar{i}_s$ increases linearly with T. This leads to an exponential decrease of $\alpha(T)$, which is faster the larger the dimensionless Stokes' losses are in the modes whose phonons are thermally excited at a given T. At low temperatures $\bar{i}_s \to 0$, but we have to keep in mind that for the acoustic modes of very low frequency the low-temperature limit is never to be reached. In practice, the contribution of very low frequency modes can be neglected for two reasons: first, the low density of states of these modes, i.e. the number of oscillators which may be excited is very small and, second, the influence of the electronic transition of the impurity on these modes may be negligible, i.e. $\bar{n}_{st}(\omega_s) \to 0$ with decreasing frequency ω_s. If non-radiative transitions are absent, the sum of the integrated intensities of PEL and the phonon sideband represents the full intensity of the electronic transition (the oscillator strength). The oscillator strength (which determines the optical T_1) is temperature independent, and the sum of these two strongly temperature dependent terms is thus independent of temperature:

$$S_0(T) + S_p(T) = \text{const} . \qquad (2.5)$$

Thus, (2.5) tells us that quite often the temperature dependence of the absorption spectrum results in transferring integrated absorption from the PEL to the phonon sideband.

Hence, the PEL not only broadens with temperature but also loses its integrated intensity. For many impurity systems the latter is the main reason for the absence of PELs in high-temperature homogeneous spectra. We shall refer to temperatures at which ZPLs are prominent in homogeneous spectra as "low temperatures". These are usually temperatures below 10 K. For most of the large organic molecules in solid matrices, including the Shpol'skii systems, the temperature limit above which the ZPLs become practically absent in homogeneous spectra lies around 20-50 K.

Figure 2.1 shows the dependence of the homogeneous absorption spectrum $\chi(\omega)$ on temperature. The decrease of DWF and the broadening of PEL are indicated. The luminescence spectrum $\phi(\omega)$ has the same features. For the basic model $\phi(\omega)$ is the mirror image of $\chi(\omega)$ if reflection is performed about the middle of the PEL [2.1], see Fig.2.6a below.

Let us note that the same formula (2.3) will describe the behavior of the Mössbauer γ-resonance line if we replace the Stokes shifts P_s by the recoil energy K_s transferred to the mode s in the process of absorption or emission of the γ photon. This is the second aspect of the remarkable analogy between PELs and the Mössbauer line caused on a fundamental level by the symmetry of the Hamiltonian of the harmonic oscillator in the coordinate and momentum variables [2.1,4].

2.2.2 Relative Width (Q-factor) of PEL. Peak Intensities

The Q-factor of a PEL may be defined as the ratio of the frequency ω_0 of the PEL to its homogeneous linewidth Γ_0. At liquid-helium temperature $Q = \omega_0/\Gamma_0 \simeq 10^8$ if $\omega_0 = 10^4$ cm^{-1} and $\Gamma_0 = 10^{-4}$ cm^{-1}. Such high Q-factors form a good starting point for high-resolution spectroscopy and make PELs sensitive indicators of subtle processes and interactions in molecules and solids. This is the third aspect of the analogy between PELs and the Mössbauer line. For convenient detection, the peak intensity of a PEL has to be well above the background. The peak intensity of PEL is proportional to the value of the dipole moment of the transition multiplied by the DWF $\alpha(T)$ and is inversely proportional to the linewidth $\Gamma_0(T)$. The latter two factors are specific for the impurity center, if at T = 0 the linewidth $\Gamma_0(0)$ is purely radiative (no non-radiative transitions), then both the integrated intensity of the PEL and the linewidth $\Gamma_0(0)$ increase in proportion to the dipole moment so that the peak intensity actually stays unaltered. The peak intensity I_{00} depends on $\alpha(T)$ whose largest possible value is 1, and on $\Gamma_0(T)$ whose smallest value is the radiative value, Γ_0. One may immediately conclude that I_{00} in homogeneous impurity spectra cannot be higher than the peak intensity of the same transition in the Doppler-free spectrum of the same atom or molecule in the gas phase. Methods exist to overcome inhomogeneous Doppler and solid-matrix broadening (e.g., Doppler-free spectroscopy with counter propagating beams and PSHB techniques, respectively). The great advantage of the latter is that the elimination of the inhomogeneity in solid matrices persists for much longer times.

The presence of phonon sidebands is one specific feature of the spectroscopy of impurity centers that is not present in gas phase spectroscopy. In particular, the phonon sidebands form the inevitable background on which the PELs must be detected. To estimate the ratio of I_{00} to the maximum intensity of the phonon sidebands, knowledge of the Debye-Waller factor (2.3) is required. The limiting values of α at low temperatures are determined by the Stokes shifts of the lattice oscillators accompanying the electronic transition, i.e. by the loss in the potential energy after the vertical electronic transition measured in the number of lattice phonons n_{st} emitted in the process of relaxation to the new equilibrium positrons. For impurity centers such as Tl$^+$ in alkali halides or F-centers with strong electron-phonon interactions, n_{st} is about 30 to 100 and $\alpha(0)$ = exp(-100) to exp(-30). Thus, there is no observable PEL in the spectra of absorption and luminscence[2] for these cases. On the other hand, transi-

---

[2] But PELs have been detected in Raman spectra [2.19] and may also be found in the spectra of nonlinear phenomena.

tions in the inner shells of the rare-earth impurity ions are very well screened from the lattice, so $n_{st} \ll 1$ and α is close to unity. In this case the phonon sideband is absent and often does not show up even at high temperatures because the processes leading to thermal broadening of the PEL cause it to merge with the phonon sideband.

Figure 2.1 displays the case for a moderate electron-phonon interaction, which illustrates the main features of the temperature dependence of impurity molecules in matrices. The DWF at $T = 0$ might be $\alpha(0) = 0.5$. The linewidths of the PELs are much narrower than those drawn in Fig.2.1.

We take, for example, $S_0 = S_p$ which gives $\alpha(T) = 0.5$. Further, we assume the simplest relation for the PEL integrated intensity $S_0 = I_{00}\Gamma_0$, (for the Lorentzian shape $S_0 = (\pi/2)I_{00}\Gamma_0$) and $S_p = I_p\Gamma_p$, I_{00} being the peak intensity of the PEL and I_p is the maximum intensity of the phonon sideband. For the ratio r of the maximum intensities we then have

$$r = \frac{I_{00}}{I_p} = \frac{S_0}{\Gamma_0} \cdot \frac{\Gamma_p}{S_p} = \frac{\Gamma_p}{\Gamma_0} . \tag{2.6}$$

If we take $\Gamma_p = 100$ cm^{-1}, $\Gamma_0 = 10^{-4}$ cm^{-1} (appropriate for an electronic transition with $T_1 = 2 \cdot 10^{-7}$ s at low temperatures), we have $r = 10^6$, which indicates that the PEL is the dominant feature in the spectrum. It should be pointed out that the bandwidth of the phonon sideband actually does not depend on the matrix element of the electronic transition, provided the electron-phonon interaction is constant. Then I_p is proportional, and r is inversely proportional, to the probability W_{if} of the electronic transition. Thus, we can expect the best ratios r(0) for partially forbidden electronic transitions.

Up to now we have considered the homogeneous spectrum, i.e. the spectrum of one single molecule or the spectrum of a number of molecules at equivalent positions of the matrix, with energy differences well within 10^{-4} cm. In order to measure the spectrum of one molecule it would be a difficult or even unrealistic task to introduce impurities with such precision. In practice, one must deal with the inhomogeneous spectra of the impurity molecules where the Q-factors and peak intensities are much less favourable. The goal then is to develop ways to suppress the influence of inhomogeneous broadening.

2.2.3 Role of Local Modes

Impurity molecules also have intramolecular vibrations, and those modes whose frequencies are out of resonance with the lattice (matrix) modes become local modes of the impurity center. Those electronic transitions which are accompanied by the creation-annihilation of the quanta of local modes but which do not change the population of lattice modes are also zero-phonon lines (ZPL), more precisely - vibronic zero-phonon lines.

In the basic model [2.1], each local mode creates a series of replicas of the PEL, and its phonon sidebands shifted in frequency by $n\Omega$, where Ω is the local-mode frequency, and $n = \pm 1, \pm 2, \ldots$. Similar vibronic ZPLs and their sidebands occur at the corresponding linear combinations of local mode frequencies if there is more than one local mode. The integrated intensities of the vibronic sidebands are determined by the Franck-Condon factors and by the population of initial vibrational levels. The essential point for high-resolution spectroscopy is that the homogeneous vibrational ZPLs are generally much broader than the PEL because at least one excited vibrational level with its associated vibrational broadening is involved in the corresponding transition. The vibrational linewidth is determined by the decay time T_{1v} of the excited state of the local mode. At $T = 0$, $T_{1v}(0) \simeq 10^{-11}$ s, which is 10^3-10^4 times shorter than the electronic decay time so that the homogeneous linewidth of the vibronic ZPL, $\Gamma_v(0)$, is correspondingly broader than PEL, and we have $\Gamma_v(0) = 0.1$-1 cm^{-1}. $T_{1v}(T)$ decreases with the increase of temperature, but in most cases rather slowly. The influence of dephasing on the vibronic ZPL is similar to that for the PEL.

The vibronic ZPLs can be and often really are quite sharp and intense peaks in the spectra. We note that their integrated intensity relative to their own phonon sideband is governed by the same Debye-Waller factor $\alpha(T)$ as in the case of the PEL, i.e. by the interaction with the lattice (matrix) modes. The part of the total electron-vibrational interaction due to shifts of the equilibrium positions of local modes governs the distribution of the intensities between the PEL and vibronic ZPLs through the Franck-Condon factors and does not change the relative intensities of a ZPL and its phonon sideband. The DWF at $T = 0$, $\alpha(0)$, is nothing else but the Franck-Condon factor for 0-0' transitions in all the lattice modes, and $\alpha(T)$ is the corresponding thermal average over all the $n = n'$ transitions, where the prime refers to the number of phonons in the same vibrational mode after the electronic transition. The remarkably sharp vibronic ZPL structure of the luminescence spectra of O_2^-, S_2^-, and NO_2^- ions in alkali halide single crystals gives good examples of strong electron-phonon interactions, the better part of which is carried by the intramolecular (local) modes [2.11].

The estimates given above show that in inhomogeneous spectra where sharp ZPLs are present the PELs have line-widths only a little narrower than those of the vibronic ZPLs. However, the hidden inner structure of the inhomogeneous PELs due to a distribution of very narrow homogeneous lines is 10^3-10^4 times more selective than that of vibronic ZPLs. This hidden structure can be studied by using high-resolution laser probing.

2.3 Inhomogeneous Broadening of the Vibronic Spectrum

2.3.1 Inhomogeneous Broadening of Purely Electronic Lines. Inhomogeneous Distribution Function

The first consequence of the high spectral resolution (high Q-factor) of PELs is that even very small differences in the influence of the matrix on the impurity cause frequency shifts of PELs that are much larger than their linewidth. This leads to a tremendous (in comparison with the PEL's linewidth) inhomogeneous broadening. The high spectral resolution of a homogeneous PEL disappears very easily in a real ensemble of absorbing centers.

In a reasonable approximation, small differences in impurity sites cause shifts of the homogeneous spectrum as a whole in frequency space without any change of the PEL lineshape. In that case, it is convenient to introduce the inhomogeneous distribution function (IDF) $\rho(\omega)$. For a certain electronic transition $\rho(\omega)\Delta\omega$ gives the fraction of impurities which have their PELs in the frequency interval $\Delta\omega$ at ω [2.20]. The IDF depends not only on the inhomogeneities in the matrix (e.g., other impurities of the same and different kind, point, linear and surface defects of the matrix, and the stress fields in it) but also on the character of the transition under consideration. If both the ground and excited electronic states undergo equal energy shifts, there will be no inhomogeneous broadening and the IDF becomes a very narrow peak. This is the case for most f-f transitions in rare-earth ions and for the intranuclear γ transitions in the Mössbauer phenomenon. There are IDFs of varied and often of quite complicated shapes. For instance, in the case of a Shpol'skii multiplet [2.20], the IDF consists not only of several sharp peaks (corresponding to the multiplet's components) but also of a broad weak background around them. The latter has to be kept in mind when interpreting the results of Raman scattering, for instance.

Of course, the inhomogeneous structure of the matrix has some influence on the frequencies of local modes, on the Debye-Waller factors, and on the electronic transition probabilities as well. The IDF may be easily generalized to describe the first of these influences (Sects.2.5 and 2.6.4). The other effects should be of importance in some special cases, e.g., for electronic transitions forbidden by spatial symmetry of the impurity center. Further, for very small samples, very low concentrations, and very high spectral selectivity the statistical fluctuations of the number of impurities have to be taken into account.

Below we shall mainly consider smooth, one-maximum IDFs with simple bell-like shapes. The IDF's bandwidth Δ is always much larger than the homogeneous $\Gamma_0(0)$ (provided strong non-radiative quenching is not present at T=0) and often narrower than the phonon sideband's Γ_p. Figure 2.2 shows an example of how the unique properties of the homogeneous PELs are destroyed by inhomogeneity. The superposition of three

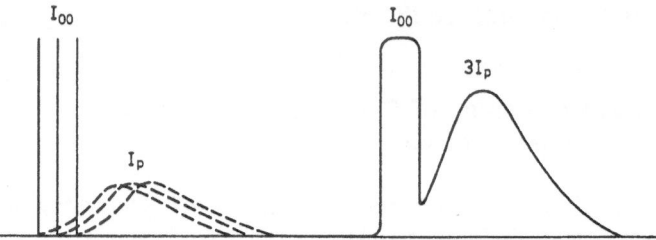

Fig.2.2. Formation of inhomogeneous zero-phonon line and phonon sideband for the case of three impurity molecules

shifted homogeneous spectra results in summing up the phonon sideband intensities while the resultant PEL becomes only broadened. The simplest calculation shows that the peak intensity of an inhomogeneous PEL increases with the number of impurities N as $(\Gamma_0/\Delta)N$, while the maximum intensity of a phonon sideband grows as $\Gamma_p N/(\Gamma_p + \Delta) \simeq N$. Thus, the ratio of the peak intensities of the PEL and the respective phonon sideband (2.6) in an inhomogeneous spectrum is reduced by a factor of $r = \Delta/\Gamma_0$. In other words, if $\Gamma_0 << \Delta << \Gamma_p$, we will see PELs in the conventionally measured inhomogeneous spectra, whose shape is governed not by the homogeneous shape but by IDFs, as well as the phonon sidebands whose shapes are distorted to some extent by IDFs. This is the case for a number of optical transitions of rare-earth ions and O_2^--type molecules[3] as impurities in ionic crystals and frozen gas matrices with $\Delta \simeq 0.1\text{-}1$ cm^{-1}, and also for Shpol'skii systems, where $\Delta \simeq 1\text{-}5$ cm^{-1}.[4]

In these cases with relatively small inhomogeneities, rather sharp (linewidth: $\simeq \Delta$) and intense PELs are present even in conventionally measured inhomogeneous spectra of absorption and luminescence. However, many solid organic impurity systems, chemically analogous (or even identical) to the Shpol'skii systems, do not have any sharp spectral lines even at low temperatures. It is well-known that the reason lies not in strong electron-phonon coupling or other specific features of these systems, but in the strong inhomogeneous broadening. In fact, the latter may be almost completely eliminated and parts of the homogeneous spectra revealed by applying quasi-monochromatic laser excitation in the proper way. One technique for achieving this is frequency-selective excitation of flourescence line-narrowing (FLN), another technique is persistent spectral hole burning (PSHB) spectroscopy.

---

[3] In the luminescence spectra of O_2^-, we have actually vibronic ZPLs with $\Gamma_v(T) \simeq \Delta$ $\simeq 1$ cm^{-1}, and the observed line-shapes are convolutions of homogeneous and inhomogeneous contours [2.21]. One of the remarkable features of these spectra is the fine-structured phonon sidebands with quite homogeneous shapes [2.11].

[4] If there is a Shpol'skii multiplet, this estimate holds for each of its components taken separately.

2.3.2 Selectivity of the Spectral Response of an Inhomogeneous Absorption Band

Phonon sidebands are always present in impurity spectra, and they also appear in selectively excited inhomogeneous spectra. The IDF itself tells nothing about the phonon sidebands. Figure 2.3 represents an inhomogeneous absorption band (curve 3) composed of homogeneous spectra with a sharp PEL and a smooth phonon sideband (curve 2), where the IDF (curve 1) is several times broader than the phonon sideband.

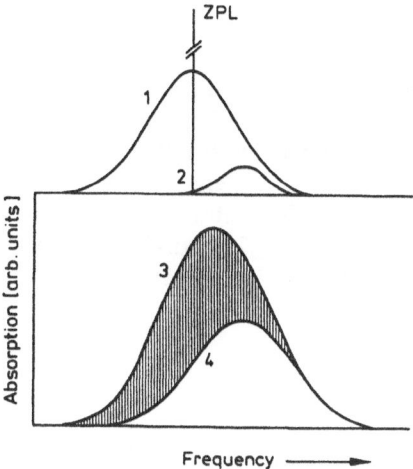

Fig.2.3. Illustrating the inhomogeneous distribution function (*1*), the homogeneous absorption spectrum of the impurity (*2*), and the inhomogenous absorption band (*3*) as the sum of two parts: a continuous contribution (*4*) formed by the overlapping phonon sidebands of spectrum *2* and a particular one (between curves *3* and *4*) formed by homogeneous zero-phonon lines distributed as given by curve *1*. The calculation is performed for the Debye-Waller factor of spectrum *2* ($\alpha = 0.5$)

An inhomogeneous impurity absorption band is actually a superposition of two distinctly different parts: a continuous band representing the sum of overlapping phonon sidebands (Fig.2.3, curve 4), and a sum of sharp, purely electronic zero-phonon lines. Only the second part leads to spectral selectivity as a result of excitation. Under narrow-line laser illumination with frequency ω_L and linewidth $\Delta\omega_L$, which is much narrower than that of the homogeneous PEL ($\Delta\omega_L \ll \Gamma(T)$), a selected subset of impurities whose PELs lie in the frequency interval $2\Gamma(T)$ about ω_L get excited. Strictly speaking, the homogeneous linewidth varies slightly in different parts of the IDF, but in most cases the dependence is very weak and may be neglected. Thermal broadening of PELs is the main reason for the loss of selectivity with increasing temperature.

It is clear from Fig.2.3 that the intensities of the two components of the inhomogeneous band depend strongly on the exact excitation frequency: the contribution of the PEL is more pronounced on the long-wavelength side of absorption, while the phonon wings prevail on the short-wavelength side. This is an obvious consequence of the shape of the homogeneous spectrum shown in Figs.2.1 and 3. If the IDF is narrow in comparison with the distance from the PEL to the maximum of the phonon sideband, the dependence on frequency becomes still more pronounced: on the short-wavelength side the inhomogeneous band consists of a sum of phonon sidebands only, whereas on the long-wavelength side the band consists of PELs. The shape of the latter is determined mostly by the IDF. Of course, if the IDF or the phonon sidebands are not smooth, the picture becomes more complicated [2.22,23].

The dependence of the PEL absorption on the excitation frequency ω may be described using the ratio of the contribution of the PEL band $\kappa_0(\omega,T)$ to the total absorption intensity at a given frequency[5]

$$s_0(\omega,T) = \frac{\kappa_0(\omega,T)}{\kappa_0(\omega,T) + \kappa_p(\omega,T)} ,$$ (2.7)

where $\kappa_p(\omega,T)$ is the contribution of the phonon sidebands to the total absorption at ω; and $s_0(\omega,T)$ describes the relative number of impurities excited via their PELS. A simple integral relation holds for this function. Since the inhomogeneous band is a sum of the homogeneous absorption spectra of single impurities, the integral of the PEL absorption over ω, $S_0(T)$ (the area between the solid lines 3 and 4 in Fig.2.3), divided by the total area of the absorption band gives the Debye-Waller factor

$$\frac{S_0(T)}{S_0(T) + S_p(T)} = \alpha(T) ,$$ (2.8)

where $S_p(T)$ is the absorption by phonon sidebands. We see that the PEL part of the total inhomogeneous absorption equals $\alpha(T)$, which decreases quite rapidly with temperature.

Equation (2.8) holds not only for the basic model, but also for more elaborate models of impurity absorption provided the influence of the inhomogeneities on $\alpha(T)$ can be neglected. Of course, in more detailed models $\alpha(T)$ does not follow (2.4), which is precisely correct only for the basic model. If the dependence of $\alpha(T)$ on inhomogeneities is to be taken into account, then $\alpha(T)$ may be replaced by a proper average $\langle \alpha(T) \rangle$. Therefore, the second reason for the loss of selectivity is the decrease of the integrated PEL absorption with increasing temperature governed by the Debye-Waller factor. It must be pointed out that the fraction of the total illumination absorbed by PELs does not depend on the PEL's homogeneous linewidth $\Gamma(T)$. This is clear because the narrower the PEL and the stronger its peak absorption, the smaller the number of impurities that have their PELs in the resonance interval $2\Gamma(T)$. The homogeneous linewidth governs the width of spectral response to excitation of an ensemble of PELs.

To describe the spectral selectivity of inhomogeneous spectra under monochromatic excitation two functions of frequency and temperature are

---

[5] Regarding the selectivity of monochromatic excitation in an inhomogeneous band, see [2.24].

needed: $\Gamma(T)$ - the homogeneous linewidth and $s_0(\omega,T)$ - the fraction of illumination absorbed by PELs. For the latter function we can write formally

$$s_0(\omega,T) = a(\omega,T) \cdot \alpha(T) \,, \tag{2.9}$$

where $a(\omega,T)$ depends on the shapes of the homogeneous absorption spectrum and the IDF; $a(\omega,T)$ may be a strongly fluctuating function if there are sharp peaks or dips in either of them. For rough estimates we can set $a(\omega,T) = 1$, which is equivalent to extrapolation of the integral relation (2.8) to each single frequency.

2.3.3 Inhomogeneous Distribution Function Under Monochromatic Laser Excitation. Site-Selection Spectroscopy

Laser illumination at ω_L in an inhomogeneous band such as that shown in Fig.2.3 is partly absorbed by the phonon sidebands, partly by PELs. The latter portion, whose relative absorption is given by $s_0(\omega,T)$, consists of selective excitation of only the PELs in a narrow interval of frequencies $\omega_L \pm \Gamma(T)$ and selects *via* excitation a group of impurities whose PELs are close to ω_L. It is reasonable to believe that a sufficiently large number of these impurities are identical not only in PEL frequency, but also in other aspects. Further, the selectively excited impurities emit a spectrum which is quite close to the homogeneous spectrum of luminescence. The excitation absorbed by phonon sidebands (or by PELs of centers different in other aspects from the main body of excited impurities) gives rise to broad background luminescence. The latter may have some structure, especially when there are sharp vibronic ZPLs present in the phonon sidebands and when the IDF is broader than the frequencies of the local modes. Thus the spectrum of luminescence under laser excitation represents a sum of a broad background with a complicated distribution of intensities and a nearly homogeneous spectrum. If the homogeneous PELs in absorption are sharp and intense and if the $s_0(\omega_L,T)$ is not too small, then the PELs and vibronic ZPLs in the total emission spectrum are pronounced and correctly positioned. Their shapes are only slightly affected, while the shapes of the phonon sidebands are usually fairly distorted. Spectra of this sort are called "site-selective" and the method to determine them - site-selection spectroscopy or fluorescence line narrowing (FLN). Note that in site-selective fluorescence spectra one may usually observe the positions and relative intensities of the vibronic ZPLs only. Nevertheless, this is a real achievement in comparison with the conventional structureless inhomogeneous spectra produced by the majority of complex molecules in frozen solutions with non-selective excitation.

Szabo [2.13] was the first to apply laser excitation to the inhomogeneously broadened ($\simeq 1$ cm^{-1} in width) PEL of ruby and to detect a PEL in luminescence whose width was reduced to 0.01 cm^{-1}. This value was still considerably larger than that expected if the width were determined by the optical lifetime. *Szabo* interpreted the difference as a result of nuclear spin interactions. Thus, the high sensitivity of FLN made it possible to obtain information on nuclear spin hyperfine interactions in a crystal by means of optical spectroscopy. However, there are strong and narrow PELs in the conventional inhomogeneous spectra of ruby, which is why *Szabo's* result was not especially striking in the sense of revealing the remarkable features of homogeneous PELs.

Personov et al. [2.25] were the first to apply laser excitation to the broad structureless spectra of complex molecules in frozen solution. For perylene in n-undecane and ethanol at 4.2 K, excitation with the helium-cadmium laser (whose frequency matches the structureless PEL absorption band of perylene) brings about the striking effect of elimination of inhomogeneous broadening: a fine-structured spectrum of luminescence was obtained in both matrices, the structure of which corresponds to that of the spectrum of perylene in Shpol'skii matrices.

The achievements of the site-selective studies of the vibrational structure of the fluorescence, phosphorescence and excitation spectra of various organic molecules have been reviewed by *Personov* [2.26]. We note that the detection of the PEL in a site-selection experiment in resonance with the laser excitation is difficult because of the strong background of scattered laser light. To eliminate scattered excitation light, specific methods have been developed, e.g. stroboscopic excitation and time-delayed recording. In these techniques one actually has to deal with time-dependent spectra of luminescence, which is an interesting topic by itself [2.27].

The main features of the fine structure of site-selective spectra may easily be understood by recalling the above considerations and accounting for the off-resonant character of the emission. There is also a partial analogy with the features of off-resonant holes (Sect.2.6.4) [2.23].

As an example let us consider the effect of multiple PELs [2.26,28]. If the IDF of a molecular impurity system is broader than the frequencies of local modes then the vibronic bands of different local modes (Ω_i) overlap. The excitation in this complicated band selects more than one group of molecules, which have their vibronic ZPLs at the frequency ω_L. After the excess vibrational energy stored via vibronic ZPLs relaxes, the molecules emit their own PELs which are shifted from ω_L by the local mode frequencies Ω_i and whose number corresponds to that of the overlapping bands.

In a site-selective spectrum, the PEL is accompanied by a broad band which looks like a phonon sideband but which, in addition to the true phonon sidebands excited by direct excitation of ZPLs, contains also a contribution from the molecules excited at ω_L via their phonon sidebands (Sect.2.3.2) [2.28,29]. To understand the temperature dependence of the site-selective spectrum of luminescence one must keep in mind that the Debye-Waller factor is involved twice: first, in excitation through $s_0(\omega,T)$, see (2.9), and second, in luminescence. If we take for $s_0(\omega,T)$ the approximation $a(\omega,T) = 1$, then we obtain that the relative integrated intensity of the ZPL in site-selective luminescence spectra is proportional to $\alpha^2(T)$, therefore its decrease with temperature is enhanced in comparison to the picture shown in Fig.2.1.

Selectively-excited luminescence spectra may be essentially affected when some hole-burning takes place [2.16,29]. For instance, a redistribution of intensities between the PEL and the sideband, resulting from different burning rates in the ZPL and phonon sideband, may occur [2.45].

2.4 Persistent Spectral Hole-Burning

2.4.1 Burning of Spectral Holes in the Inhomogeneous Distribution Function

In site-selective spectroscopy, the impurity molecules excited with near-monochromatic excitation usually luminesce or return back to the ground state via non-radiative transitions, and the initial IDF is restored. In general, however, a fraction of the selected group of molecules may be subject to phototransformations and thus not return back to exactly the same ground state. The new ground states may have very long lifetimes if they are formed and fixed by new equilibrium positions of the atoms of the molecule or matrix. As has been emphasized in Sects.2.2,3 even a slight change in the molecule or its surroundings is strong enough to move the PEL's frequency far away from ω_L in comparison to the PEL's low-temperature homogeneous linewidth. When this occurs, a fraction of the excited molecules is absent at ω_L after the excitation-deexcitation cycle is over, and a spectral hole (dip) is left in IDF as a long-lived label of the selectively excited body of molecules. The hole shape (depth, width, asymmetry) and its dependence on temperature, irradiation time, etc. (Sect.2.5) provide a wealth of high-resolution spectral information about homogeneous PELs. The amazing fact is that the information is revealed by the molecules that are *absent* from the ensemble which formed the initial IDF.

Processes leading to spectral hole formation may be chemical reactions in the molecule itself or various rearrangements of the particles of the matrix. Some mechanisms of hole formation have been discussed in Chap.1. Here we provide a list of the main features of persistent spectral hole burning (PSHB).

1) A hole in the IDF has different manifestations in different spectra. Usually a hole is detected by recording the absorption (transmission), luminescence or excitation spectrum. But a hole in the IDF may be detected also in a vibrational band or in spectra of other kinds where inhomogeneous broadening of the transition is essential. We shall see further that excitation with one single frequency can create holes in several different overlapping IDFs and more than one hole in one IDF, if vibronic ZPLs are involved. One single hole in an IDF may give rise to several holes in the observed spectra.

2) An essential advantage of PSHB in comparison with site-selection spectroscopy is that the PELs are studied in the absence of any background of strong excitation resonant with the PELs.

3) In many cases of interest the holes are persistent, i.e. they may have extremely long lifetimes at low temperatures. For instance, in [2.32] a half-decay time of 22,000 years is reported for holes in the spectra of quinizarin molecules in deuterated alcohol glass (C_2D_5OD/CD_3OD 3:1). On the other hand, the process of PSHB itself must involve relatively brief excitation: if the excitation lasts endlessly, all the inhomogeneous band will be burned away (i.e., transformed into an entirely different absorption spectrum) without any spectral selectivity left that is correlated with ω_L. The selectively burned holes are created as a result of the fact that PSHB develops along three different time scales. First, the molecules absorbing resonantly through their PELs at ω_L are removed from ω_L. Second, the molecules having their PELs at frequencies lower than ω_L are transformed due to absorption in phonon sidebands; third, the impurity molecules having their PELs at frequencies higher than ω_L absorb weakly through their anti-Stokes phonon sidebands. Of course, the last process actually has to be neglected at low temperatures and for PEL frequencies well above ω_L. However, the absorption via phonon sidebands is often competitive with direct PEL absoption, especially if there are vibronic ZPLs or other sharp peaks present. Thus, the hole shape is a time-varying function depending on the irradiation dose, and on the illumination time.

4) PSHB allows a piece of solid matter activated with impurities, having an IDF shown in Fig.2.3, to serve as a medium for photochemistry with high spectral selectivity. The sample may actually be regarded as a broad-band high-resolution spectral device for persistent recording and study of the spectrum of the incident light. Such spectral recording may also allow storage of the phase relations between the Fourier components of the incident light, which is equivalent to the recording of the time dependence of light pulses. In this fashion, space-time domain holography of fast processes of pico- and nanosecond duration becomes possible (Sects.2.8 and 7.4).

5) All methods of PSHB involve the preparation of persistent spectral filters and the study or utilization of their properties. The accuracy of the method in high-resolution spectroscopy depends fundamentally upon the understanding of the relation of the hole shape to the characteristics of the homogeneous spectral line. To achieve narrow holes, the holes should not be "overburned", i.e. the irradiation doses must be rather modest. The same applies to the fluence used for reading. Further, the characteristic time τ_{HB}, i.e. the hole formation time plus the measurement time, must not be too long. The dependence of the hole shape on τ_{HB} is usually weak, but for some materials τ_{HB} plays a central role in determining the hole shape. This may be the case for glassy matrices even at very low temperatures, where the processes of spectral diffusion seem to still be present (Chap.3).

6) PSHB may be performed for modestly photosensitive impurity centers. In fact, holes can almost always be observed if one waits long enough: because of their usually large absorption cross-section the molecules whose PELs are in resonance with the laser frequency ω_L are being excited so frequently (in comparison with the molecules absorbing via their phonon sidebands) that something must happen to them in the end. The portion of exciting energy absorbed at ω_L in various phonon sidebands and the microscopic events leading to PSHB are distributed over a large number of impurity molecules covering an interval of the IDF which is much broader than $\Gamma(T)$.

Of course, the quantum efficiency of PSHB is important in considering the time and laser power needed to burn a given hole and the actual observed parameters of holes - width, depth, stability. High photosensitivity results in a decrease of the excited state lifetime, as compared to optical or energy-transfer-limited values, and leads to broad energy levels, preventing the creation of narrow holes.

7) The dose of energy delivered to the sample to create a detectable hole may be small even if the hole-burning efficiency is very low. First, only a small fraction β of impurities has to be transformed to create even a deep hole. For $\Delta = 100$ cm^{-1} and $\Gamma_0 = 0.001$ cm^{-1} we have $\beta = 10^{-5}$. For an impurity concentration of 10^{-4} moles/mole, only one molecule out of 10^9 in the sample has to be phototransformed to produce a narrow, 100 % deep hole. Second, with sensitive techniques (Sect.2.7.2 and Chap.7) even very shallow holes having a relative depth of 10^{-6} may be detected. In that case only one out of 10^{15} molecules need be transformed. Regarding further minimization of the number of phototransformations, we must not forget about the limits set by the fluctuations of the number of the impurities in the frequency interval of one hole, especially when we want to create and detect narrow holes in small spatial spots (Sect.7.3.4).

2.4.2 Early Observations of Persistent Spectral Hole-Burning

Spectrally selective bleaching of impurity molecules under excitation by a few pulses of 694-nm ruby-laser light was reported by *Gorokhovskii* et al. in 1974 for a frozen solution of free-base phthalocyanine (H$_2$Pc) in n-octane (Shpol'skii system) at 5 K [2.33]. H$_2$Pc molecules undergo phototransformation when excited to the first excited singlet state. The transformation involves a 90° rotation of the proton-pair orientation within the core of the molecule (tautomerization) [2.34], which causes a change of the ZPL frequency by tens of wavenumbers due to interactions between the molecule and the matrix. The free molecule is of high symmetry and no changes are to be expected in the absence of the host matrix. The selectivity is due to the high PEL excitation rate of the molecules whose PELs are in resonance with the laser frequency.

Figure 2.4 shows the results of the first hole-burning experiments. A narrow hole (dip) appears in the inhomogeneous band at the ruby-laser

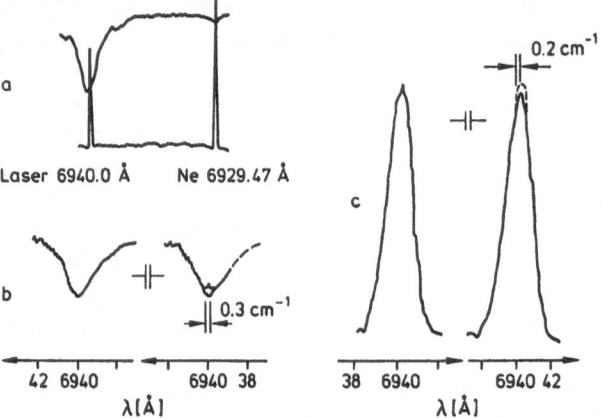

Laser 6940.0 Å   Ne 6929.47 Å

0.2 cm⁻¹

0.3 cm⁻¹

42 6940   6940 38   38 6940   6940 42
λ[Å]   λ[Å]

Fig.2.4. (*a*) Transmission Shpol'skii band of 3 cm⁻¹ width of the molecule H₂Pc in n-octane at 5 K along with a neon line and the ruby laser spectrum. (*b*) The same transmission band before (on the left) and after (on the right) irradiation of the sample with 10 laser pulses. The hole at the laser frequency is clearly displayed. (*c*) Spectral hole measured in the luminescence band of the same origin before (on the left) and after (on the right) irradiation with 5 laser pulses. The dashed lines show the recreation of the band shapes after nonselective irradiation of the sample

frequency after irradiating the sample with the laser pulses at helium temperature (10~20 pulses of 100 mJ, 0.5 ms duration). At helium temperature and in darkness the hole persists for many hours after the excitation is over. This remarkable feature enables spectral measurements of the hole resonant with ω_L to be performed free from background caused by the scattered exciting radiation that plagues FLN studies. The spectral hole in resonance with the excitation frequency was detected in the 0-0 absorption as well as in the luminescence bands. In these experiments, a hole-width of $\delta_H = 0.2$ cm⁻¹ was measured, which was limited by the spectral resolution of the apparatus. The detected holewidth was by one order of magnitude narrower than the bandwidth of the inhomogeneously broadened Shpol'skii component in conventionally measured spectra (1-8 cm⁻¹, depending on the preparation of the sample). This confirmed the inhomogeneous nature of the "narrow" Shpol'skii lines and indicated the presence of extremely narrow homogeneous purely electronic zero-phonon lines in the spectra of large molecules in organic matrices.

Another version of PSHB was demonstrated independently and at the same time by *Kharlamov* at al. [2.35]. A hole of 0.7 cm⁻¹ width was detected in the broad-band absorption of the photochemically stable molecule perylene in ethanol glass after 30 min irradiation of the sample at 4.2 K with a He-Cd laser ($\lambda_L = 441.56$ nm, laser power: 15 mW). The formation of the hole in this case was suggested to arise from some non-photochemical process, i.e. processes taking place in the environment of the impurity molecule.

As described in Chap.1, persistent spectral holes induced by photochemical transformations in the molecule itself are referred to as photochemical hole-burning (PHB), whereas hole-burning in the spectra of more stable guest molecules, when phototransformation of some kind takes place in the matrix, is called photophysical or non-photochemical hole-burning (NPHB) [2.36]. Of course, there are many different processes leading to hole-burning, some of which are intermediate between photochemical and photophysical situations. To be general, we call the entire phenomenon persistent spectral hole-burning (PSHB), which emphasizes that the most distinguishing feature is the very long lifetime of the holes.

In succeeding experiments performed with an improved apparatus by a number of researchers, it was shown that hole-burning spectroscopy enables efficient elimination of inhomogeneous broadening (at least the trivial part of it) and attainment of the electronic-state-lifetime-limited theoretical linewidth at the lowest temperatures [2.16,37]. PSHB thus allows extremely-high-resolution spectroscopy of large molecules in matrices at low temperatures.

2.5 Kinetics of Persistent Spectral Hole-Burning

As was mentioned above, since hole-burning is an inherently time-dependent process, the kinetics of hole formation is of central importance. Hole-burning kinetics has been theoretically considered by a number of researchers [2.16,38-49] on the basis of models of increasing complexity. The basic features, however, can be understood with a simple model proceeding from the following assumptions:

1) Weak excitation ensuring the absence of optical saturation;

2) low impurity concentration ensuring the absence of impurity-impurity interactions, e.g. energy transfer;

3) small optical thickness of the sample, providing uniform irradiation by monochromatic light of frequency ν_0 and intensity I_0;

4) one-dimensional inhomogeneity: only the frequency of purely electronic transitions, ω, varies for different impurity sites. The homogeneous spectra of absorption $\kappa(\nu-\omega)$ and luminescence $\phi(\nu-\omega)$ of an impurity center are supposed to be shifted by the local field differences without any distortion of their shapes and without changes in their integrated intensity. The latter implies a similar orientation of all impurities and equal values of the absorption cross section σ, as well as quantum yields η and η_f of photochemistry and fluorescence, respectively. The spectral properties of such a system are then fully described by the generally time-dependent inhomogeneous distribution function (IDF) $\rho(\omega,t)$ [2.16,46].

5) one-photon mechanism of photochemistry;

6) stability of photoproducts that guarantees the irreversibility of the hole-burning process;

7) large photochromic shift as compared to the width of the homogeneous spectrum. This prevents the disturbance of the hole by the spectrum of the photoproduct.

These assumptions are reasonable and can be realized, to a good approximation, for a majority of systems by decreasing the excitation intensity and utilizing samples with sufficiently low impurity concentrations. Of course, Assumption 5 does not work at all for a relatively small but very important new class of materials - systems providing two-photon (gated) hole-burning (Sect.7.3.3) [2.50,51].

The kinetics of hole-burning is governed then by the simple equation

$$\dot{\rho}(\omega,t) = -I_0\sigma\kappa(\nu_0-\omega)\eta\rho(\omega,t) \ , \tag{2.10}$$

which yields

$$\rho(\omega,t) = \rho(\omega,0)\ \exp[-I_0\sigma\kappa(\nu_0-\omega)\eta t] \ . \tag{2.11}$$

The spectra of the excitation $I_e(\nu,t)$, and the fluorescence (with broadband excitation) $I_f(\nu,t)$, depend on the burning time t and are given by

$$I_e(\nu,t) = I\sigma\eta_f \int_{-\infty}^{\infty} \kappa(\nu-\omega)\rho(\omega,t)d\omega \tag{2.12}$$

and

$$I_f(\nu,t) = I\sigma\eta_f \int_{-\infty}^{\infty} \phi(\nu-\omega)\rho(\omega,t)d\omega \ , \tag{2.13}$$

respectively, with I being the intensity of monochromatic excitation and I the spectral density of broad-band excitation. Here and in what follows the homogeneous spectra κ and ϕ are assumed to be normalized to unity.

For realistic shapes of the honogeneous spectra κ and ϕ, (2.12,13) cannot be solved analytically; below we shall present the results of numerical calculations. However, the situation at small burning times t $\ll$ $I_0\sigma\kappa_{max}\eta$ can easily be analysed. The difference spectra of excitation and fluorescence before and after hole-burning for a time t are given by

$$\Delta I_e(\nu,t) = I_e(\nu,0) - I_e(\nu,t) \simeq I_0I\sigma^2\eta\eta_f t \int_{-\infty}^{+\infty} \kappa(\nu-\omega)\kappa(\nu_0-\omega)\rho(\omega,0)d\omega \tag{2.14}$$

and

$$\Delta I_f(\nu,t) = I_f(\nu,0) - I_f(\nu,t) \simeq I_0I\sigma^2\eta\eta_f t \int_{-\infty}^{+\infty} \phi(\nu-\omega)\kappa(\nu_0-\omega)\rho(\omega,0)d\omega \ , \tag{2.15}$$

respectively.

Omitting the factor $I\sigma\eta t$, the right-hand side of (2.15) coincides with the formula for the selectively-excited fluorescence spectrum, as it is

38

observed in fluorescence line-narrowing experiments. The analysis given for FLN [2.52,53] applies here as well.

Both (2.14 and 15) contain contributions from burning via the PEL and phonon sidebands. However, because of the large difference in the PEL and phonon sideband peak intensities, phonon side-holes become observable only when no-phonon holes (especially the purely electronic one) are already essentially saturated. Note also that for an initial IDF $\rho(\omega,0)$ that is constant in the region of the hole, the excitation spectrum (2.14) is symmetric even for an asymmetric κ. Any information about the asymmetry of κ reveals itself only when the saturation effects become important.

For model calculations [2.16,29,38] homogeneous spectra are assumed to consist of narrow Lorentzian-shaped PELs κ_0, ϕ_0 and broad phonon-sideband profiles κ_p, ϕ_p of the form

$$\kappa_p(\nu-\omega) = \begin{cases} (1-\alpha)\Gamma_p^{-2}(\nu-\omega)\exp[-(\nu-\omega)/\Gamma_p] \ , & (\nu-\omega) > 0 \\ 0, & (\nu-\omega) < 0 \end{cases}$$

$$\phi_p(\nu-\omega) = \kappa_p(\omega-\nu), \tag{2.16}$$

where α is the Debye-Waller factor (DWF), and Γ_p characterizes the width of the phonon sideband. The evolution of the zero-phonon hole in the excitation spectrum is depicted in Fig.2.5a. In accordance with (2.14) the initial hole in the excitation spectrum also has a Lorentzian shape with width $\delta_0 = 2\Gamma$ (twice the homogeneous width of the absorption line; the

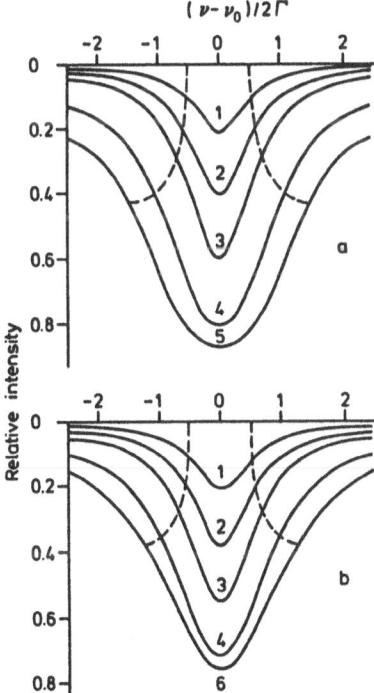

Fig.2.5a,b. Evolution of zero-phonon hole in the absence (a, W=∞) and presence (b, W=15) of reverse processes. Burning times t = 0.5(1), 1.2(2), 2.5(3), 9(4), 20(5) and ∞(6) in the units of $\pi\Gamma/2I_0\sigma\eta$

excitation linewidth is assumed to be zero), and the depth H is proportional to the burning time. In the process of burning a hole, broadening takes place gradually and the hole loses its Lorentzian shape. In the initial stage of burning, the ratio of the relative rates of broadening and deepening of the hole is 0.5 for a Lorentzian ZPL. The time dependence of the holewidth δ at small burning times is given by

$$\delta(t) \simeq 2\Gamma[1 + \dot{H}(0)t/2I_e^0] , \quad H(t) = \Delta I_e(\nu_0, t) , \qquad (2.17)$$

where I_e^0 is the ZPL contribution to the initial excitation spectrum. The evolution of the phonon side holes in the excitation (b) and flourescence (c) spectra is illustrated in Fig.2.6.

In several papers, the influence of the alteration of the various assumptions 1) - 7) was considered:

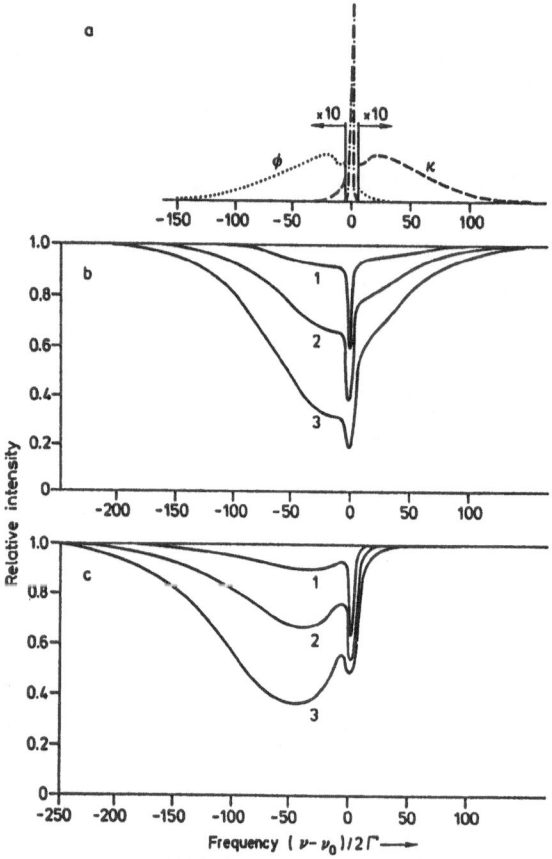

Fig.2.6. Evolution of the phonon sidehole in the spectra of excitation (b) and luminescence (c) for model shapes of homogeneous spectra (a). Debye-Waller factor $\alpha = 0.5$: phonon sideband width $\Gamma_\rho = 25\Gamma$. Burning times t = 10(1), 50(2), 150(3) in the units of $\pi\Gamma/2I_0\sigma\eta$

40

a) At higher excitation intensities, the population of metastable triplet states of molecular impurities and optical saturation of the $S_1 \leftarrow S_0$ transition become important. The influence of metastable states on spectral hole burning was studied in [2.38,40,46] and results mainly in broadening of the holes even at the very initial stage of burning.

b) To consider PSHB in samples with high impurity concentration it is necessary to take into account impurity-impurity interactions. The influence of energy transfer on selectively excited luminescence was considered in [2.54]. A similar analysis can be applied to the hole spectrum in the initial stage of burning as well. Another possible, but less apparent, effect of impurity-impurity interactions on PSHB was considered in [2.16]. The photoreaction of a group of impurities is effectively equivalent to the creation of new defects for the impurities involved earlier in the hole formation process. The new defects act as a new source of inhomogeneous broadening. Considering impurities as interacting elastic dipoles, an equation was derived to describe such a process

$$\frac{\partial \rho}{\partial t} = - \Delta BQ \frac{\partial \rho}{\partial \omega} - \frac{|\Delta A|}{\pi} Q \frac{\partial}{\partial \omega} P \int_{-\infty}^{+\infty} \frac{\rho(\omega')}{\omega - \omega'} \, d\omega' - K\rho \, , \qquad (2.18)$$

where

$$Q = \int_{-\infty}^{+\infty} K(\omega)\rho(\omega,t)d\omega \, ,$$

and

$$K(\omega) = I_0 \sigma\eta \cdot \kappa(\nu_0 - \omega) \, .$$

ΔA, ΔB in (2.18) are proportional to the change of the elastic momentum of an impurity in the photochemical reaction. The effect of such quasi-static impurity-impurity interactions is to increase the relative rate of hole broadening and to shift the hole maximum away from the burning frequency. In particular, when photoactive impurities themselves are the source of inhomogeneous broadening, (2.17) is to be modified to

$$\delta(t) \simeq 2\Gamma[1 + (1/2 + |\Delta A/A|)\dot{H}(0)t/I_e^0] \, . \qquad (2.19)$$

Such hole broadening effects may be considerable, if a large number of holes are burned at different frequencies (optical spectral memory).

c) Hole-burning in the spectra of optically dense samples under quasi-monochromatic excitation with spectral width $\Delta\omega$ was considered in [2.16]. It was demonstrated to result in the narrowing of holes in the transmission spectrum compared to the widths in optically thin samples. In [2.45,49], the case of monochromatic excitation was investigated.

d) There exists experimental evidence that not only the frequencies of purely electronic transitions but most of the parameters of homogeneous spectra (vibrational frequencies [2.55], linewidths [2.56], Debye–Waller factors [2.29], etc.) are affected, to some extent, by the matrix inhomogeneity. In general, only a partial correlation exists between the variations of different parameters [2.57]. To take this into account, a higher-dimensional IDF of actual parameters must be introduced. Formulae (2.12,14) would then include multiple integration over these parameters. For example, hole–burning kinetics with an inhomogeneous distribution of DWFs was considered in [2.29].

In [2.43,58], an inhomogeneous distribution of homogeneous widths uncorrelated with the frequency of purely electronic transitions was proposed. This leads to a peculiar hole shape that deviates from a Lorentzian even in the initial stage of burning. For the model distribution $\rho(\omega,\Gamma) = \rho(\omega)\rho_\Gamma(\omega)$, where

$$\rho_\Gamma(\Gamma) = \begin{cases} 1/(\Gamma_2 - \Gamma_1), & \Gamma_1 < \Gamma < \Gamma_2 \\ 0, & \Gamma < \Gamma_1, \ \Gamma > \Gamma_2 \end{cases} \tag{2.20}$$

the resulting hole shape at small burning times is

$$\Delta I_e \simeq \ln \frac{\Gamma_2^2 + (\nu-\nu_0)^2}{\Gamma_1^2 + (\nu-\nu_0)^2} \tag{2.21}$$

with the width $\delta = 2\sqrt{\Gamma_1\Gamma_2}$.

Concerning the formation of an antihole, it is also important to know how the transition frequencies of the initial form of the impurity and the photoproduct are correlated. A model for describing the hole–burning in a system with two metastable impurity configurations mutually transformable via photo- or dark processes was proposed in [2.38], where a phenomenological two-dimensional IDF, $\rho(\omega_1,\omega_2)$, was introduced (ω_1,ω_2 being the transition frequencies in the respective impurity configurations). In [2.59], the form of $\rho(\omega_1,\omega_2)$ was specified for axially symmetric impurities in a cubic lattice, where photoinduced transitions between different geometrically equivalent impurity orientations occur (e.g., O_2^--type molecular ions in alkali halides [2.60]). In Fig.2.7, the results of these calculations are presented, which demonstrate the existence of only a partial correlation between transition frequencies in different orientations of the impurity.

e) The kinetics of two-photon hole burning was theoretically considered in [2.42,47]. The overall burning efficiency in this case is proportional to the product of the light intensities applied to the first and second steps of excitation, but no major differences in hole shape and its evolution are expected for continuous-wave PSHB. However, interesting effects of hole-narrowing are to be expected for pulsed excitation [2.48].

f) Proper consideration of reverse processes [2.38] is essential when the photoproduct is thermally unstable at the burning temperature or when it

42

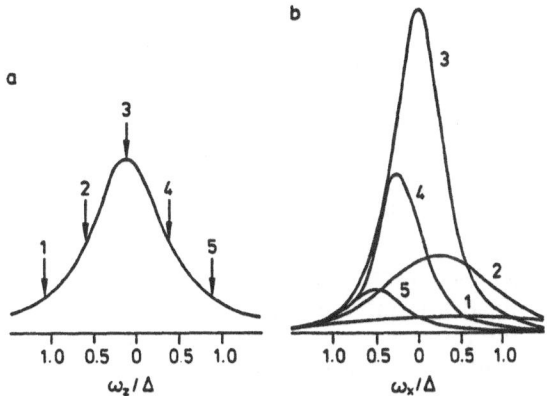

Fig.2.7a,b. Inhomogeneous distribution of the transition frequencies ω_g in an ensemble of axially symmetric impurities, fixed along a certain C_4-axis of a cubic crystal (a) and frequency distributions after a 90^0 reorientation for fixed values of the initial frequency (b). Curves *1-5* correspond to the arrows *1-5* in (a)

is photochemically unstable and the photochromic shift is insufficient to avoid the absorption of the burning light by the photoproduct. The latter includes the case of relatively strong phonon-sideband absorption.

The main effect of reverse reactions consists of saturation of the shape of the hole at large burning times (compare a and b in Fig.2.5). For the Lorentzian absorption line the stationary hole is also Lorentzian with the width

$$\delta(\infty) = \Gamma(1 + \sqrt{1+W}) , \qquad (2.22)$$

determined by the ratio W of the rates of selectively-excited (via ZPL) and non-selective (via phonon sideband and thermal) reverse processes.

g) The situations where the product and educt absorption bands strongly overlap were considered in [2.41,46]. In the case of weak correlation between educt and product transition energies the redistribution of impurities during the burning process does not noticeably affect narrow zero-phonon holes.

2.6 Spectroscopic Applications

2.6.1 Homogeneous Zero-Phonon Line Broadening and Dephasing in Crystals

As discussed above (Sect.2.2), homogeneous ZPLs (especially PELs) are very narrow at the liquid-helium temperature and therefore extremely sensitive to various impurity-matrix interactions. In particular, the measurements of the homogeneous broadening of ZPLs with increasing temperature give information about relaxation processes of electronic and

43

vibrational excitations in impurity molecules. In addition to radiative and non-radiative decay of the excited electronic level, homogeneous widths are controlled by pure dephasing of the excited electronic state and the decay and dephasing of the excited molecular vibrations.

To a reasonable approximation, the theoretical PEL shape is Lorentzian with the total homogeneous linewidth (HLW) Γ depending on the excited-state population decay time T_1, and on the pure dephasing time T_2^*, see (2.2). For a dipole-allowed purely electronic transition the depopulation time T_1 may be considered as temperature-independent at the liquid-helium temperature, whereas the dephasing time T_2^* is increasing with decreasing temperature due to the disappearance of thermally excited lattice modes. As the temperature approaches zero, Γ approaches its limiting value determined by T_1: $\Gamma(0) = 1/(2\pi c T_1)$. By studying the temperature dependent-component of the linewidth, $\Gamma(T)-\Gamma(0)$, one can study the pure dephasing processes. The low-temperature dephasing of the impurity electronic excitation in crystalline matrices is due to the Raman-like quadratic terms in the electron-phonon coupling with phonons or other matrix excitations, thus it is very sensitive to the density of low-frequency vibrational states of the matrix. Because of this the study of PEL broadening at low temperatures is of particular interest. Measurements of the temperature dependence of linewidth may serve as a useful probe of the low-frequency dynamics of matrices in the vicinity of the impurity center.

The most accurate measurements of PEL HLW have been made by burning narrow non-saturated zero-phonon holes in the inhomogeneous band of the corresponding transition. The PEL HLW low-temperature broadening process was first studied in 1977 [2.61-63] for the S_1-S_0 transition in H_2-tetra-4-tert-butylphthalocyanine molecules in tetradecane (H_2-Pc*-C_{14}) at temperatures of 1.8-30 K. A 692.947 nm Ne-discharge line and scanning with a Fabry-Perot interferometer were used to burn and measure holes, respectively, in the Shpol'skii component at 693.0 nm. The HLW was determined as the zero-burning-time limit of the holewidths δ_t produced at different irradiation times t. As follows from (2.17), after making a correction for the instrumental width, this method yields the limiting holewidth of $\delta_0 = 2\Gamma(T)$. Data for $\Gamma(T)$ are presented in Fig.2.8a. At 1.8 K the HLW was found to be $\Gamma = 0.002$ cm^{-1}. The life-time-limited value, calculated from the fluorescence decay time ($T_1 = 5.2$ ns), was $\Gamma(0) = 0.001$ cm^{-1}. The dephasing component at 1.8 K is then $1/(\pi c T_2^*) = 0.001$ cm^{-1} and the corresponding pure dephasing time, $T_2^* = 10$ ns. The pure dephasing broadening $\Gamma(T)-\Gamma(0)$ is shown in Fig.2.8b on a double logarithmic scale. Curve 1 represents the theoretical dephasing rate caused by Raman scattering of Debye phonons. The characteristic Debye temperature T_D is taken to be 50 K. Curve 1 fits the experimental data well not only at high temperatures $T > T_D$, where theory [2.64] tells us that a simple T^2 dependence should be expected, but the T^2 law approximates the experimental data fairly well also at lower temperatures 10 K$<T<T_D$. However, for $T < 10$ K the behavior of the experimental data is entirely different from that predicted considering only the Debye

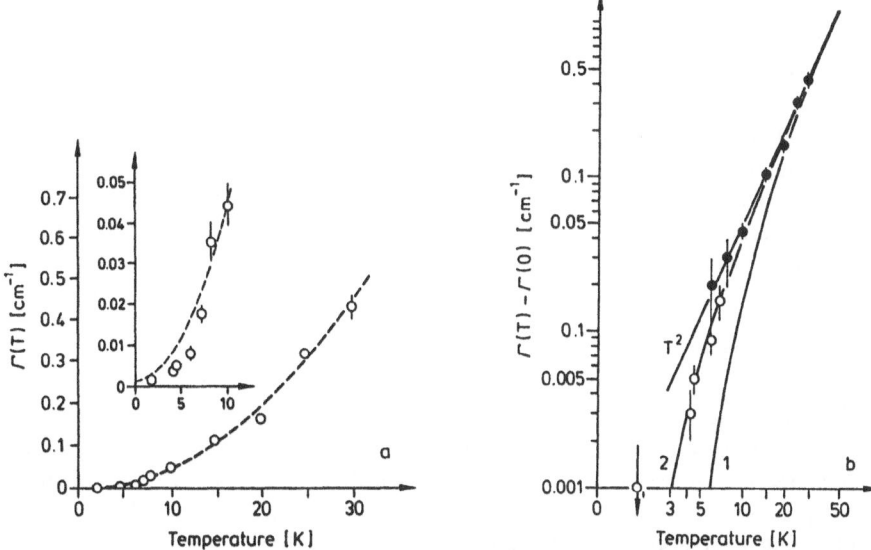

Fig.2.8a,b. Temperature dependence of the linewidth for the purely electronic S_1-S_0 transition in H_2Pc^*-C_{14} in an ordinary (a) and a double-logarithmic (b) plot. Curves *1* and *2* are calculated by using $\Gamma(T)-\Gamma(0) = 6.42(T/T_D)\int_0^{T/T_D} x^6 e^x(e^x-1)^{-1}dx$, $T_d = 50$ K for (*1*) and $\Gamma(T)-\Gamma(0) = 0.098n(\omega)[n(\omega)+1]$ with $\omega = 10$ cm^{-1} for (*2*)

phonons: instead of $\Gamma(T) \simeq T^7$ the experiment shows that $\Gamma(T)$ decreases exponentially when $T \rightarrow 0$. All experimental data in this regime can be described by curve 2 taking into account one kind of phonon only, presumably pseudolocalized phonons of frequency ω_0. For this single-mode model $\Gamma(T)$ is expressed as [2.65]

$$\Gamma(T) = bn(\omega_0)\,[n(\omega_0) + 1], \quad n(\omega_0) = 1/(e^{\hbar\omega_0/\kappa T} - 1)\,. \tag{2.23}$$

The fitting parameters for curve 2 are: the pseudo-local mode frequency $\omega_0 = 8$-12 cm^{-1} and the quadratic electron-phonon coupling constant $b = 0.06$-0.19 cm^{-1}. Therefore in this case, the perturbation of the lattice modes by the impurity molecule is strong enough to create a peak in the density of the low-frequency vibrational states at the impurity - the pseudolocal mode. This mode is responsible for the better part of the dephasing process at low temperatures. The calculated total HLW $\Gamma(T)$ is presented in Fig.2.8a (dashed line). Its extrapolated limit at $T = 0$ coincides, within the experimental accuracy, with the independently measured decay-limited $\Gamma(0)$. This was a result of basic importance for the theoretical understanding, because it proved that the PEL HLW is actually limited by the radiative lifetime at the lowest temperatures. At the same time, the resolution of PSHB spectroscopy approaches here the $\Gamma(0)$ limit. This result was confirmed later [2.66], with more accurate measurements, by using single-mode tunable cw dye laser with frequency stability better than 10 MHz.

45

In the time period since these early measurements, PEL broadening in crystals has been studied by PSHB [2.39,66,67] as well as by photon echo experiments [2.40,68-70] for a number of impurity molecules; the crucial role of pseudolocal modes in optical dephasing at low temperatures has been confirmed. For example, for the S_1-S_0 PEL of free-base porphin in n-decane, $\Gamma(T)$ was measured by PSHB at temperatures 1.2-4.2 K, which yielded an exponential dependence on temperature [2.67]. The data were analyzed assuming optical dephasing due to exchange interactions with a pseudolocal mode of frequency ω_0. In the slow exchange limit,

$$\Gamma(T) - \Gamma(0) = \frac{e^{-\hbar\omega_0/\kappa T}}{\pi\tau} , \qquad (2.24)$$

and the pseudolocal mode frequency was determined independently by measurements of phonon sidebands (ω_0 = 7.0 and 5.6 cm^{-1} in S_0 and S_1 states, respectively). The homogeneous width of the pseudolocal mode in the S_1 state was obtained by hole-burning in the phonon sideband, from which the corresponding lifetime was estimated to be τ = 115 ps. These parameters fit rather well the broadening caused by the exchange interaction. Recently, the data of [2.67] were reconsidered [2.71] on the basis of the newly developed nonperturbative theory of the ZPL width of impurities in crystals [2.3], but no satisfactory agreement with the theory was obtained.

The temperature-dependent optical dephasing (measured by picosecond photon echoes) for tetracene and pentacene monomer and dimer molecules in a p-terphenyl crystal shows exponential dependence over many orders of magnitude of T_2 [2.69]. The activation energies (pseudolocal mode frequencies) obtained from the exponential fit varied (from 10 to 30 cm^{-1}) for different impurity molecules in the same crystal host, excluding the coupling to regular acoustic host phonons as a relevant dephasing process. An exponential dependence of dephasing at low temperatures with an activation energy of 13.8 cm^{-1} has also been detected (using the photon echo) for a spin-forbidden T_1-S_0 transition of dibrombenzophenone in dibromdiphenyl ether [2.70]. In this case, the situation is more complicated because the molecules excited to T_{1x}, T_{1y}, and T_{1z} states all participate in the echo signal.

These experiments have underscored the dominant role of pseudolocal vibrations for the molecular impurity dephasing in crystal hosts at low temperatures. It seems reasonable that a well-pronounced pseudolocal peak will arise in the phonon density of states when a large molecule is imbedded in the crystal and strong perturbations are created. In the frequency region of such local modes, the acoustic phonon density of states is relatively low, and therefore at temperatures $\kappa T < \hbar\omega_0$ the pseudolocal mode predominates over acoustic phonons in dephasing processes. At higher temperatures acoustic phonons will dominate because the higher-frequency phonons with higher density of states become thermally populated. However, experimental studies of the HLW are difficult at temperatures above 40 K because of the broadening and damping (decreasing Debye-Waller factors with increasing temperature) of the PELs. From this point of view the observations of homogeneous broadening of purely vibrational transitions are important because sometimes they allow one to follow $\Gamma(T)$ up to higher temperatures. Interesting results were obtained by burning holes in the infrared absorption band of the molecular vibration ν_3 of the ReO_4^- impurity molecule in alkali halide crystals (see also Chap.6 for details) [2.72,73]. The temperature dependence of the HLW, measured over 4 orders of magnitude using transient and persistent hole-burning, indicates the dominant role of acoustic lattice phonons in the dephasing of the vibrational excitation in the ground electronic state. At T < 10 K the HLW was independent of temperature, showing the limiting value determined by the decay time T_1 of the vibrational excited state. In this simple case, the molecule is in a site of cubic symmetry in the crystal and the formation of persistent holes is probably caused by the reorientation of the molecule during vibrational de-excitation [2.73] (see also Chap.6).

2.6.2 Photochemical Hole-Burning in Glassy Matrices

A different situation is encountered when the host is a glass or polymer rather than a crystal or polycrystal. In this case, it is reasonable to expect specific effects which make both the measurement of true homogeneous linewidths and the approach to lifetime-limited width more difficult. First, the presence of two-level systems (TLS) [2.74a] provides a high density of very low-frequency lattice states, which cause pure dephasing broadening of homogeneous lines at considerably lower temperatures than in the case of crystals with pseudolocal modes. Actually, the manifestation of a TLS is to some extent analogous to that of a pseudolocal mode. Essential differences arise from the extremely wide distribution of TLSs down to very low energies, and, on the other hand, from different distances between an impurity and its closest TLS neighbors. Second, the high-entropy, relatively soft structure of glasses provides possibilities for very low excitation energy rearrangements of the structure, leading to diffusion and spectral diffusion [2.32,105]. The broadening of spectral holes by the latter is stronger for larger values of τ_{HB} - the characteristic hole-burning plus measurement time.

PSHB spectroscopy is now an effective tool for studying the structure of glasses and dynamic processes in them at low temperatures (Chaps.3 and 5). In this subsection, we shall refer only to the results obtained in our laboratory on glassy systems in which molecules showing photo-chemical hole-burning were used.

The PSHB experiments immediately reveal essential differences between the widths and temperature broadening of holes in crystalline versus glassy matrices [2.31,58,63]. Figure 2.9 shows the data obtained for the S_1-S_0 electronic transition of the H_2-Pc^* molecule in isopropylene-ether glass (GL) and in a polycrystalline matrix of tetradecane (C_{14}). Two

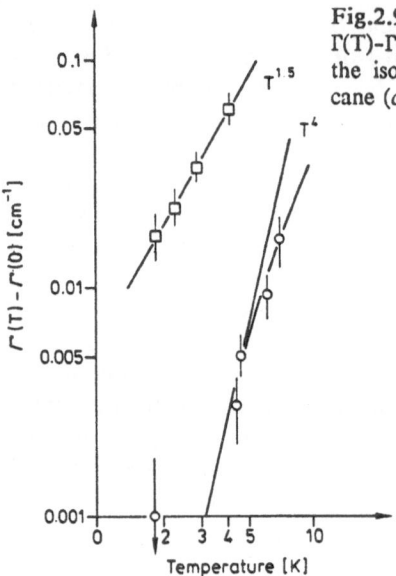

Fig.2.9. Temperature broadening of the linewidth $\Gamma(T)-\Gamma(0)$ for the purely electronic S_1-S_0 transition in the isopropylene-ether glass (*squares*) and in tetradecane (*circles*)

differences between glassy and crystalline hosts were observed. First, the holewidths are strongly dependent on the matrix: holes are the narrowest in C_{14} (a Shpol'skii system), the largest in GL, and intermediate in n-nonane (C_9) where the inhomogeneous distribution of the transition energies is about as large as in GL. The hole halfwidths at 1.8 K were 0.017, 0.0035 and 0.002 cm^{-1} in GL, C_9, and C_{14}, respectively. Second, the temperature dependence of the holewidth in the glass is distinctly different from that in the polycrystals. A rather slow broadening which follows the law $T^{1.5\pm0.3}$ between temperatures 1.8 and 4.2 K was observed. The difference between the holewidth and shape for polycrystalline and GL matrices is shown in Fig.2.10 for the molecule H_2-tetra(tert-butyl) porphirazine (TAP*). The hole in the S_1-S_0 transition at 1.8 K is Lorentzian

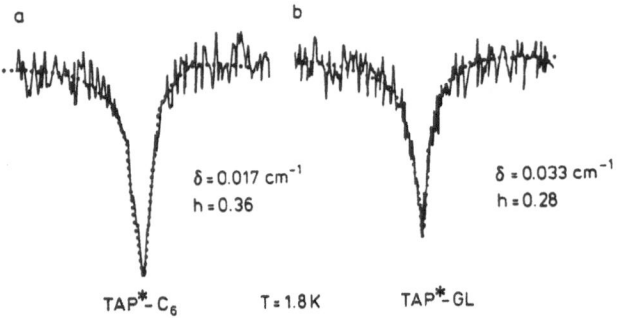

Fig.2.10a,b. Comparison of holes in the excitation spectra of a TAP* molecule at 1.8 K in matrices of polycrystalline n-hexane (a) and glass (b). Burning wavelengths are 619.7 nm (a) and 618.3 nm (b); burning intensity and time are $2 \cdot 10^{-6}$ W·cm^{-2} and 0.3 s, respectively. The dots represent Lorentzian approximations

48

for the n-hexane matrix (δ_H = 0.017 cm^{-1} and the corresponding Γ = 0.005 cm^{-1}). In the case of GL, the holewidth is two times larger, and the hole contour in the wings clearly shows a rather slow decrease in comparison with a Lorentzian profile. The peculiar hole shape in a GL matrix is in accordance with model calculations (Sect.2.5) [2.58] where a non-correlated inhomogeneous distribution of HLW, Γ, is taken into account (in this case Γ varies between 0.005 and 0.05 cm^{-1} [2.58]).

These features are obviously connected with the specific structure and dynamics of glassy matrices. No changes in the holewidth were noticed in these first experiments when the time of the burning-recording cycle τ_{HB} was varied from 10^2 to 10^{-2} s. Because of the softness of the structure some rearrangements in the impurity surroundings are to be expected, which produces spectral diffusion [2.75,76], i.e., additional broadening of the hole during the time of observation.

Actually there were two questions to answer in the PHSB studies of glassy systems: does the measured holewidth δ_H in glass correspond to the HLW Γ in the same way as in crystals, or is there any additional broadening? Further, how does the hole broadening depend on the time of the burning-recording cycle and on the temperature?

To address the second question, a method for fast profile scanning by using the Doppler effect was developed (Sect.2.7.1) [2.77] and applied to shorten τ_{HB} down to microseconds [2.78]. For these measurements, metastable (MS) holes were utilized. MS holes are created in the IDF of a S_1-S_0 transition when the electronic excitation inter-system crosses from the S_1 state to the metastable triplet state T_1 [2.78-80]. The MS holes disappear when the T_1-state population returns to the S_0 state, i.e. during the T_1-state lifetime. MS holes are not fixed by long-lived changes in the positions of the atoms, and actually they are not persistent holes. A small fraction of T_1 states undergo photochemistry and so a persistent photochemical hole can also be formed under S_1 excitation.

Holes measured with the Doppler spectrometer are shown in Fig.2.11 for the S_1-S_0 transition of the impurity molecule II_2-octaethylporphin in

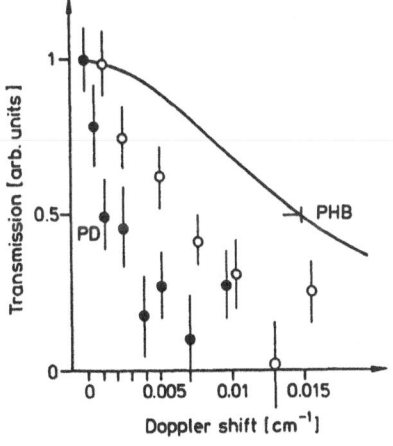

Fig.2.11. Decrease of the holewidth δ_H in the transmission spectrum of an OEP molecule in PS at 1.45 K with the shortening of (burning + measuring) cycle time τ_{HB} by measurements with the Doppler spectrometer. δ_H = 0.03 cm^{-1} at τ_{HB} = 3 s (solid line); δ_H = 0.004 cm^{-1} at τ_{HB} = 10^{-4} s (open circles); and δ_H = 0.0014 cm^{-1} at τ_{HB} = 10^{-5} s (filled circles)

polystyrene (OEP-PS) at 1.45 K. The sample was prepared by polymerizing a 10^{-4} M/ℓ solution of OEP in styrene. The experimental points correspond to the sample transmission measured at different frequencies corresponding to the different Doppler velocities. The MS holewidths δ_H = 0.004 and 0.0014 cm^{-1} were obtained for pulse durations τ = 10^{-4} and 10^{-5} s, respectively. The persistent hole measured under the same conditions but at τ_{HB} = 3 s was much broader, δ_H = 0.03 cm^{-1}, while the radiative linewidth was $\Gamma(0)$ = 0.0002 cm^{-1}. Thus, in this case (at 1.45 K) the holewidth in glass depends on the hole-burning and the measurement time τ_{HB}, and decreases considerably with decreasing τ_{HB} into the microsecond time domain.

Returning to persistent hole-burning the first measurements at sub-Kelvin temperatures (down to 0.3 K) by *Thijssen* et al. [2.81] showed that the holewidths (and the corresponding HLW derived as $2\Gamma = \delta_H$) follow a $T^{1.3}$ temperature dependence for a number of impurities in glassy hosts. Recently, spectral holes were studied down to 0.05 K by *Gorokhovskii* et al. [2.82]. The holewidths versus temperature display two crossovers, from one power law to another. The persistent holes were produced in the S_1-S_0 absorption (transmission) of an OEP-PS sample in the temperature region of 1.5–0.05 K. The sample was placed in a superfluid ^{3}He-^{4}He solution whose temperature was measured with the 10% accuracy below 0.1K and with 1% accuracy at temperatures above 0.1 K. A laser beam from a single-mode CR-699-21 laser with an intensity $5 \cdot 10^{-7}$ W·cm^{-2} was directed to penetrate the sample twice. No hole saturation effects were observed at the hole-burning times used (2–10 s). The holes at 0.05 K are

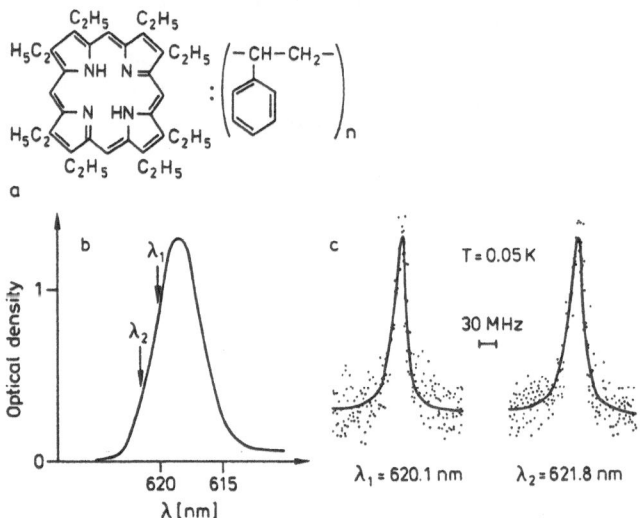

Fig.2.12a,b. Spectral holes in S_1-S_0 transition of OEP molecules in PS burned and measured at 0.05 K. (a) The structure of the molecules involved; (b) the position of burning wavelengths, λ_1 and λ_2 in the inhomogeneous 0-0 absorption band; (c) spectral holes, the solid lines represent Lorentzians of widths (FWHM) δ_1 = 26 MHz and δ_2 = 34MHz

shown in Fig.2.12. The holes were burned in two spectral regions, at λ_1 = 620.1 nm and λ_2 = 621.8 nm on the long-wavelength side of the inhomogeneous absorption band (Fig.2.12b) in an interval about 2 cm^{-1} wide around both wavelengths for all the measurement temperatures. The holes at 0.05 K for the two spectral regions (Fig.2.12c) were Lorentzian in shape with different hole widths: δ_1 = 26±1.5 MHz at λ_1 and δ_2 = 34±1.5 MHz at λ_2, both being considerably broader than the radiative-lifetime-limited value $\delta(0) = 2\Gamma(0) = 1/\pi T_1$ = 10.8 MHz. (The S_1-state lifetime T_1 = 29.5±1 ns was measured from the fluorescence decay at 4.2 K).

The temperature dependence of the holewidths $[\delta(T)-\delta(0)]$ for holes burned at λ_2 is shown in Fig.2.13. The experimental dots form a rather complicated curve with two turning points, at T = 0.09 and 0.36 K. At higher temperatures, 0.36 K<T<1.5 K, the hole broadening may be approximated with the power dependence $T^{1.25\pm0.14}$, which is close to the $T^{1.3}$ dependence observed by *Thijssen* et al. [2.81]. A crossover to the law $T^{0.84\pm0.06}$ ocurred in the region of 0.09 K < T < 0.36 K (the law $T^{1.0\pm0.1}$ was also observed below about 1K in [2.81]). At still lower temperatures, T < 0.09 K, another crossover to a rather fast T dependence takes place with the law $T^{1.52\pm0.24}$. The extrapolation of this dependence indicates that the $2\Gamma(0)$ limit will be reached only at temperatures below 0.01 K. One must conclude that a complicated body of very low-frequency excitations of the matrix will be frozen out only at these very low temperatures.

There seems to be no theory for the temperature dependence of the holewidth that is able to interpret a complicated experimental curve $\delta_H(T)$ with two crossover points. We must not forget that the holewidth in

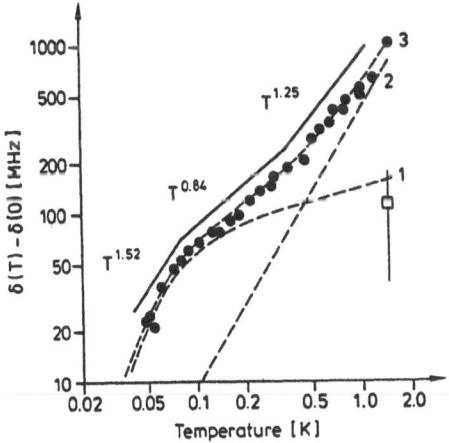

Fig.2.13. Temperature broadening of spectral holes in the region 0.05-1.5 K for OEP-PS for burning at λ_2 (Fig.2.12) shown in log-log coordinates. $\delta(0) = 2\Gamma(0) = 11$ MHz (T_1 = 30 ns). Curve *1* represents the homogeneous part with the law $2\Gamma(T) = A\exp(\hbar\omega_0/\kappa T)[\exp(\hbar\omega_0/\kappa T)+1]^{-1.75} \cdot [\exp(\hbar\omega_0/\kappa T)-1]^{-0.25}$; $\hbar\omega_0/\kappa$ = 0.12 K and A = 286.9 MHz. Curve *2* represents the spectral diffusion contribution with the law $\delta_D = BT^\alpha$, α =1.66 and B = 0.027 MHz. Curve *3* is the sum of curves *1* and *2*. The square corresponds to the holewidth measured with Doppler-spectrometer at 1.45 K and τ_{HB} = 10^{-5} s (Fig.2.11)

glasses contains some inhomogeneous part which was shown to depend on the time τ_{HB} spent on the burning and detecting of the hole, i.e. spectral diffusion is present (Chap.3). The full holewidth may be given as follows

$$\delta_H(T) = 2\Gamma(T) + \delta_D(T, \tau_{HB}) , \tag{2.25}$$

where δ_D is the contribution by the spectral diffusion.

The TLS may take part in both processes of hole broadening: in dephasing [2.83] and in spectral diffusion [2.76]. To fit the experimental data *Gorokhovskii* et al. [2.82] have used the form for $2\Gamma(T)$ proposed by *Lyo* [2.85], see also [2.106], for the dipole-quadrupole interaction between impurity and TLS in the form

$$2\Gamma(T) = A e^{\hbar\omega_0/\kappa T} (e^{\hbar\omega_0/\kappa T} + 1)^{-1.75} (e^{\hbar\omega_0/\kappa T} - 1)^{-0.25} , \tag{2.26}$$

and the form for $\delta_D(T, \tau_{HB}) \simeq T^\alpha$ proposed in [2.76,86]. Curve 1 in Fig.2.13 represents the homogeneous part of $\delta_H(t)$ with the pseudolocal mode frequency $\omega_0 = 0.08$ cm^{-1}. Curve 2 exhibits the contribution of spectral diffusion with the exponent $\alpha = 1.66$. The spectral diffusion contribution becomes negligible below 0.1 K. It is interesting to compare curve 1 with the holewidth measured with the Doppler spectrometer. The experimental point at T = 1.45 K with the measurement time $\tau_{HB} = 10^{-5}$ s is shown in Fig.2.13; and it is in agreement with the homogeneous holewidth predicted by curve 1.

From the experimental data available for spectral hole broadening in glasses at low temperatures one can conclude that holewidths measured slowly (i.e., at $\tau_{HB} \gg T_1$) are affected by a substantial inhomogeneous broadening, caused by a diffusion-like process of rearrangement in the matrix resulting in spectral diffusion. Diffusional broadening of the hole was also observed when holes were burned in a T_1-T_0 transition (pyrene in butylbromide [2.105]). The inhomogeneous contribution may be decreased by shortening the measurement time τ_{HB}. The processes of spectral diffusion become inactive at very low temperatures (for OEP-PS – below 0.1 K), where inhomogeneous broadening of the holewidth caused by spectral diffusion becomes negligible. Therefore, it seems reasonable to suppose that a cross-over from the law $\delta_H(T) \simeq T^{1.3}$ to an approximately linear T dependence takes place because of a competition between inhomogeneous and homogeneous effects on the holewidth.

The behavior of $\delta_H(T)$ at very low temperatures (at T < 0.1 K for OEP-PS) is due to the homogeneous broadening $\Gamma(T)$ of the purely electronic zero-phonon line of the impurity molecule. Holewidths close to the electronic lifetime-limited value $\Gamma(0) = (2\pi c T_1)^{-1}$ are possible at still lower temperatures (for OEP-PS at T < 0.01 K). The difference $\delta_H(T)-2\Gamma(0)$ at higher temperatures is due to dephasing processes. For OEP-PS, $\delta_H(0.05) \simeq 4\Gamma(0)$.

The results for OEP-PS also show that dephasing in glasses at very low temperatures is sensitive to the details of the local structure of the matrix in the vicinity of impurity. The same microscopic processes are, to some extent, responsible for the position of ZPLs in the inhomogeneously broadened band and, therefore, a correlation arises between the parameters of $\Gamma(T)$ and the hole-burning frequency. On the other hand, at some higher temperature the $\Gamma(T)$ dependence on the hole-burning frequency vanishes (for OEP-PS the curve $\delta_1(T)$ and $\delta_2(T)$ merge at $T > 1$ K). This indicates that the processes causing spectral diffusion are relatively insensitive to the local structure of the impurity centers.

2.6.3 Homogeneous Linewidths of Vibronic Transitions and Relaxation

Hole-burning measurements on vibronic (electron-vibrational) transitions in molecular impurity absorptions enable determination of their homogeneous linewidths and thus extraction of information about the lifetimes and relaxation mechanisms of the vibrational levels of the excited electronic state.

As an example, we consider here the relaxation of the vibrational state 515 cm^{-1} above the S_1-S_0 origin in the molecule H_2Pc^* in n-nonane (C_9) at temperatures 1.8 to 60 K [2.84]. The C_9 matrix provides a large inhomogeneous broadening ($\Delta \simeq 150$ cm^{-1}), which is necessary to observe the relatively broad holes characteristic of vibronic transitions with homogeneous widths in excess of 0.1 cm^{-1}. The temperature broadening $\Gamma_v(T)$ of the vibronic ZPL is shown in Fig.2.14. The experimental data fit the

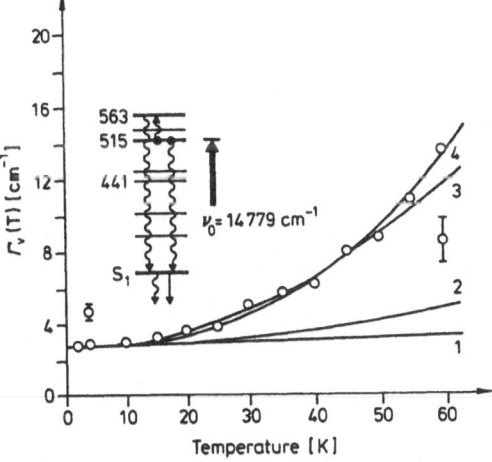

Fig.2.14. Temperature dependence of HLW Γ_v of the vibronic transition $\nu_{00}+515$ cm^{-1} for H_2Pc^* molecules in n-nonane. Insert shows the scheme of actual vibronic levels and the channels of anharmonic decay. The experimental data are shown by empty circles. Curves 1-4 are calculated by using:
(1) $\Gamma_v(T) = 2.8[n(\omega_1)+1]$, $\omega_1 = 74$ cm^{-1};
(2) $\Gamma_v(T) = 2.8[n(\omega_1)+1]+2.8n(\omega_2)$, $\omega_2 = 48$ cm^{-1}
(3) $\Gamma_v(T) = 2.8[n(\omega_1)+1]+2.8n(\omega_2)+1.5n(\omega_3)\cdot[n(\omega_3)+1]$, $\omega_3 = 20$ cm^{-1}
(4) $\Gamma_v(T) = 2.8+0.56\cdot10^{-3}\cdot T^{2.4}$

53

approximation $\Gamma_\upsilon(T) = \Gamma_\upsilon(0)+\alpha T^{2.4\pm0.2}$ well. An extrapolation to zero temperature gives $\Gamma_\upsilon(0) = 2.8$ cm^{-1}, yielding the lifetime of the 515 cm^{-1} molecular vibration, $T_1 = 1.9$ ps. It is natural to suppose that the decay of an internal vibration in an impurity molecule is caused by anharmonic interactions with crystal phonons and other molecular vibrations.

The structure of the lower vibronic levels of the H_2Pc^* molecule (see Fig.2.14, insert) allows one to discuss possible schemes for energy relaxation. There are two channels of anharmonic decay which involve only one crystal phonon: (a) the transition to the nearby 441 cm^{-1} level with the creation of a 74 cm^{-1} phonon, and (b) the transition to the higher 563 cm^{-1} level with the annihilation of a 48 cm^{-1} phonon. The calculated temperature dependence of the respective decay rates (probabilities) is shown in Fig.2.14, where curve 1 corresponds to the decay channel (a) and curve 2 results from the contributions of both channels, (a and b), assuming equal anharmonic coupling constants with the 441 and 563 cm^{-1} vibrations. Both curves demonstrate a weak dependence of the decay probability on temperature for T < 60 K owing to the high frequency of the phonons involved. It is obvious that, in addition to (a) and (b), a dephasing process which is of the same order of perturbation as the two-phonon decay should also be taken into account. Since there are many possibile combinations of phonons and local vibrations that can produce decay, we consider the broadening by two-phonon Raman scattering only. Curve 3 summarizes the broadening of the vibronic transition by the anharmonic decay processes (a and b) and by single-mode dephasing due to the scattering of a pseudolocal vibration of frequency 20 cm^{-1}. Curve 3 gives a rather good approximation for the HLW measured at T<50 K. The deviations at T>50 K indicate that other crystalline and molecular vibrations should be included.

The homogeneous linewidths of 14 vibronic transitions up to 1600 cm^{-1} above the S_1-S_0 origin have been measured by PSHB at 4.2 K for free-base porphin (H_2P) in n-octane [2.87]. The widths of individual vibrational states were found to vary strongly, from 4 to 180 GHz, whereas the inhomogeneous linewidth varied only by a factor of 3. These measurements were performed on vibronic lines associated with the A site of the Shpol'skii component for which the corresponding 0-0 line was shown to be only weakly temperature broadened between 1.6 and 4.2 K [2.39], indicating a small dephasing effect on Γ_υ. It is reasonable to relate the differences in the vibronic linewidths to the different relaxation times T_1 of the various vibrational levels. However, no correlation between Γ_υ and the energy of the vibronic level was observed.

For H_2P in different n-alkanes it was also shown [2.90] that vibronic HLWs and relaxation times depend strongly on the nature of the site in which the molecule is incorporated. Relaxation is faster in the A site than in the B site, which was interpreted as the result of a tighter fit of the H_2P molecule in the A site as opposed to the B site.

2.6.4 Off-Resonance Hole-Burning and Non-Correlation Effects

In this subsection we shall discuss holes which appear in the IDF and in the inhomogeneous vibronic bands far from resonance with the laser excitation frequency (off-resonance holes). As was mentioned in Sect.2.2, the selective bleaching during PSHB actually makes a hole in the inhomogeneous distribution function (IDF) of the purely electronic transition (or in several IDFs if the inhomogeneous bands of different purely electronic transitions are overlapping).

In the simplest case, when hole-burning takes place at the laser frequency ω_L in the single IDF of a S_1-S_0 purely electronic transition, a sequence of hole-replicas can also appear at the frequencies of all vibronic transitions of the bleached molecules. The hole-burning absorption spectra reveal a vibronic structure in the excited electronic state of the impurity molecules that is otherwise hidden under the inhomogeneously broadened structureless absorption profile (Chap.5).

The hole-burning absorption spectra were measured and the energies of vibronic transitions were obtained for many aromatic molecules in glasses [2.89] by using a two-channel apparatus developed to detect changes in the optical density of the order of 10^{-4}. Figure 2.15 represents two examples of hole-burning absorption spectra where fine vibronic structures with many vibronic zero-phonon transitions are revealed as a set of off-resonant holes after single-frequency hole-burning in the inhomogeneous 0-0 absorption band.

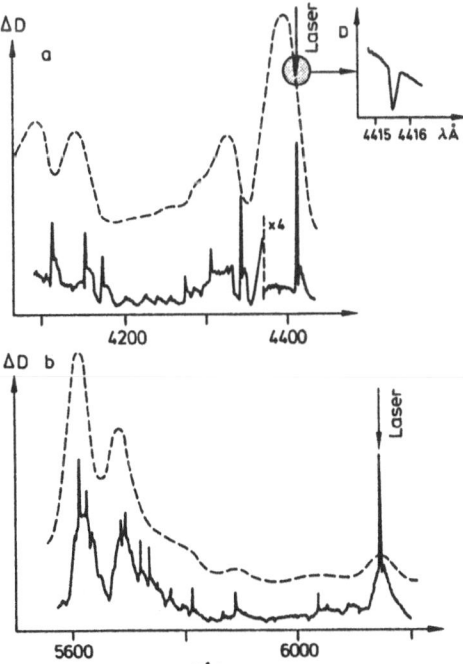

Fig.2.15a,b. Hole-burning spectra of off-resonant holes at 4.2 K for molecules of perylene in ethanol (a) and porphin in polystyrene (b). Insert shows the resonant hole in the 0-0 absorption band. Dashed lines represent the absorption spectra of the samples before hole-burning [2.92]

The properties of vibrational off-resonant zero-phonon holes in the hole-burning absorption spectrum are similar to those of the zero-phonon lines in the spectrum of selectively excited luminescence; thus, off-resonant holes have most of the advantages and complications of site-selective spectroscopy (Sect.2.3). Like the multiplets of ZPLs in [2.25], the complex multiplets of holes may occur in the region of the 0-0 transition when the hole-burning is performed in the region of vibronic transitions with considerable inhomogeneous broadening. The appearance of multiplets of holes was first observed in absorption by *Kharlamov* et al. [2.89].

The hole-replicas are usually broader than the resonant hole as a result of the lack of absolute correlation between different transitions in the impurity molecules selected by the laser excitation [2.38,57]. A comparison of the widths of resonant holes with those of hole replicas has revealed, in addition to the inhomogeneous distribution of purely electronic frequencies, the inhomogeneous distribution of molecular vibrational frequencies in the spectra of impurity molecules. We discuss hole multiplets and non-correlation effects, following [2.93] for the molecule of H_2-P_c^* in C_9 at 4.2 K (Fig.2.16). For this system the 0-0 absorption band has an inhomogeneous bandwidth $\Delta = 170$ cm^{-1}, and the first vibronic band actually consists of two overlapping vibronic bands for the transitions $\nu_{00}+515$ cm^{-1} and $\nu_{00} + 441$ cm^{-1}, as shown in Fig.2.16a. Laser hole-burning at

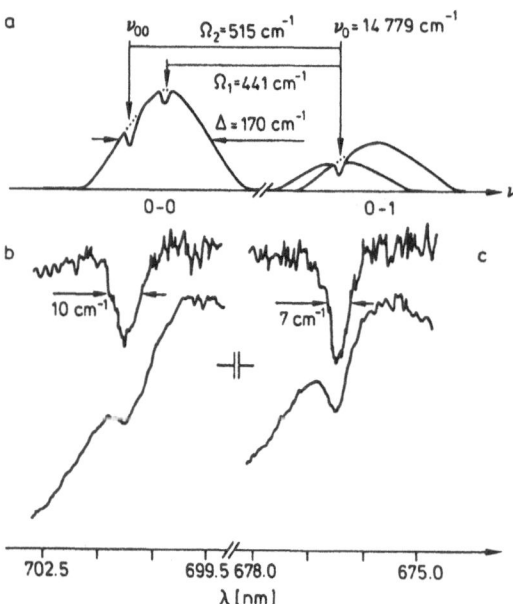

Fig.2.16a–c. Off-resonant holes in the excitation spectra of H_2Pc^* in n-nonane at 5 K. (a) Schematic inhomogeneous absorption bands of 0-0 and two overlapping vibronic 0-1 transitions with one resonance hole at ν_0 and two hole replicas in 0-0 band. (b) Off-resonant hole due to the PELs at the frequency $\nu_{00} = \nu_0 - \Omega$, $\Omega = 515$ cm^{-1}. (c) Resonant hole due to the vibronic ZPLs. Luminescence is recorded at 736.5 nm (b) and 700.9 nm (c)

the frequency ω_L (ν_0 in the figure) creates two replica holes in the 0-0 band, because two types of bleached molecules are available whose burning was performed via the two different vibronic transitions. An additional inhomogeneous distribution of the vibrational frequency Ω_i follows from the fact that the width of the replica hole was found to be broader than that of resonant hole (Fig.2.16 b and c). The non-saturated resonant holewidth in the vibronic transition is $\delta_\nu = 2\Gamma_\nu$; assuming a Lorentzian shape and width K for the vibrational frequency distribution, the replica's holewidth in the electronic origin transition may be obtained as $\delta_0' = \Gamma_0 + \Gamma_\nu + K$. The holewidths δ_0' and δ_ν were found to be 10 and 7 cm^{-1}, respectively. Neglecting the small value of Γ_0 and taking into account the instrumental width of 1 cm^{-1}, one obtains $\Gamma_\nu = 3$ cm^{-1} and K = 5.6±0.5 cm^{-1}. When the resonant hole was burned in the 0-0 band, the hole-replica in the vibronic transition was obtained with the width $\delta_\nu' = 10$ cm^{-1}. In this case, $\delta_\nu' = \Gamma_0 + \Gamma_\nu + K'$, where K' is the spread of vibrational frequencies for molecules with fixed frequency ν_{00}. For this system K' = 6.3±1 cm^{-1}.

Thus, the non-correlation effect between the origin and vibronic transitions may be schematically represented, as shown in Fig.2.17a. Monochromatic excitation in the 0-0 band fixes the energy difference between the adiabatic potentials of the ground and excited states of the impurity but does not fix their curvatures. It is interesting to compare the variation of molecular vibrational frequencies for molecules at a fixed burning frequency with that for different burning frequencies in the inhomogeneous line. The latter value $\Delta\Omega$ was obtained from the positions of the vibronic

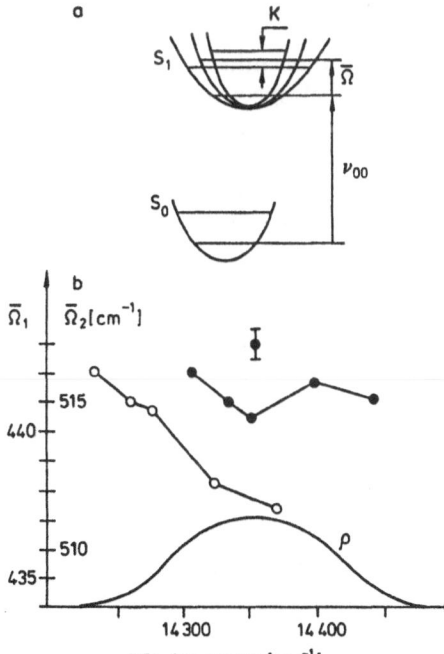

Fig.2.17. (a) Origin of vibrational inhomogeneity: the fixing of the PEL frequency ν_{00} does not fix the intermolecular vibrational frequency Ω. (b) The variation of the mean vibrational frequencies $\bar{\Omega}_1 = 440$ cm^{-1} (*filled circles*) and $\bar{\Omega}_2 = 515$ cm^{-1} (*open circles*) for molecules of $H_2P_c^*$ in C_9 at 5 K having different PEL frequencies fixed inside the IDF [2.93]

57

ZPL in luminescence measured under selective excitation within the inhomogeneous band. The results of these experiments for molecules of $H_2P_c^*$ in C_9 are depicted in Fig.2.17b. For the 515 cm^{-1} vibration a monotonic decrease of Ω with the increase of ν_{00} was observed over a range $\Delta\Omega = 5$ cm^{-1}, i.e. of the order of 7 K.

For the 515 cm^{-1} vibration the relative magnitude of both the inhomogeneous vibrational dispersion K/Ω and the vibrational frequency variation $\Delta\Omega/\Omega$ are close to the ratio of the vibrational and electronic frequencies: $K/\Omega \simeq \Delta\Omega/\Omega \simeq \Omega/\nu_{00} \simeq 1\%$. This indicates that vibrations may be as sensitive to the inhomogeneity of matrix as the electronic transition is (for more information on this, see Chap.6). For the 441 cm^{-1} vibration the variation of Ω within the inhomogeneous band is smaller ($\Delta\Omega = 1$ cm^{-1}), which indicates a different influence of matrix imperfections on different impurity vibrations.

A further study of correlation effects on off-resonant holes was performed by *Friedrich* et al. [2.92]. The shape of the hole replica in the 0-0 transition initiated by hole-burning in a vibronic transition was calculated taking into account two correlated Gaussian inhomogeneous distributions, one for the PELs and one for the vibronic ZPLs. For quinizarin molecules in an ethanol-methanol glass this correlation was shown to be rather high and the hole replicas observed were rather narrow. However, even in the case of full correlation, the replica's hole-width in the 0-0 transition cannot be narrower than the linewidth Γ_v of the corresponding vibronic transition. This fact and the lack of absolute correlation limit the spectral resolution possible in off-resonant hole measurements.

2.6.5 Hole-Burning in the Spectra of Chlorophyll-like Molecules

The chlorophyll-like molecules important in photosynthesis and other biological processes have broad structureless optical spectra even when isolated in low-temperature solid matrices, if the spectra are acquired with methods of conventional spectroscopy. Quite narrow vibronic structure was revealed in the luminescence and excitation spectra when the site-selection technique was applied to these systems, yielding vibrational frequencies in S_0 and S_1 states for chlorophyll *a* and some of its derivatives in various matrices (Chap.5) [2.94-97,107]. Hole-burning methods have provided two additional pieces of information about the mechanisms of photoinduced transformations in the chlorophyll-like molecules and about the relaxation rates of the vibronic states.

The first (rather broad) holes were burned in the S_1-S_0 band of chlorophyll *a* (Chl*a*) and protochlorophyll (PChl) in ether glass at 4.2 K by *Avarmaa* et al. [2.98]. The hole lifetime and dark-recovery kinetics correspond to the triplet-state decay parameters ($\tau_T = 5$ ms) that indicated an intramolecular metastable (MS) mechanism of hole formation (Sect.2.4).

Besides the MS transient hole, an additional permanent burning process was observed for Chl*a* and PChl, which gave a long time constant of tens of seconds in the hole recovery kinetics. A feature typical of NPHB - the

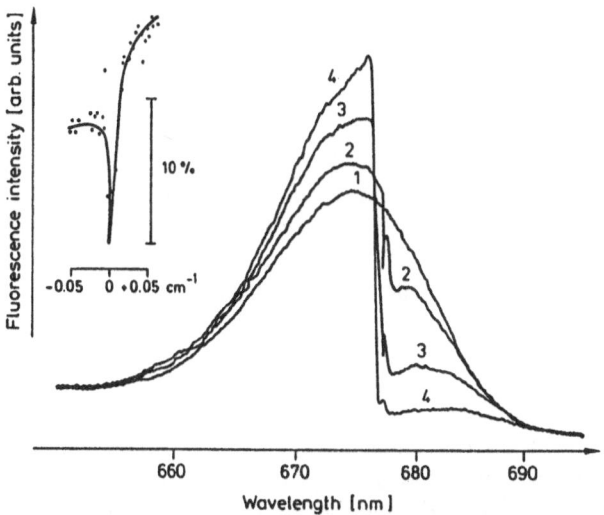

Fig.2.18. Hole-burning in the excitation spectrum of Chlorophyll a in ether-butanol at different doses before (*1*) and after irradiation for 0.5 min (*2*), 5 min (*3*) and 30 min (*4*) with 0.2 W·cm^{-2} at 676.4 nm. Insert shows an unsaturated hole shape at T = 1.8 K. Fluorescence recording in the vibronic region near 715 nm [2.103]

redistribution of the molecules over the sites within the broad IDF - was found also for these molecules in glasses, as well as in polycrystalline n-alkanes [2.99,100]. Figure 2.18 shows the site redistribution measured in the 0-0 excitation band of Chl*a* after burning with different irradiation doses.

As seen in Fig.2.18, after prolonged irradiation the molecules absorbing in the long-wavelength part of IDF, $\nu < \nu_L$, are transferred (with an efficiency of $\simeq 10^{-5}$-10^{-4}) to the region of $\nu > \nu_L$ as a result of burning via phonon wings. The sites, once on the short-wavelength side of the laser line, will not transform further since they are unable to absorb ν_L. It can be seen that the centers which have undergone site interconversion are distributed over a broad region roughly proportional to the width of the initial IDF. Therefore, it is concluded that for every Chl*a* molecule there exists a wide distribution of possible matrix-molecule configurations responsible for the formation of the IDF. Further, from the behaviour of the hole-filling processes on thermal annealing, one may infer [2.101] that the ground-state activation barriers for spontaneous site interconversions are also spread over a wide interval of energies. The unsaturated holes obtained were very narrow, 0.01 cm^{-1} [6] at 1.8 K (Fig.2.18b), which up to now is the highest Q-factor ($\omega/\delta\omega = 1.5\cdot10^6$) observed in the spectroscopy of chlorophyll-like molecules.

A new band caused by a photoproduct was observed in the spectra of pheophytin *a* (Pheo*a*) after irradiation in the 0-0 absorption and was

---

[6] The holewidth still exceeds the lifetime-determined limit by one order of magnitude (T_1= 7 ns).

ascribed to the tautomeric rotation of inner hydrogens in the distorted tet-rapyrrole ring [2.102]. The narrow holes were burned in the 0-0 bands of the original and tautomeric forms. The efficiency of the photoconversion of the tautomer was estimated to be $\simeq 10^{-3}$, almost two orders of magnitude higher than the quantum yield of the tautomer formation. The potential barrier between the two forms in the ground state, $\Delta E \simeq 50$ cm^{-1}, was estimated from the temperature dependence of the decomposition of the photoproduct in the dark. The barrier was found to have a considerable inhomogeneous distribution. A concurrent spontaneous site-redistribution mechanism is active for Pheo$a$, too, but its efficiency is several times lower.

Hole-burning was utilized to measure the HLW Γ_ν of the vibronic levels of the S_1 electronic state for Pheo$a$ and Chl$a$ in ether at 5 K [2.103,104], the results of which are collected in Table 2.1. The homogeneous widths of 0.8-4.9 cm^{-1} obtained are significantly smaller than the linewidths of 7-10 cm^{-1} observed in the site-selection spectra under excitation in resonance with the 0-0 transition, i.e. the latter actually are in-

Table 2.1. Homogeneous widths (Γ_ν) and coherence lifetimes (T_2) of the vibrational sublevels of S_1 excited electronic state for Pheo$a$ and Chl$a$ in ether at T = 5 K determined by the hole-burning technique [2.104]

ω [cm^{-1}]	Γ_ν [cm^{-1}]	T_2 [ps]	ω [cm^{-1}]	Γ_ν [cm^{-1}]	T_2 [ps]
Pheophytin a					
			1240	2.1	2.5
513	1.1	4.8	1265	2.1	2.4
564	1.6	3.3	1295	2.4	2.2
584	1.5	3.5	1360	3.5	1.5
600	1.6	3.3	1380	2.4	2.2
670	1.1	5.0	1425	2.5	2.1
698	1.3	4.1	1472	2.5	2.1
709	2.8	1.9	1510	4.9	1.1
735	1.2	4.4	1540	3.1	1.7
770	0.9	6.6	1575	4.5	1.2
797	1.3	4.2			
870	2.3	2.3	Chlorophyll a		
886	4.1	1.3			
905	1.3	4.1	740	1.4	3.8
982	1.5	3.5	925	2.0	2.7
1060	2.5	2.1	984	2.8	1.9
1075	1.8	2.9	1075	3.8	1.4
1098	2.6	2.0	1250	2.4	2.2
1120	1.8	1.1	1345	2.2	2.4
1150	2.6	2.0	1510	4.6	1.2

homogeneously broadened as a result of non-correlation effects (Sect.2.6.4). Regarding vibrational relaxation, an overall trend can be seen in the increase of its rate with vibrational energy. Still, the slowest decaying modes are at 743 cm^{-1} for Chl$a$ and 770 cm^{-1} for Pheo$a$. A majority of the vibrational sublevels decay at the rate of $(2-5) \cdot 10^{11}$ s^{-1}, while the fastest decaying mode is at 1510 cm^{-1}, having a decay rate $9 \cdot 10^{11}$ s^{-1}. On the average, the relaxation times are comparable to those for H_2-porphin in n-octane, where values of T_2 = 1-40 ps were found [2.87,90]. The results for Pheo$a$ agree reasonably well with the estimate $T_2 \simeq$ 1-3 ps for vibrations in the 1200 cm^{-1} region made on grounds of preliminary hot-luminescence measurements. Therefore, in spite of a higher complexity of the Chl-type molecules, the vibrational relaxation is not notably faster than in the parent tetrapyrrole compounds.

2.7 Special Methods of Hole-Burning and Detection

2.7.1 Detection of Holes by Doppler Scanning

The purely electronic ZPL is an optical analog of the Mössbauer line, and it is natural to think about studying the ZPL in a similar way as well – via Doppler scanning of the laser frequency [2.77,78]. The PEL linewidths are in the range of Γ = 0.0001-0.01 cm^{-1}. To obtain a Doppler shift $\Delta\nu$ of the same order of magnitude we need velocities of the source relative to the absorber to be v = 1-100 m·s^{-1}, which is easily achievable.

Applied to the PEL itself this novel idea encounters a serious difficulty: the tremenous inhomogeneous broadening Δ is orders of magnitude larger than the Doppler shift $\Delta\nu \simeq$ 0.01 cm^{-1}. We overcome this difficulty by replacing the PEL by its PSHB image. Then we only have to perform measurements on a narrow-bandwidth spectral filter. To create the Doppler shifts in spectra of samples positioned in a liquid-IIe cryostat, it is convenient to use a moving mirror outside the cryostat. The setup of such a Doppler spectrometer is shown in Fig.2.19.

Using PSHB, the sample (OEP-PS, 10^{-4} M/ℓ) transparency was increased by a factor of 5-10 at the hole maximum. To avoid hole broadening small burning intensities (below 10^{-6} W·mm^{-2}) were used. The beam from an argon-ion-laser-pumped single-frequency power-stabilized ring dye laser was first passed twice through the sample at 1.5 K. The laser beam was therefore attenuated by the transparency spectrum of the sample. The beam was further directed to reflect twice from a pair of corner prisms mounted on a rotating disc. The double reflection provides the Doppler shift $\Delta\nu$ = 4ν(v/c), where v is the tangential velocity of the rotating prisms. The Doppler-shifted probe beam was finally passed through the sample once again and its total intensity was detected. The influence of any scattered light was suppressed by optoacoustical modulation of the probe beam.

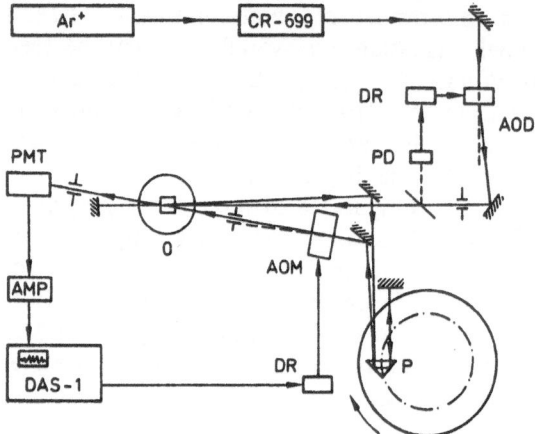

Fig.2.19. Set up of an optical Doppler-spectrometer with a single-frequency dye laser CR-699. Acousto-optical deflector AOD, its driver DR and photodiode PD form the feedback loop for laser output power stabilization; twelve prisms P are mounted on the rotating disk; acousto-optical modulator AOM directs the Doppler shifted beam through the sample in the cryostat O; the photon-counting system consists of the photomultiplier PMT, gated amplifier AMP and pulse analyzer DAS-1

The intensity of the signal as a function of the Doppler shift is Lorentzian with a maximum at $\Delta\nu = 0$ and with halfwidth $\delta_D = \delta_0(2) + \delta_0$, where $\delta_0(2) \simeq 0.65\delta_0$ signifies the width of the line formed after two passes through the filter. The results in Fig.2.20 give for the holewidth $\delta_0 = 0.017 \pm 0.004$ cm^{-1}. The method of Doppler-scanning has been applied to perform very fast hole-burning experiments with τ_{HB} down to 10^{-4} s (Sect.2.6.2) [2.74b,78].

The method can be quite useful when one must measure holes narrower than the narrowest laser linewidth available in the laboratory. That situation occurs in optically thick samples where the holewidth in transparency can be narrower than the linewidth of the hole-burning laser, the limit $\Gamma(T)$ being given by the purely-electronic homogeneous linewidth.

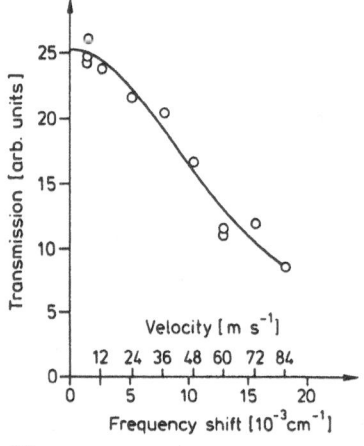

Fig.2.20. Hole in the transmission spectrum of OEP-PS at 1.5 K measured with the Doppler spectrometer shown in Fig.2.19. After laser irradiation at $\lambda_L = 620.8$ nm, $\delta_L = 10$ MHz, 10^{-6} W·mm^{-2} the sample's transmission is measured as a function of the tangential velocity of the rotating prisms (*empty circles*). Solid line is Lorentzian curve with $\delta = 0.023$ cm^{-1}

Thus, Doppler scanning provides one more - an instrumental cost - advantage of PSHB spectroscopy: to probe very sharp holes, very-narrow-line tunable lasers are not always required.

2.7.2 Holographic Detection of Spectral Holes

In [2.108,109] use of the (static) holographic technique for writing and reading spectral holes has been reported (Sect.7.4). Holography provides a zero-background method of performing high-resolution PSHB spectroscopy. The high detection sensitivity permits the hole-burning intensities to be maintained at an extremely low level (about 10^{-6} W·cm^{-2}). This holographic method is also of interest to investigations at very low temperatures where burning energies as well as readout powers have to be low in order to avoid local heating effects.

In a field of two intersecting laser beams (object and reference beams) the sample is exposed to a spatially modulated interference pattern of illumination intensities. This pattern is recorded in the corresponding spatially modulated structure of spectral holes, i.e. the interference pattern is "photographed" by the spatial modulation of the absoption coefficient and refractive index of the sample. For readout, the sample is illuminated with the reference beam with attenuated intensity. Two signal beams then emerge from the sample. One of them continues in the direction of the reference beam and represents the transmittance signal. The other, the holographic signal beam, propagates in the original direction of the object beam and respresents the diffracted response produced by the structure of the burnt-in holes. Both the transmitted and the holographic intensities depend on the frequency of the reference beam. Theory [2.109] shows that in a rather good approximation the dependence of the holographic efficiency on the reading frequency is a Lorentzian function with the same shape and spectral width as the hole in the absorption. If holes are absent, the holographic signal is zero (while the transmittance one is not). Thus, holographic detection is performed on a low-level background comprising only the stray light and the photomultiplier dark current.

The usefulness of the method was demonstrated for hole detection in the absorption spectrum of chlorin molecules imbedded in polyvinylbutyral film [2.108] and was later used for studies of the hole shapes of the same molecular system as a function of the temperature and applied electric field [2.109].

2.7.3 Creation of Sharp Antiholes

The creation of a sharp peak in absorption under monochromatic illumination may be considered as the production of an antihole. A rather sharp peak may be formed when the photoproduct has well-pronounced ZPLs in its spectra (Chaps.5,6). However, the antihole is usually much broader than its parent hole because of the inhomogeneous broadening of the ZPLs of the photoproduct. The latter is caused by differences in the initial structure of the selected impurities undergoing hole-burning: although their ZPL frequencies are the same, differences in structure near the centers appear in the product spectrum because the product energy levels interact differently with the local environment.

In this subsection, we shall consider another method of creating a sharp peak in absorption - by means of spectrally selective damping of the hole-burning efficiency [2.110]. The key feature of the method is the prevention of hole formation by narrow-band stimulated emission. The method has been called reversed hole-burning and the resulting peak - the antihole. Of course, the meaning of the term "antihole" is different from that used in the preceeding paragraph. A sample containing photo-active molecules is simultaneously irradiated by two laser beams. The first provides a broad-band intense illumination in the region of impurity absorption producing nonselective bleaching. The second beam is in resonance with a vibronic luminescence band and is spectrally much narrower. Its purpose is to restrict the photo-transformations of the molecules whose PELs are in resonance with this beam via shortening of their excited electronic state lifetime through stimulated emission.

As we know, the process of hole formation often takes place in the excited electronic state. In this case, the shortening of the lifetime of the excited state results in a decrease in the frequency of phototransformations for the molecules undergoing stimulated emission. In other words, the selected body of impurities who have their PELs in resonance with the spectrally narrow second beam receive protection from hole-burning, while the rest of the absorption band is bleached away by the first beam. This procedure may lead to a rather narrow antihole in the new IDF, which actually is a considerably bleached initial IDF.

This form of antihole burning was demonstrated for OEP-PS ($5 \cdot 10^{-4}$ M/ℓ) [2.110]. The sample of 2.5 mm thickness was illuminated with two pulsed dye lasers at 5 K. The first laser (rhodamine 6G dye, peak intensity of $2 \cdot 10^8$ W·cm^{-2} at 585 nm) populated nonselectively the S_1 state via vibronic absorption. The second, narrow-line laser (DCM dye, $\lambda_L = 650.33$ nm, $\Delta\omega_L = 1$ cm^{-1}, peak intensity of 10^8 W·cm^{-2}) was in resonance with a vibronic ZPL of the impurity fluorescence and thus, performed a selective depopulation of the S_1 state by stimulated emission.

Figure 2.21 displays the transmission spectra of the sample before and after a 1000 s illumination and the difference absorption spectrum with two stable antiholes. More than one hole is created because λ_L stimulates two different zero-phonon transitions. The antiholes are about 6 cm^{-1} wide, while the inhomogeneous broadening of the 0-0 S_1-S_0 absorption band is 60 cm^{-1}. Antiholes of this kind may be useful in studies of spectrally selective photochemistry and in manufacturing narrow-band absorption filters.

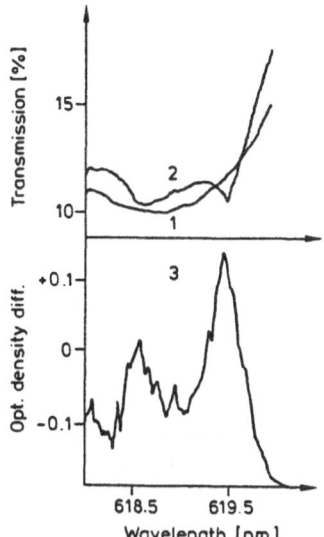

Fig.2.21. Two antiholes in the 0-0 absorption band of OEP-PS at 5 K created by simultaneous irradiation with two lasers: the laser $\lambda_1 = 585$ nm with $\Delta\lambda_1 = 7$nm gives nonselective population of the S_1 state, and the laser $\lambda_2 = 650.33$ nm with $\Delta\lambda_2 = 5\cdot10^{-2}$ nm gives selected stimulated S_1 state depopulation. Curves *1* and *2* show inhomogeneous transmission of the sample before and after the irradiation by two lasers for 10^3 s; curve *3* is the difference absorption spectrum

2.8 Hole-Burning Time-and-Space-Domain Holography

2.8.1 Hole-Burning by Picosecond Pulses

PSHB by light pulses presents new possibilites for optical data storage and signal processing on the nano- and picosecond time scale. These applications bring PSHB close to photon-echo phenomena. (See Chap.7 for additional discussion of such applications). The conventional method of performing high-resolution spectroscopy and optical data storage is to first burn spectral holes and then read (or detect) them by means of tunable narrow-line CW lasers. When dealing with large numbers of bits (10^{10}-10^{12} bits·cm^{-2}), parallel processing can become useful (if not a must). From the basic principles of the physics of oscillations it follows that the detection of a spectral hole with accuracy $\delta\nu$ requires a measurement time at least as long as $1/\delta\nu$. This means that the probing of a hole of width $\delta\nu_H = 10^{-3}$ cm^{-1} takes at least about 10^{-7} s. The burning time of such a narrow hole is at least as long or longer, depending on the quantum efficiency of the process.[7] Therefore, single-beam readout of 10^{12} bits stored on a 1x1 cm^2 area takes at least 10^5 seconds, i.e. one day. Of course, working with poorer spectral resolution enables one to read faster. The loss of spectral storage density may be compensated by increasing the spatial storage area and the number of spatial storage spots.

---

[7] Hole-burning and reading time of 30 ns (corresponding to a holewidth of 10^{-3} cm^{-1}) have actually been achieved already by means of semiconductor heterojunction lasers. With the optimal materials, high-density optical memories could be manufactured [2.17,111,112] (Chap.7).

A complementary technique to conventional quasi-monochromatic processing is the excitation and reading of the PSHB medium by means of short light pulses. A pulse of 2–3 ps duration covers about 5 cm^{-1} of frequency space comprising tens of thousands of homogeneous PEL lines. The problem is then how these resonances will respond when excited by a picosecond pulse where the phases of all the Fourier components are strictly correlated. May we expect that the unique properties of PELs determined by the excited electronic-state lifetime τ_e will be displayed at all since $\tau_e (\simeq 10^{-8}$ s) is 10^4-fold longer than the excitation time? *Mossberg* [2.113] proposed that the answer is yes by virtue of Fourier-transform considerations, provided the interaction processes could be considered as linear. This is the case in the experiments to be described below. Actually a proper understanding of the situation needs a bit of theory of time-dependent spectra. It has beeen shown for two-level systems in [2.114] that the Fourier components of the pulse can, in fact, be stored with the accuracy of the homogeneous linewidth provided the light-matter interaction is linear, and that the results of PSHB are not detected earlier than τ_e after the excitation. There is another useful feature of this approach: one can utilize the very high spectral resolution provided by PELs without having to use very narrow-line lasers. Thus one can write information in tens of thousands of spectral channels at each spatial spot with the dimension of about 10^{-8} cm^2 over an area of several cm^2 simultaneously.

The storage capabilities of an inhomogeneous spectral band of Δ cm^{-1} width are fully utilized by a light pulse of $t_p = (\Delta \cdot c)^{-1}$ s duration. For typical inhomogeneous bandwidths (for organics) $\Delta = 100$ cm^{-1}, so $t_p = 0.3 \cdot 10^{-12}$ s. To store all the intensities of the Fourier components of a pulse lasting t_p seconds, an inhomogeneous bandwidth Δ broader than t_p^{-1} is needed, i.e. 100 cm^{-1} is a sufficient bandwidth for picosecond pulses. For femtosecond pulses $\Delta \geq 5000$ cm^{-1} is required. This is clearly an application where large inhomogeneous broadening turns out to be a useful feature of the material.

The homogeneous linewidth $\Gamma(T)$ determines the ultimate spectral resolution in the frequency domain. As we know, as $T \rightarrow 0$, $\Gamma(T) \rightarrow \Gamma(0) = (T_1)^{-1}$, where T_1 is the excited electronic-state decay time. For impurities in a crystalline matrix, the homogeneous linewidth $\Gamma(T)$ approaches the $\Gamma(0)$ limit and, consequently, the holewidths $\delta_H(T) \simeq 2\Gamma(0)$ for temperatures less than $\simeq 2$ K. In glassy matrices the holewidths $\delta_H(T)$ at these temperatures are usually 10 or more times larger than $2\Gamma(0)$. As described in Chaps.2–4, the origin of hole broadening at these temperatures is far from being well understood. In any case, the inverse holewidth δ_H^{-1} places an upper limit on the temporal duration of the pulse sequence to be stored.

2.8.2 Theory of Time-and-Space-Domain Holographic Recording and Playback

Following [2.115] (see also the reviews [2.116,117] and Sect.7.4), let us consider a plate with dimensions $(2x_{max}, 2y_{max}, d)$ containing photochemi-

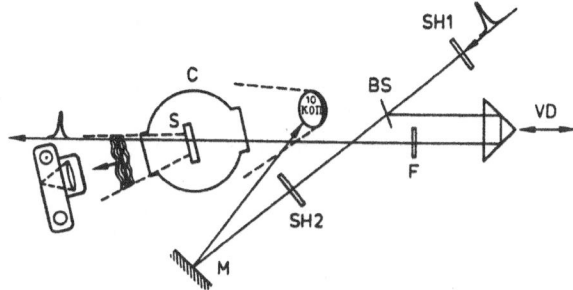

Fig.2.22. Schematic diagram of experimental set-up for recording time-and-space-domain holograms. Beamsplitter BS divides the expanded picosecond laser beam between reference and object channels; light from the object scene (a picosecond pulse scattered by a coin) travels through the windows of the cryostat C to the sample S; reference pulses pass the delay VD and strike the sample at 10° with respect to the object beam; to reproduce the holographic image of the scene the reference beam is attenuated by a filter F and the input beam illuminating the object is blocked by a shutter SH2

cally active dye molecules with zero-phonon absorption lines (ZPL) with homogeneous width $1/T_2$. The ZPL frequencies are distributed over an inhomogeneous band around ω_0 with the bandwidth Δ. Let the plate be illuminated, as shown in Fig.2.23, by a light pulse scattered by the object under study (object pulse)

$$S(r,t) = s(x,y,t-z/c) \exp[i\omega_0(t-z/c)] \; , \tag{2.27}$$

with the spectral width of $\Delta\omega_s$ and the duration $t_s \ll T_2$. If the front edge of the object pulse reaches the plate at the moment $t = 0$, the trailing edge of the pulse will leave the plate at $t = t_s$.

Let us further suppose that, with the delay t_R ($|t_R|<T_2$) relative to the front edge of the object pulse, the plate is also illuminated by a short plane-wave reference pulse, which is passed through the plate at an angle Θ with respect to the Z-axis. If the duration of the reference pulse is short enough compared to the object pulse, it can be considered to be a δ-function pulse and written as

$$R(r,t) = R_0\delta(t-n_R \cdot r/c-t_R) \exp[i\omega_0(t - n_R \cdot r/c - t_R)] \; , \tag{2.28}$$

where the unit vector in the direction of the reference pulse is $n_R = (-\sin\Theta,0,\cos\Theta)$, which for small angles Θ may be approximated as $(-\Theta,0,1)$.

The frequency spectrum of the pair consisting of one reference pulse and an object pulse, acting jointly on the medium, is given by

$$I(x,y,\omega) = R_0^2 + \left| s_F(x,y,\omega-\omega_0) \right|^2 + R_0 s_F(x,y,\omega-\omega_0) \cdot \exp[-i\omega(x\Theta/c - t_R)]$$

$$+ R_0 s_F^*(x,y,\omega-\omega_0) \cdot \exp[i\omega(x\Theta/c - t_R)] \; , \tag{2.29}$$

where s_F stands for the Fourier transform. We note here that (because $t_s \ll T_2$ and $|t_R| < T_2$) the total response is coherent, therefore the amplitudes have to be summed. The reference and object pulses need not spatially overlap at all. The phase memory of the excited electronic state of the molecules remembers the phases of the Fourier components of the first pulse until the second pulse arrives.

The optical properties of the medium are described by the dielectric permittivity

$$\epsilon(\mathbf{r},\omega) = \epsilon_0 + (\sigma c/2\pi\omega) \int \frac{d\omega' \rho(\mathbf{r},\omega')}{\omega' - \omega + i/T_2} , \qquad (2.30)$$

where σ is the integrated absorption cross section of the ZPL, and $\rho(\mathbf{r},\omega)$ is the inhomogeneous distribution function defined in Sect.2.2. The latter is altered in the course of recording by means of PSHB. The constant part of the permittivity will be taken as $\epsilon_0 = 1$, for simplicity.

Since T_2 is much longer than the overall duration of the pair of object and reference pulses, the permittivity (2.30) can be written as

$$\epsilon(\mathbf{r},\omega) = 1 - (\sigma c/2\omega) [i\rho(\mathbf{r},\omega) - \hat{H}\{\rho(\mathbf{r},\omega)\}] , \qquad (2.31)$$

where the $\hat{H}$ denotes the Hilbert transformation

$$\hat{H}\{f(\omega)\} \equiv (1/\pi) \int d\omega' \cdot f(\omega')/(\omega'-\omega) . \qquad (2.32)$$

Further, the intensities of the object and reference pulses are assumed limited so that nonlinear saturation and power broadening effects can be neglected. In the case of moderate intensities the PSHB bleaching effect due to resonant excitation is proportional to the intensity (2.29). If we further assume that the plate has a considerable resonant optical density and that the spectrally selective bleaching does not change it significantly, the altered inhomogeneous distribution function may be expressed as

$$\rho(\mathbf{r},\omega) = \rho_0[1 - m\sigma\eta \, I(x,y,\omega) \, \exp(-\rho_0 \sigma\eta)] , \qquad (2.33)$$

where ρ_0 is the density of centers before hole-burning, η is the PSHB quantum efficiency, and m is an integer representing the number of applied identical pairs of reference and object pulses. Use of multiple pulse-pairs enables one to achieve the spectral contrasts required.

Using (2.31,33), we can write the complex linear transmittance function of the plate as follows:

$$K(x,y,\omega) = e^{-d(i\omega/c - \rho_0 \sigma/2)}[1 + (\kappa/2)(1 + i\hat{H})\{I(x,y,\omega)\}] , \qquad (2.34)$$

where $\kappa = m\sigma\eta$, and the exponential factor describes the attenuation and phase shift of the output signal after crossing the plate of thickness d. One can see that the second term in (2.34) contains exhaustive information about the object pulse.

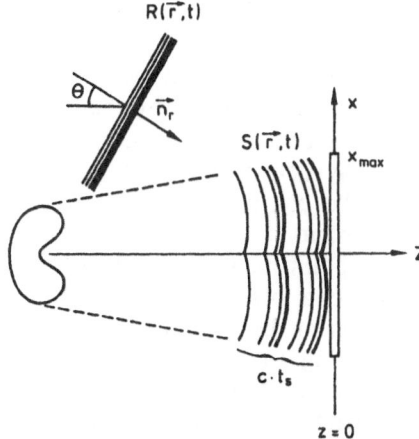

R(r̄,t)

θ

$\vec{n}_r$

S(r̄,t)

x

x_{max}

Z

c·t_s

z = 0

Fig.2.23. Diagram of the holographic recording procedure for wave-front conjugation. Object pulse S(r,t) reached the recording plate at the moment $t = 0$; reference pulse R(r,t) is shown to be delayed

Now let us calculate the response of the plate for a δ-function readout
pulse. The resulting output signal from the hologram is

$$F^{out}(r,t) = f_0(r,t) \exp[i\omega_0(t + x\Theta/c)] + f_s(r,t)\cdot\exp[i\omega_0(t + t_R)]$$
$$+ f_s^*(r,t)\cdot\exp[i\omega_0(t - t_R + 2\Theta x/c)] , \qquad (2.35)$$

where

$$f_0(r,t) = (1/R_0\kappa + R_0/2)\cdot\delta(t+\Theta x/c)$$
$$+ R_0^{-1}Y(t+\Theta x/c) \int s(x,y,\tau)\cdot s^*(x,y,\tau-t-\Theta x/c)d\tau ,$$
$$f_s(r,t) = Y(t+\Theta x/c)\cdot s(x,y,t+t_R) ,$$
$$f_s^*(r,t) = Y(t+\Theta x/c)\cdot s^*(x,y,t_R-t-2\Theta x/c) . \qquad (2.36)$$

In writing (2.35), we have dropped the constant factor $R_0^2\kappa\cdot\exp(-\sigma\rho_0 d/2)$
which is common to all three terms, and we have assumed that $Z = +0$.
In (2.36) $Y(\tau)$ is the Heaviside unit step function.

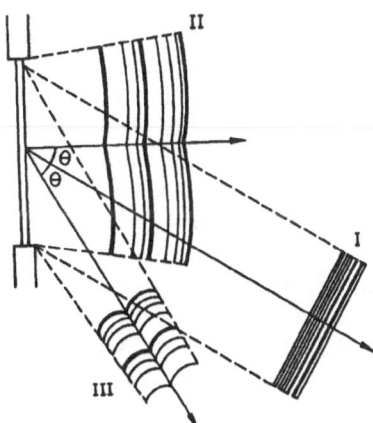

II

θ
θ

I

III

Fig.2.24. Diagram showing three different
kinds of pulses appearing in the output of the
hologram: transmitted readout probe pulse (*I*),
reproduced object pulse (*II*) and conjugated
object pulse which occurs only if the readout
probe pulse is applied in the direction opposite to that of the recording reference pulse

69

Equations (2.35,36) show that at the output plane of the spectral hologram there will appear three different light waves (pulses) (Fig.2.24). The
first term with $f_0(r,t)$ propagates in the direction n_R and describes a
transmitted read-out pulse with a distorted shape determined by the auto-
correlation function of the object pulse. The second term $f_1(r,t)$ propagates along the Z axis and represents the playback of the stored object
signal as a virtual image. The last term is the conjugated image of the
object with reversed time behaviour. The causality function $Y(\tau)$ guarantees that no output signal appears at any point behind the hologram
before the readout pulse has arrived. It also cuts off part of the reproduced object pulse according to the time ordering between the reference
and object pulses during the process of writing the hologram. If $t_R \leq$
$-|\Theta x_{max}|$, i.e. if the reference pulse came before the object pulse, the
latter is completely reproduced but the conjugated object pulse is not pre-
sent; on the contrary, if $t_R \geq |\Theta x_{max}|+t_s$, the original object pulse is
absent; finally, in the intermediate case of overlapping object and refer-
ence pulses, $-|\Theta x_{max}| \leq t_R \leq |\Theta x_{max}|+t_s$, both conjugated and nonconju-
gated object pulses are partially reproduced.

In [2.115] it was shown that if the two-dimensional (thin) hologram
condition ($\Theta^2 d \leq 2\pi c/\omega_0$) is violated so that the synchronism of writing
and readout pulses must be considered, the conjugated object pulse can be
reproduced only if a counter-propagating probe is applied in the direction
$-n_R$. In the experiments [2.118-123], both thin and thick holograms were
produced by changing the angle Θ.

2.8.3 Experimental Results and Discussion

In the experiments, an actively mode-locked Ar-ion laser (Spectra-Physics
model 171) was utilized to pump a Rhodamine 6G picosecond dye laser,
which provided 2-3 ps duration pulses (5-6 cm^{-1} spectral FWHM) at 82
MHz repetition rate with an average output power of 100 mW.

As a spectrally selective medium for recording time-and-space holograms by PSHB, polystyrene doped with octaethylporphin (OEP-PS) at
concentrations of $10^{-4} \simeq 10^{-3}$ M was used. The inhomogeneously broadened 0-0 impurity absorption band was 200 cm^{-1} (FWHM) centered at the
wavelength of 617 nm. Homogeneous ZPL widths were less than 0.05
cm^{-1} at 1.8 K. Samples were prepared in blocks with thickness 0.3-1.0 cm
and optical density 1-3 cm^{-1} and were placed in a liquid He cryostat
with optical windows. To record high-contrast spectral holograms, PSHB
exposures of at least 0.1 mJ·cm^{-2} were needed. Depending on the average
intensity of the incident light this exposure contains 10^9-10^{11} identical
sequences of writing pulses.

In Fig.2.22, a typical experimental set-up [2.117,123] is presented. An
object - a ten kopeck coin - was positioned in front of the entrance
window of the cryostat containing an OEP-PS block at 2 K. One beam
was directed at the coin to provide a scattered object beam, and a second

beam of plane-wave reference pulses was passed through the sample at a 10^0 angle, 20 ps ahead of the object pulses scattered from the coin. Ten minutes of PSHB exposure was enough to store a high-contrast spatial holographic image of the coin. The image was clearly visible from behind the cryostat when the illumination of the original coin was blocked and only the plane-wave beam was passed through the sample to act as a readout reference pulse. The holographic images, which persisted at least for several hours (until the liquid helium boiled away), were either photographed through the output window of the cryostat (Fig.2.25) or temporally analyzed with the help of a synchroscan streak camera system (Fig.2.27). As usual, the photographs and streak camera images cannot reproduce, to a full extent, the rich information contained in the holographic images of the picosecond events (such as the spatial depth of the image, for example).

Fig.2.25. Image of a coin photographed from the hologram

In a slightly different set-up [2.117-123], the original object image incident on the recording sample was first spoiled by a distorter (a fragment of a broken glass jar) and then recorded. During the writing of the hologram the reference delay was adjusted so that reference pulses arrived at the sample some tens of picoseconds later than the object's distorted wave front. Afterward the recorded holograms were illuminated by plane wave readout pulses travelling in the opposite direction compared to the original reference pulses (compare with the last term of (2.36)). This procedure resulted in fully conjugated (i.e., time-and-space reversed) object pulses. These pulses passed through the distorter in the opposite direction, the distortion of the opposite sign compensated the initial temporal and spatial distortions, and a nondistorted object image emerged from the distorter (Fig.2.26). Another example of a holographic time-reversal of picosecond signals, visualized with the help of the streak camera is presented in Fig.2.28.

The following applications of time-and-space domain holograms in optical processing have been suggested (Sect.7.4) [2.115].

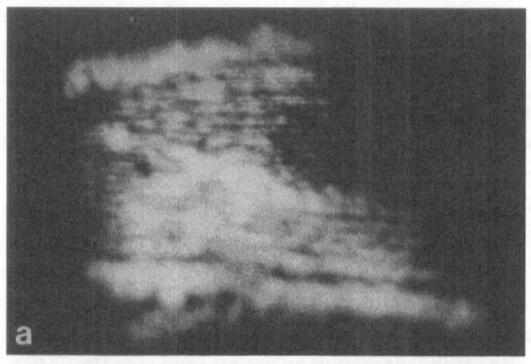

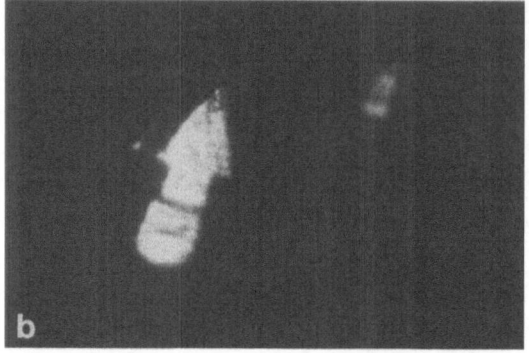

Fig.2.26. (a) Distorted object image photographed at the input of the cryostat, after passing through a distorter. (b) Restored image of the object photographed at the output of the cryostat as the signal from a conjugated hologram passed the distorter in the opposite direction

1) Multifrequency parallel recording and readout of an optical spectral memory. Hole-burning and detection of 1600 holes in an absorption band using picosecond pulses have been reported [2.124].

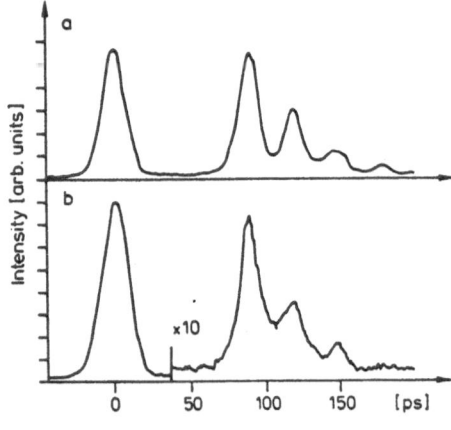

Fig.2.27. Streak-camera images of the applied sequence of hole-burning pulses (*a*) and of the resulting accumulated stimulated photon echo signal (*b*). Note that in the hole burning sequence the asymmetrical pulse appears after the single symmetrical reference pulse

2)   Reversal of both the wave front and the time dependence of light pulses via detection of the conjugated probe wave [2.117,123].

3)   Processing of picosecond signals: a) to sum and subtract signals (the "minus" sign may be created by using a half-wave plate between the object and the hole-burning medium); b) space-and-time convolution of pulses using one pulse as the probe to read the hologram of the other pulse; c) filtration of the spatial and temporal frequencies of signals by using a hologram filter with specific structure.

4)   Synthesis of picosecond signals via transformation of the incident light pulses in holograms with predetermined spatial-spectral hole structures, created by means of tunable single-mode lasers or other controlled light sources.

5)   Recognition of picosecond events as a generalization of the standard holographic methods of pattern recognition. If a hologram is illuminated by a signal which coincides with one of the signals stored in that hologram and if the scattered light is focused, then a δ-function pulse at a particular spot of the focal plane will be created [2.118,119].

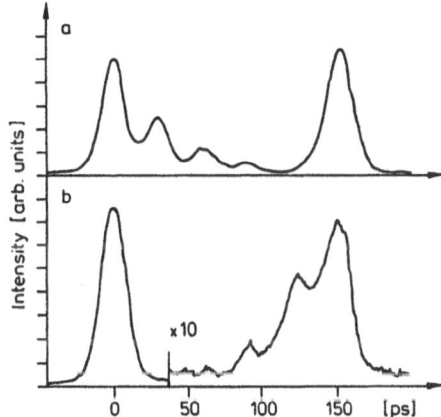

Fig.2.28. Streak-camera images of the applied hole-burning sequence (*a*) and that of the time-reversed holographic signal (*b*). Note that in this case for hole burning the asymmetrical pulse is applied first

2.9 Concluding Remarks

Persistent spectral hole-burning (PSHB) is a rapidly growing field of molecular and solid-state spectroscopy and photochemistry providing high-resolution spectra, information about photochemical reactions, and the possibility of various applications.

The objectives of high-resolution spectroscopy are the understanding of vibronic structure, perturbations of external fields, and the low-energy tunneling states in amorphous matrices. A possible highlight of the latter studies is the search for the manifestations of the fractal structure of disordered solid matter. Studies of hole-burning mechanisms are actually studies of *photochemical reactions* in molecules in the solid state and at low temperatures. *Applications* in science and technology are based on the production of spectral filters of different designs. Another interesting application is encountered in hole-burning holography. The time-and-space domain holography of pico- and nanosecond events suggests intriguing possibilities for data and image processing, and for the experimental modeling of some features of associative memories [2.125].

Considering the richness and diversity of PSHB science and applications, the prospects for continued progress in this field are promising.

References

2.1   K.K. Rebane: *Impurity Spectra of Solids* (Plenum, New York 1970)
2.2   I.S. Osad'ko: Theory of Light Absorption and Emission by Organic Impurity Centers, in *Spectroscopy and Excitation Dynamics of Condensed Molecular Systems*, Vol.4, ed. by V.M. Agranovich, R.M. Hochstrasser (North-Holland, Amsterdam 1983) p.437
2.3   D. Hsu, K. Skinner: J. Chem. Phys. **81**, 1604 and 5471 (1984)
2.4   E.D. Trifonov: Dokl. Acad. Nauk SSSR 147, 826 (1962) (in Russian)
2.5   E.F. Gross, B.S. Razbirin, S.A. Permogorov: Dokl. Akad. Nauk SSSR 154, 1306 (1964) (in Russian)
2.6   K.K. Rebane, V.V. Hiznyakov: Opt. Spektrosk. 14, 491 (1963) (in Russian)
2.7   E.V. Shpol'skii: Usp. Fiz. Nauk 71, 215 (1960); and 80, 255 (1963) (in Russian)
2.8   M.A. Krivoglaz, S.I. Pekar: Trudy Inst. Fiziki Adad. Nauk Ukr. SSR 4, 37 (1953) (in Russian)
2.9   J.J. Rolfe: J. Chem. Phys. **40**, 1664 (1964)
2.10   K.K. Rebane, L.A. Rebane: Izv. Akad. Nauk Eston. SSR 14, 309 (1965) (in Russian)
2.11   K.K. Rebane, L.A. Rebane: Pure and Appl. Chem. 37, 161 (1974);
       L.A. Rebane: Vibronic Spectra of Molecular Centers in Alkali Halides and Local Lattice Dynamics, in *Physics of Impurity Centres in Crystals*, ed. by G.S. Zavt (Tallinn 1972) p.353;
       L.A. Rebane, O.I.Sild: Spectroscopy of Molecular Impurities in Alkali Halides, in *Defects in Insulating Crystals*, ed. by V.M. Tuchkevich and K.K. Shvarts (Zinatne Riga, Springer Berlin, Heidelberg 1981) p.619
2.12   K. Rebane, P. Saari, T. Tamm: Izv. Akad. Nauk Eston. SSR 19, 251 (1970) (in Russian)
2.13   A. Szabo: Phys. Rev. Lett. 25, 924 (1970)
2.14   A.A. Gorokhovskii, R. Kaarli, L.A. Rebane: Opt. Commun. 16, 282 (1976)
2.15   W.H. Hesselink, D.A. Wiersma: Theory and Expermental Aspects of Photon Echoes in Molecular Solids, in *Spectroscopy and Excitation Dynamics of Condensed Molecular Systems*, Vol.4, ed. by V.M. Agranovich, R.M. Hochstrasser (North-Holland, Amsterdam 1983) p.249
2.16   L.A. Rebane, A.A. Gorokhovskii, J.V. Kikas: Appl. Phys. B 29, 235 (1982)
2.17   A. Szabo: US Patent No 3, 896, 420 (July 22, 1975);
       G. Castro, D. Haarer, R.M. Macfarlane, H.P. Trommsdorff: "Frequency Selective Optical Data Storage System", U.S. Patent No 4, 101, 976 (July 18, 1978)
2.18   L. Allen, J.H. Eberly: *Optical Resonance and Two-Level Atoms* (Wiley-Interscience, New York, 1975)
2.19   C.J. Buchenauer, D.B. Fitchen, J.B. Page: Experimental Study of Resonant Raman Scattering by Impurities, in *Light Scattering Spectra in Solids*, ed. by R.F. Wallis (Plenum, New York 1968) p.496;
       L.A. Rebane, T.Y. Haldre: Pis'ma JETP 26, 674 (1977) (in Russian)

2.20   T.B. Tamm, J.V. Kikas, A.E. Sirk: Zh. Prikl. Spektrosk. 24, 315 (1976) (in Russian)

2.21   A.M. Freiberg, L.A. Rebane: Fiz. Tverd. Tela 16, 2626-2631 (1974) (in Russian); and Strain Broadening and Shape of No-Phonon Lines in the Spectra of the O_2^- Impurity in KCl, in *Molecular Spectroscopy of Dense Phases*, ed. by M. Grosmann, S.G. Elkomass, J. Ringeissen (North-Holland, Amsterdam 1976) p.495

2.22   R.I. Personov: Izv. Akad. Nauk SSSR 42, 232 (1978)

2.23   K.K. Rebane: Zh. Prikl. Spektrosk. 37 906 (1982); L.A. Rebane: Zh. Prikl. Spektrosk. 34, 1023 (1981)

2.24   M.N. Sapozhnikov, V.I. Alekseev: Phys. Status Solidi (b) 120, 435 (1983)

2.25   R.I. Personov, E.I. Al'shits, L.A. Bykovskaya: Opt. Commun. 6, 169 (1972); Zh. Eksp. Teor. Fiz. Pis'ma 15, 609 (1972)

2.26   R.I. Personov: Site Selection Spectroscopy of Complex Molecules in Solutions and its Applications, in *Spectroscopy and Excitation Dynamics of Condensed Molecular Systems*, Vol.4, ed. by V.M. Agranovich, R.M. Hochstrasser (North-Holland, Amsterdam 1983) p.555

2.27   V.V. Hizhnyakov, I.K. Rebane: Zh. Eksp. Teor. Fiz. 74, 885 (1978) (Engl. transl. Sov. Phys. JETP 47, 463 (1978); I.K. Rebane, A.L. Tuul, V.V. Hizhnyakov: Zh. Eksp. Teor. Fiz. 77, 1302 (1979) (Engl. transl. Sov. Phys. JETP 50, 655 (1979) V.V. Hizhnyakov, I.K. Rebane: Proc. Estonian SSR Acad. Sci. phys. math. 35, 406 (1986)

2.28   K.K. Rebane, R.A. Avarmaa, A.A. Gorokhovskii: Izv. Akad. Nauk SSSR 3, 1794 (1975) (in Russian)

2.29   J. Friedrich, J.D. Swalen, D. Haarer: J. Chem. Phys. 73, 705 (1980)

2.30   J. Friedrich, D. Haarer: J. Chem. Phys. 76, 61 (1982)

2.31   A.A. Gorokhovskii, J.V. Kikas, V.V. Palm, L.A. Rebane: Fiz. Tverd. Tela 23, 1040 (1981); Engl. transl.: Sov. Phys. - Solid State 23, 602 (1981)

2.32   W. Breinl, J. Friedrich, D. Haarer: J. Chem. Phys. 80, 3496 (1984); and Optical spectral diffusion processes in organic glasses on a logarithmic time scale, in Proc. Int'l. Conf. Luminescence (Madison, WI 1984)

2.33   A.A. Gorokhovskii, R.K. Kaarli, L.A. Rebane: JETP Lett. 20, 216 (1974); Opt. Commun. 16, 282 (1976)

2.34   K.N. Solovjev, E.I. Zalesskii, V.N. Kotlo, S.F. Shkirman: JETP Lett. 17, 463 (1973)

2.35   B.M. Kharlamov, R.I. Personov, L.A. Bykovskaya: Opt. Commun. 12, 191 (1974)

2.36   J.M. Hayes, G.J. Small: Chem. Phys. 27, 151 (1978); Chem. Phys. Lett. 5, 435 (1978)

2.37   J. Friedrich, D. Haarer: Angew. Chemie 23, 113 (1984)

2.38   A.A. Gorokhovskii, J.V. Kikas: Zh. Prikl. Spektrosk. 28, 832 (1978)

2.39   S. Voelker, R.M. Macfarlane, A.Z. Genack, H.P. Trommsdorff, J.H. van der Waals: J. Chem. Phys. 67, 1759 (1977)

2.40   H. de Vries, D.A. Wiersma: J. Chem. Phys. 72, 1852 (1980)

2.41   R. Janiso, J. Kikas: Izv. Akad. Nauk Eston. SSR 30, 247 (1981)

2.42   D.M. Burland, D. Haarer: IBM J. Res. Dev. 23, 534 (1979)

2.43   J.M. Hayes, R.P. Stout, G.J. Small: J. Chem. Phys. 74, 4266 (1981)

2.44   B.M. Kharlamov, R.I. Personov, L.A. Bykovskaya: Opt. Spektrosk. 39, 137 (1975) (in Russian)

2.45   J. Kikas, J. Malkin: Proc. Acad. Sci. Estonian SSR 36, 62 (1987)

2.46   A.U. Jalmukhambetov, I.S. Osad'ko: J. Chem. Phys. 77, 247 (1983)

2.47   O.N. Korotaev, E.I. Donskoi, V.I. Gladkovskii: Opt. Spektrosk. 59, 492 (1985)

2.48   I. Rebane: Proc. Acad. Sci. of Eston. SSR 35, 296 and 400 (1986), 36, 204 (1987) (in Russian)

2.49   W.E. Moerner, F.M. Schellenberg, G.C. Bjorklund, P. Kaipa, F. Lüty: Phys. Rev. B 32, 1270 (1985)

2.50   A. Winnacker, R.M. Shelby, R.M. Macfarlane: Opt. Lett. 10, 350 (1985)

2.51   H.W.H. Lee, M. Gehrtz, E.E. Marinero, W.E. Moerner: Chem. Phys. Lett. 118, 611 (1985)

2.52   R. Avarmaa: Izv. Akad. Nauk Eston: SSR 23, 238 (1974)

2.53   J. Kikas: Chem. Phys. Lett. 57, 511 (1978)

2.54   D.L. Huber, D.S. Hamilton, B. Barnett: Phys. Rev. B 16, 4642 (1977)

2.55   A.A. Gorokhovskii, J.V. Kikas: Opt. Commun. 21, 272 (1977)

2.56   P. Avouris, A. Campion, M.A. El-Sayed: J. Chem. Phys. 67, 3397 (1977)

2.57   T.B. Tamm, P.M. Saari: Chem. Phys. Lett. 30, 219 (1975)

2.58   A.A. Gorokhovskii, J.V. Kikas, V.V. Palm, L.A. Rebane: Izv. Akad. Nauk SSSR 46, 952 (1982)

2.59   J. Kikas, M. Ratsep: Phys. Status Solidi (b) 112, 409 (1982)

2.60   A.B. Treshchalov, L.A. Rebane: Fiz. Tverd. Tela 20, 469 (1978)

2.61   A.A. Gorokhovskii, L.A. Rebane: Opt. Commun. 20, 144 (1977)

2.62   L.A. Rebane: Relaxation of Electron-Vibrational Excitations in Impurity Molecules in Solid Matrices, in *Ultrafast Relaxation and Secondary Emission*, Proc. Int'l. Symp. UPS-78, Tallin (1978) p.89

2.63   A.A. Gorokhovskii, L.A. Rebane: Izv. Akad. Nauk SSSR 44, 859 (1980)

2.64   D.E. McCumber: J. Math. Phys. 5, 221 (1964)

2.65   M.A. Krivoglaz: Fiz. Tverd. Tela 6, 1707 (1964)

2.66   S. Völker, R.M. Macfarlane, J.H. van der Waals: Chem. Phys. Lett 53, 8 (1978)

2.67   A.I.M. Dicker, J. Dobkowski, S. Völker: Chem. Phys. Lett. 84, 415 (1981)

2.68   T.I. Aartma, D.A. Wiersma: Chem. Phys. Lett. 42, 520 (1976)

2.69   F.G. Patterson, W.L. Wilson, H.W.H. Lee, M.D. Fayer: Chem. Phys. Lett. 110, 7 (1984)

2.70   G. Wäckerle, H. Zimmermann, K.P. Dinse: Chem. Phys. Lett. 110, 107 (1984)

2.71   D. Hsu, J. Skinner: J. Chem. Phys. 83, 2097 (1985)

2.72   A.R. Chraplyvy, W.E. Moerner, A.J. Sievers: Opt. Lett. 6, 431 (1981)

2.73   W.E. Moerner, A.J. Sievers, R.H. Silsbee, A.R. Chraplyvy, D.K. Lambert: Phys. Rev. Lett. 49, 398 (1982)

2.74   R.C. Zeller, R.O. Pohl: Phys. Rev. 134, 2029 (1971)
       K.K. Rebane, A.A. Gorokhovski: J. Luminescence 36, 237 (1986)

2.75   V. Hiznyakov, I. Tehver: Phys. Status Solidi (b) 95, 65 (1979)

2.76   T.L. Reinecke: Sol. State Commun. 32, 1103 (1979)

2.77   K.K. Rebane, V.V. Palm: Opt. Spektrosk. 57, 381 (1984)

2.78   K.K. Rebane: J. Luminescence 31/32, 744 (1984); Cryst. Latt. Def. and Amorph. Mat. 12, 427 (1985)

2.79   R.A. Avarmaa, K.H. Mauring: Opt. Spektrosk. 41, 670 (1976)

2.80   R.M. Shelby, R.M. Macfarlane: Chem. Phys. Lett. 64, 545 (1979)

2.81   H.P.H. Thijssen, R.E. van der Berg, S. Völker: Chem. Phys. Lett. 103, 23 (1983)

2.82   A.A. Gorokhovskii, V.H. Korrovits, V.V. Palm, M.A. Trummal: JETP Lett. 42, 249 (1985); Chem. Phys. Lett. 125, 355 (1986)

2.83   J.M. Hayes, R.P. Stout, G.J. Small: J. Chem. Phys. 74, 4266 (1981)

2.84   A.A. Gorokhovskii, L.A. Rebane: Fiz. Tverd. Tela 19, 3417 (1977)

2.85   S.L. Lyo: Phys. Rev. Lett. 48, 688 (1982)

2.86   S. Hunklinger, M. Schmidt: Z. Physik B 5, 93 (1984)

2.87   S. Völker, R.M. Macfarlane: IBM J. Res. Devel. 23, 547 (1979)

2.88   A.A. Gorokhovskii, V.V. Palm: JETP Lett. 37, 201 (1983)

2.89   B.M. Kharlamov, L.A. Bykovskaya, R.I. Personov: Opt. Spektrosk. 42, 755 (1977); Chem. Phys. Lett. 50, 407 (1978); Zh. Priklad. Spektrosk. 28, 840 (1978)

2.90   A.I.M. Dicker, S. Völker: Chem. Phys. Lett. 87, 481 (1982)
2.91   R.I. Personov, B.M. Kharlamov: Photochemical and Photophysical Hole Burn-
       ing in Electronic Spectra of Complex Organic Molecules, in *Lasers Application*
       (Vilnus 1984)
2.92   J. Friedrich, D. Haarer: J. Chem. Phys. 79, 1612 (1983)
2.93   A.A. Gorokhovskii, J. Kikas: Opt. Commun. 21, 272-274 (1977)
2.94   R. Avarmaa: Izv. Akad. Nauk Eston. SSR 23, 93 (1974)
2.95   J. Fünfschilling, D.F. Williams: Photochem. Photoviol. 22, 151 (1975)
2.96   L.A. Bykovskaya, F.F. Litvin, R.I. Personov, Y.V. Romanovskii: Biofizika 25,
       13 (1980)
2.97   K.K. Rebane, R.A. Avarmaa: J. Chem. Phys. 68, 191 (1982)
2.98   R. Avarmaa, K. Mauring, A. Suisalu: Chem. Phys. Lett. 77, 88 (1981)
2.99   R. Tamkivi: Izv. Akad. Nauk Eston. SSR 31, 447 (1982)
2.100 K.K. Rebane, R.A. Avarmaa: J. Photochem. 17, 311 (1981)
2.101 R.V. Jaaniso: Izv. Akad. Nauk Eston. SSR 31, 161 (1982)
2.102 K. Mauring, R. Avarmaa: Chem. Phys. Lett. 81, 446 (1981)
2.103 K.K. Rebane, R.A. Avarmaa, K.H. Mauring, R.V. Jaaniso: In Proc. of Symp
       on Modern Aspects of Fine-Line and Selective Spectroscopy (Moscow 1984) p.
       63
2.104 R.A. Avarmaa, K.K. Rebane: Spectrochim. Acta 41A, 1365 (1985)
2.105 A.A. Gorokhovskii, V.V. Palm: JETP Lett. 37, 201 (1983)
2.106 M.A. Krivoglas: Zh. Eksp. Teor. Fiz. 88, 2171 (1985)
2.107 R. Avarmaa, K. Rebane: Stud. Biophys. (Berlin) 48, 209 (1975)
2.108 A. Renn, A.J. Meixner, U.P. Wild, F.A. Burkhalter: Chem. Phys. 92, 157
       (1985)
2.109 A. Renn, A.J. Meixner, U.P. Wild: *Holographischer Nachweis Photochemischer
       Absorptionloecher in Photoreaktive Festkörper*, ed. by H. Sixl (Wahl-Verlag,
       Karlsruhe 1984)
2.110 J. Kikas, I. Sildos: Chem. Phys. Lett. 114, 44 (1985)
2.111 K.K. Rebane: Laser Study of Inhomogeneous Spectra of Molecules in Solids, in
       Proc. Int'l. Conf. Lasers'82 (New Orleans, LA 1982)
2.112 A. Szabo: In Proc. Int'l. Conf. Lasers'80 (New Orleans, LA 1980) p.374
2.113 T. Mossberg: Opt. Lett 7, 77 (1982)
2.114 I. Rebane: Izv. Akad. Nauk Eston. SSR 34, 438 (1985)
2.115 P. Saari, A. Rebane: Izv. Akad. Nauk Eston. SSR 33, 322 (1984)
2.116 P. M. Saari, R.K. Kaarli, A.K. Rebane: Kvant. Elektron. 12, 672 (1985)
2.117 P. Saari, R. Kaarli, A. Rebane: J. Opt. Soc. Am. B 3, 527 (1986)
2.118 A.K. Rebane, R.K. Kaarli, P.M. Saari: Opt. Spektrosk. 55, 405 (1983)
2.119 A.K. Rebane, R.K. Kaarli, P.M. Saari: JETP Lett. 38, 383 (1983)
2.120 A. Rebane, R. Kaarli, P. Saari, A. Anijalg, K. Timpmann: Opt. Commun. 27,
       170 (1983)
2.121 A. Rebane, R. Kaarli: Chem. Phys. Lett. 101, 317 (1983)
2.122 A.K. Rebane, R.K. Kaarli, P.M. Saari: J. Molec. Struct. 114, 343 (1984)
2.123 A. Rebane, R. Kaarli, P.M. Saari: Izv. Akad. Nauk. Eston. SSR 34, 444 (1985)
2.124 J. Kikas, R. Kaarli, A. Rebane: Opt. Spektrosk. 56, 387 (1984)
2.125 T. Kohonen: *Self-Organization and Associative Memory*, 2nd ed., Springer Ser.
       Inf. Sci., Vol.8 (Springer, Berlin, Heidelberg 1988)

3. Photochemical Hole-Burning in Electronic Transitions

D. Haarer

With 25 Figures

In this chapter the basic aspects of the chemistry and spectroscopy of photochemical hole-burning (PHB) are developed. A comparison with the techniques of transient hole-burning in the field of nuclear magnetic resonance (NMR) and optical spectroscopy clearly illustrates that the chemical nature of the PHB method has far-reaching consequences: Not only fast relaxation processes in the time regime between femtoseconds and nanoseconds can be investigated, but also very slow relaxation processes which occur on time scales of seconds to years. This is due to the irreversible and stable nature of the photochemical change which is induced via laser irradiation. This irradiation creates a photochemical "memory" which is sensitive to slow matrix changes in the long time domain. These matrix changes, which generally lead to changes of the observed PHB spectra and which are discussed in conjunction with the spectroscopic term of "spectral diffusion", are of special relevance for the understanding of amorphous solids, a class of materials that is character- ized by a non-equilibrum state of the pertinent microscopic parameters, describing the multidimensional configurational space of a glass. At pre- sent, and in the near future, the main challenge for the spectroscopist will be to correlate microscopic and macroscopic matrix parameters with the observed spectral features of PHB spectroscopy.

3.1 Photochemical, Photophysical, and Spin Hole-Burning

3.1.1 Historic Survey

The phrase "hole-burning" was first used in the early days of nuclear magnetic resonance (1948) by *Bloembergen* et al. [3.1]. These authors doc- umented, for the first time, that the spectroscopic resolution of an NMR experiment could be carried beyond the "inhomogeneous linewidth limit" by saturating one "spin packet" of an inhomogenously broadened NMR line. (The term "spin packet" will be defined below.) Figure 3.1 shows an inhomogeneous NMR-line before (Fig.3.1a) and after (Fig.3.1b) local saturation at the frequency ω_0.

The NMR experiments and optical experiments reported in this book have been performed in a similar fashion:

a) A fixed-frequency "infinitely narrow" light source with frequency ω_L $= \omega_0$ and with high intensity (NMR or optical), creates a depletion of

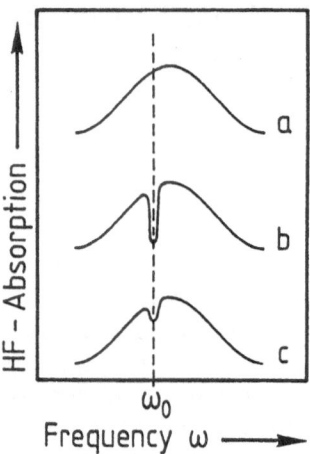

Fig.3.1. Local saturation of an NMR line at the frequency ω_0 (a) before, (b) during, and (c) after irradiation at ω_0 [3.1]

the ground state of a two-level system. The phrase "infinitely narrow" is to be understood in the following sense: The light source has a width $\Delta\omega_L$ which obeys the following inequality

$$\Delta\omega_L \ll \Delta\tilde{\omega}_h \simeq \frac{1}{T_1} + \frac{1}{T_2} \tag{3.1}$$

where T_1 and T_2 are the longitudinal and transverse relaxation times. In the above equation $\Delta\tilde{\omega}_h$ is an estimate of the homogeneous linewidth which also defines the approximate width of a "spin packet". $\Delta\tilde{\omega}_h$ should be taken as an approximate value for the real homogeneous linewidth $\Delta\omega_h$. The explicit relation between $\Delta\omega_h$ and the related relaxation times T_1 and T_2 depends on the nature of the two-level system under consideration (see below).

b)   In addition to the fixed-frequency light source, a variable-frequency, low-intensity light source is used to measure the hole, which is centered at ω_0 in the absorption line profile. Figures 3.1a and b show the transmission signal of a sample as measured by the low-intensity light source before (Fig.3.1a) and during "local saturation" at the frequency ω_L (Fig.3.1b). The "hole" at the frequency ω_L and its linewidth and line shape become the central object of any hole-burning experiment. In the early days of NMR the main parameters measured during a hole-burning experiment were the relaxation times T_1 and T_2. Today, as the content of this book will document, a hole-burning experiment can be used to measure a variety of experimental parameters characterizing both the matrix (host) and the dopant (guest) which constitute the two-level system.

3.1.2 Radiation-Induced Saturation Versus Chemical Depletion

a) *Transient Saturation*

In the previous subsection we described the first NMR hole-burning experiment, which was a true "saturation experiment". Here the depletion of the ground state is maintained by the incident radiation field at the frequency $\omega_L = \omega_0$. The macroscopic spin system is characterized by the probabilities of the spins of the ensemble being in the ground state $|0\rangle$ or in the excited state $|1\rangle$, and by their mutual phase relation. In this case the total wave function is given as

$$\Psi_{tot} = C_0|\Psi_0\rangle + C_1|\Psi_1\rangle \qquad \text{with}$$

$$|C_0|^2 + |C_1|^2 = 1 . \tag{3.2}$$

The matrix of complex coefficients C_0, C_1 (with C_0^*, C_1^* the complex conjugates) will therefore describe the evolution of the spin system. To treat the ensemble of spin systems, the relevant matrix is

$$\{\rho_{ij}\} = \begin{bmatrix} |C_0|^2 & C_0 C_1^* \\ C_1 C_0^* & |C_1|^2 \end{bmatrix} , \tag{3.3}$$

ρ_{ij} is the density matrix. (For details of the formalism see, for instance, [3.2] as here only the principle results will be summarized). The density matrix formalism can be used to express the the radiation-induced macroscopic polarization Π of an ensemble of spins. For brevity it is convenient to use the set of Pauli matrices

$$\sigma_z = \begin{bmatrix} 1 & 0 \\ 0 & -1 \end{bmatrix} ; \quad \sigma_x = \begin{bmatrix} 0 & 1 \\ 1 & 0 \end{bmatrix} ; \quad \sigma_y = \begin{bmatrix} 0 & -i \\ i & 0 \end{bmatrix} . \tag{3.4}$$

One can evaluate the expectation value of each spin component with the aid of a density matrix in this fashion

$$\langle \sigma_i \rangle = \text{Tr}\{\rho\sigma_i\} . \tag{3.5}$$

If we define the polarization Π as the average spin orientation, and use vector notation we have

$$\Pi = \langle \sigma \rangle . \tag{3.6}$$

The above definitions (3.3,4) yield the following values for the polarization Π.

$$\Pi_z = (\rho_{11} - \rho_{22})$$

$$\Pi_x = (\rho_{12} + \rho_{21})$$

$$\Pi_y = i(\rho_{12} - \rho_{21}) \, . \tag{3.7}$$

Using the polarization, the density matrix can be written in a way which allows a more pictorial interpretation

$$\{\rho_{ij}\} = \frac{1}{2} \begin{vmatrix} 1 + \Pi_z & \Pi_x - i\Pi_y \\ \Pi_x + i\Pi_y & 1 - \Pi_z \end{vmatrix} . \tag{3.8}$$

In (3.8) the diagonal terms are related to the z-polarization, i.e., the difference between excited and ground-state populations. The non-diagonal parts are related to the x and y components of the polarization.

So far we have treated a two-level system in a completely general way without distinguishing between spin systems and optical systems. Empirically, however, there is an important difference between the two cases. In spin systems, generally, both the $|0\rangle$ level and the $|1\rangle$ level are populated to a certain extent even in the absence of coherent exciting radiation. This is due to the fact that the thermal energy is sufficient to induce $|1\rangle \leftarrow |0\rangle$ transitions and, hence, maintain a finite polarization $\Pi_z < 1$.

In an optical system one can make the assumption that no thermal transitions between the $|0\rangle$ and $|1\rangle$ level will be induced. Thus in the absence of light the system is completely polarized and $\Pi_z = 1$. therefore the density matrix for this special case will have the following form

$$\{\rho_{ij}\}_{\text{no light}} = \begin{vmatrix} 1 & 0 \\ 0 & 0 \end{vmatrix} \tag{3.9}$$

since $\text{Tr}\{\rho^2\} = 1$, one can refer to this state as to a "pure" state.

Typical optical experiments (absorption, fluorescence, phosphorescence etc.) under incoherent excitation conditions with weak light sources do not cause the density matrix to depart much from that described in (3.9). Transient experiments, however, can be performed in a state of the system, in which the absorption is zero for a certain time window. Let us assume, for instance, that we have performed a $\pi/2$ excitation by irradiation with a coherent light pulse. At the end of this pulse the polarization has been rotated into the x direction, and the system can be described by the following density matrix [3.2]·

$$\{\rho_{ij}\}_{\pi/2} = \begin{vmatrix} 1/2 & 1/2 \\ 1/2 & 1/2 \end{vmatrix} . \tag{3.10}$$

We see by comparing with (3.8) that in (3.10) the Π_z polarization is zero, i.e., the number of absorbers in the ground state and in the excited state is equal. The system has no net absorption and is therefore transparent to an external probing light beam.

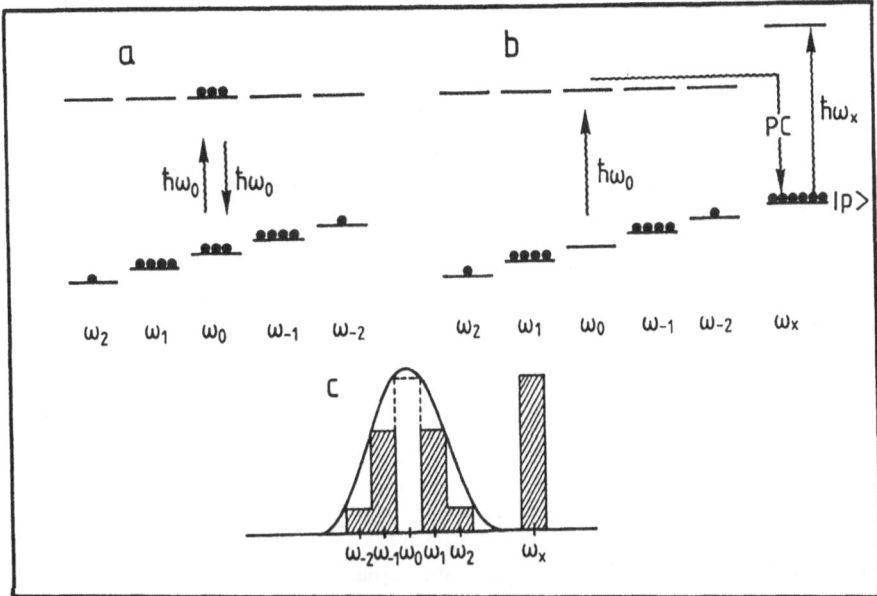

Fig.3.2. (a)Radiation saturation of an ensemble of five optical two-level systems at the center frequency ω_0. (b) Photochemical bleaching of an ensemble of five two-level systems at the center frequency ω_0. Creation of a photoproduct state absorbing at the frequency ω_x. (c) Absorption after photochemical bleaching as described in Fig.3.2b. The photoproduct shows up at the frequency ω_x

Figure 3.2a shows an example of (five) two-level systems in which the absorbers, centered at ω_0, are characterized by a polarization $\Pi_z = 0$ at frequency ω_0 (saturation). Therefore the "inhomogeneous" absorption line, as characterized by the various frequencies ω_{-2} to ω_{+2}, has a spectral hole in its center at ω_0. The width of the hole is on the order of $\Delta\omega_h$ (see below). It is obvious that, after the saturating light beam at ω_0 is removed, this state of the system will decay to the ground state as defined in (3.9) and the absorption band will return to its original shape. It should be noted, however, that a straightforward absorption experiment will only monitor the Π_z part of the polarization. By definition, the latter decays with the T_1 decay time of the $|1\rangle$ state T_{11}. We can therefore write, assuming the saturating beam is removed at $t = 0$,

$$|C_1(t)|^2 = |C_1(0)|^2 \exp(- t/T_{11})$$

and hence

$$C_1(t) = |C_1(0)| \exp(- t/2T_{11}) . \tag{3.11}$$

So far we have only dealt with the occupation numbers of the wave functions $|0\rangle$ and $|1\rangle$. They decay with a T_1 lifetime which is called the longitudinal relaxation time. We know, however, through the decay of the non-diagonal parts of the density matrix, that the phase relation characterized by the $\Pi_{x,y}$-polarization decays with a typical coherence time T_2^*, the transverse relaxation time. Note that the decay of the transverse relaxation does not require any energy relaxation, because no lattice heat bath is involved in the decay mechanism. If we insert the above empirical relaxation times into the density matrix, we obtain with the shorthand $1/\tau = 1/T_{10} + 1/T_{11}$

$$\rho_{ij}(t) = \begin{pmatrix} |C_0(0)|^2 \exp\left(-\dfrac{t}{T_{10}}\right) & C_0(0)C_1^*(0)\exp\left(-\dfrac{t}{2\tau} - \dfrac{t}{T_2^*}\right) \\[4mm] C_1(0)C_0^*\exp\left(-\dfrac{t}{2\tau} - \dfrac{t}{T_2^*}\right) & |C_1(0)|^2\exp\left(-\dfrac{t}{T_{11}}\right) \end{pmatrix} . \tag{3.12}$$

In the optical regime we know that the lifetime of the ground state, T_{10}, can be looked upon as being infinite and that the excited state decays with the lifetime T_1. We can therefore replace T_{11} by T_1 and can assume that $1/T_{10}$ is negligibly small.

By inspecting (3.12) we see again that the Π_z polarization decays with the decay constant T_1, whereas the transverse polarizations Π_x and Π_y have an additional decay process, which occurs on a time scale of T_2^*.

The homogeneous width of a "spin packet", as described in the introduction, is given by the decay of the transverse polarizations Π_x or Π_y and thus determines the spectral width of a transient hole, as shown in Fig.3.2c. By inspecting (3.12) we obatin

$$\Delta\omega_h = \frac{1}{2T_1} + \frac{1}{T_2^*} . \tag{3.13}$$

b) *Chemical Depletion*

So far we have treated the two-level system as a system which returns to the "thermal" population with a typical T_1 relaxation time. During equilibration the population numbers $|C_1|^2$ and $|C_0|^2$ generally establish a Boltzmann distribution and we have

$$|C_1|^2/|C_0|^2 = \exp(-\Delta E/kT) . \tag{3.14}$$

For optical systems the energy gap $\Delta E = E_1 - E_0$ is very large compared to kT and therefore the following condition holds

$$|C_1|^2 \simeq 0$$

and

$$|C_0|^2 = 1 . \tag{3.15}$$

This means the system is in its ground state, yielding the largest possible absorption signal.

For transient hole-burning, T_1 is the characteristic time with which a fully saturated state, such as that given by (3.10), returns to its full absorption strength. This time can be as long as minutes or hours if we are dealing with nuclear spin systems. However, T_1 is only on the order of 10^{-8} s if we are considering an optical two-level system in which $|1\rangle \leftarrow |0\rangle$ is an allowed transition (for instance the lowest optical transition of a dye molecule). If one were to perform optical transient hole-burning experiments, a typical time window for the experiment would therefore be on the order of 10^{-8} s. The first transient optical hole-burning experiment in the solid state was performed by *Szabo* in ruby [3.3]. Here a partially forbidden transition (the R_1 band) was used to facilitate the experiment through the longer lifetime of the excited state. Transient hole-burning has also been performed on vibrational modes of molecular impurities in alkali halides [3.4].

A comparatively small variation on the optical two-level system, as shown in Fig.3.3, converts the rather difficult saturation experiment into a simple chemical depletion experiment. If we introduce a three-level system, in which the $|0\rangle$ and $|1\rangle$ levels are defined as above (with a T_1 of typically 10^{-8} s) and where a third level $|p\rangle$ is connected to the $|1\rangle$ state in such a way, that it receives a fraction of the optically excited states which return with a rate $k_1 \simeq 1/T_1$ to the ground state $|0\rangle$ [3.5]. The branching ratio $p = k_{p/k_1}$ usually obeys the following inequality

$$p = k_p/k_1 < 1 \ . \tag{3.16}$$

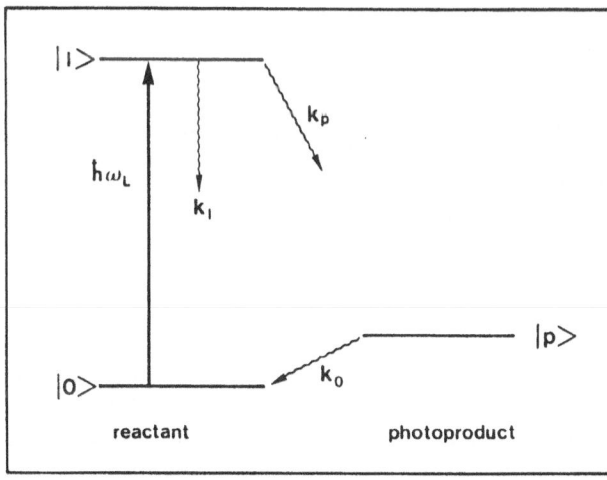

Fig.3.3. Three-level system with the reactant levels $|0\rangle$ and $|1\rangle$ and the photoproduct level $|p\rangle$. K_1, K_p and K_0 are the decay rate constants of the various states. ω_L is the hole-burning frequency

If we assume that the third level $|p\rangle$ is a photochemically generated state whose depletion rate k_0 to the original ground state is zero, then we have the ideal photochemical system, in which the laser-induced ground state depletion is permanent. For small values of k_0 the ground state absorption recovers within a time interval of roughly $t_0 = 1/k_0$.

The attractive feature of a photochemical scheme is the fact that even weak laser radiation will lead to ground state depletion through a "photochemical" channel, in which one needs n_c light quanta to create one photochemically generated state. n_c is roughly given by the ratio of the two rate constants k_1/k_p. This means that for small photochemical quantum yields $\eta \simeq 1/n_c$ one has to pump the system for sufficiently long times to convert a detectable fraction of the molecules from the $|0\rangle$ state to the $|p\rangle$ state. By introducing the photochemically induced state $|p\rangle$, the hole burning scheme of Fig.3.2a needs to be modified as is shown in Fig.3.2b. Here the presence of the state $|p\rangle$ leads to a depletion of molecular sites absorbing at ω_0 and thus produces a new photoproduct absorption" at the frequency ω_x (again the lineshapes are approximated by rectangular curves; the question of real lineshapes will be discussed below). Note, that the photochemical hole-burning process (Fig.3.2b) can be initiated under low intensity irradiation conditions, because the "bottleneck state" $|p\rangle$ can be accumulated infinitely slowly. The latter is not possible in the case of "excited-state hole-burning" (Fig.3.2a), where optical saturation has to be continuously maintained via light irradiation at the frequency ω_0. If we assume that the photoproduct absorbing at the frequency ω_x has the same oscillator strength as the original molecule, then in the photochemical scheme (Fig.3.2b) the total absorption integrated over the whole frequency range is maintained. In the case of the transient hole-burning scheme (Fig.3.2a) the total integrated absorption is reduced, because the band at ω_x does not exist.

The first optical experiments demonstrating the feasibility of a photochemical (or photophysical; see definition below) hole-burning scheme were published in 1974 by *Personov* and coworkers [3.6] and by *L.Rebane* and coworkers [3.7]. Both groups discovered the appearance of "holes" or "gaps" in the spectra of organic molecules in a host matrix after laser irradiation. In the next subsection we will discuss the photochemical requirements that are necessary for optical hole-burning to occur.

3.1.3 Photochemical Systems and Mechanisms

As has been pointed out in Chaps.1 and 2, the advantage of the method of photochemical hole-burning (PHB) is its high spectral resolution. The gain in resolution, compared to straightforward optical spectroscopy, is given by the ratio of the inhomogeneous linewidth $\Delta\omega_i$ and the homogeneous linewidth $\Delta\omega_h$. This criterion, however, can only be applied if the optical spectrum is dominated by the zero-phonon part of the optical transition involved. Since the zero-phonon criteria have been discussed in

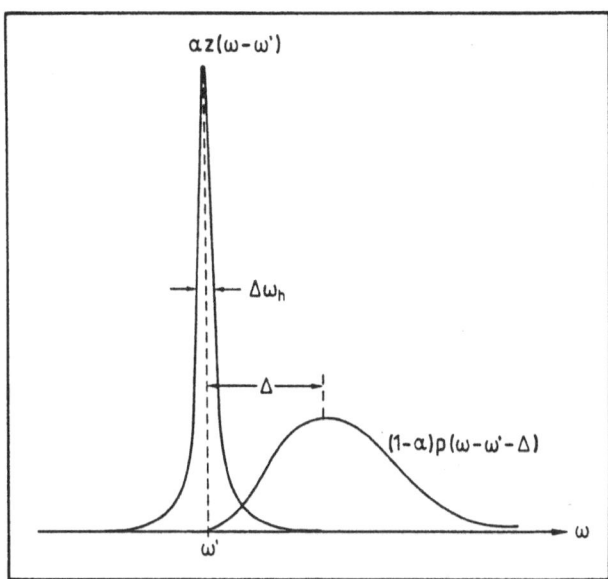

Fig.3.4. General lineshape function of an electronic excitation of a guest molecule in a solid host matrix: Schematic view of a zero-phonon line and its phonon sideband. The phonon sideband has an intensity of $(1-\alpha)$ and is displaced by the lineshift parameter Δ. The latter corresponds to half the "Stokes shift"

Chap.2, we will, at this point, only discuss the salient features of the required zero-phonon photochemistry.

The general lineshape of a molecular transition can be described by its zero-phonon part with the linewidth $\Delta\omega_h$ and a broad phonon sideband as depicted in Fig.3.4 (the width of the phonon sideband is usually several tens of wave numbers). The zero-phonon-line and the phonon sideband can be described by a normalized lineshape function $g(\omega)$, which has the following form (see for instance [3.5])

$$g(\omega-\omega') = \alpha z(\omega-\omega') + (1-\alpha)p(\omega-\omega'-\Delta) . \qquad (3.17)$$

Here the functions $g(\omega-\omega')$, $z(\omega-\omega')$ and $p(\omega-\omega'-\Delta)$ are normalized lineshape functions, α is the Debye-Waller factor which characterizes the fraction of the oscillator strength belonging to the zero-phonon part of the optical transition, and Δ is the displacement of the phonon sideband; the quantity 2Δ is sometimes referred to as the "Stokes shift" between the absorption and emission maxima of broad band transitions. In hole-burning spectroscopy one needs a situation in which α is close to one and, hence, most of the line intensity is centered at the frequency ω_0 of the laser which performs the photochemistry. For the remainder of this chapter we will assume that $\alpha \simeq 1$ and we will further assume a Lorentzian lineshape function $z(\omega-\omega')$ with

$$z(\omega-\omega') = \frac{1}{\pi} \frac{\gamma/2}{(\omega-\omega')^2 - \gamma^2/4} \,. \tag{3.18}$$

Here γ characterizes the full width at half maximum (FWHM).

From the electron-phonon coupling theory (see, for instance, [3.8]), it is well known that geometry changes in the excited state of molecules (or centers) lead to an exponential decrease of the Debye-Waller factor α and, hence, the intensity of the zero-phonon line. If we assume a coupling to a harmonic oscillator of frequency Ω_i and mass m_i and if we allow an excited state displacement q_{i0} (Fig.3.5), then the Debye-Waller factor α can be easily calculated for $T = 0$; it is, in a single-mode picture (see also [3.5]):

$$\alpha(T=0) = \exp\left[-\frac{1}{2} \frac{m_i \, \Omega_i^2}{\hbar} \, q_{i0}^2 \right] \tag{3.19}$$

Equation (3.19) shows that an excited state displacement q_{i0} suppresses the appearance of a zero-phonon line exponentially.

For photochemical reactions with repulsive excited state coordinates as shown in Fig.3.6a, the above arguments, which are based on the simple

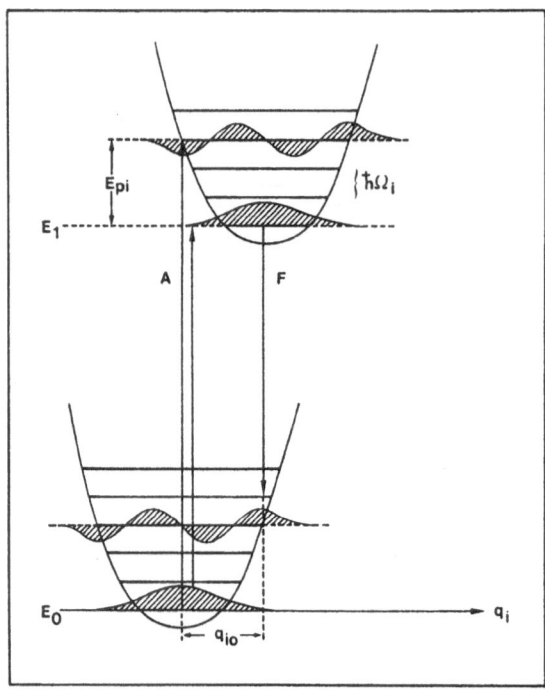

Fig.3.5. Energy diagram for interpreting phonon side-band spectra in terms of the excited state displacement q_{i0}. The phonon frequency is Ω_i The figure shows a zero-phonon and a three-phonon transition in absorption (A) and emission (F), respectively

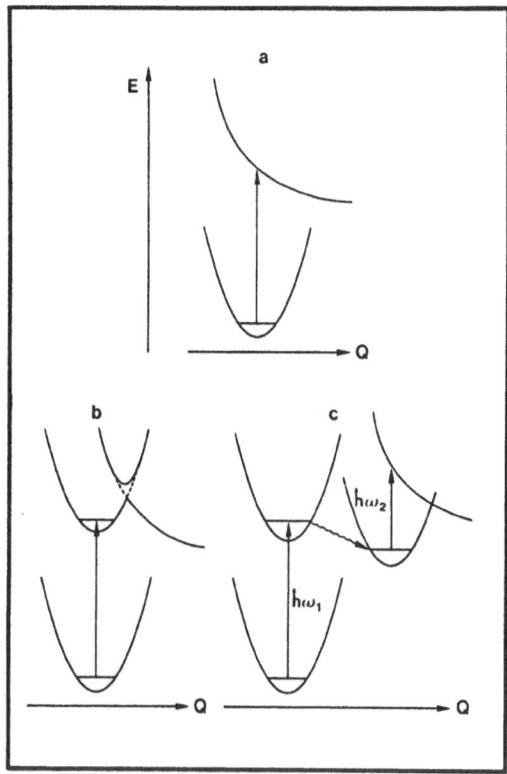

Fig.3.6a-c. Photochemical reaction schemes plotted as a function of one configurational coordinate Q. (a) Repulsive exciplex-like state. (b) Level-crossing from a "quasi-stable" excited state. (c) Two-photon reaction with photon energies $\hbar\omega_1$ and $\hbar\omega_2$. The level which absorbs the photon $\hbar\omega_2$ is a long-lived intermediate state

picture of a harmonic oscillator, can still be applied on a qualitative level. It is well known, for instance, that proton transfer processes (i.e., large geometry changes) are characterized by large Stokes shifts [3.9,10]. Such processes, or processes with large geometry changes in the excited state are therefore not compatible with the appearance of zero phonon lines and cannot be used for PHB spectroscopy. As one example we refer to the keto-enol proton transfer processes in the solid state, which show large spectroscopic shifts and broad optical spectra [3.11]. PHB experiments on such systems were therefore unsuccessful [3.12].

The question as to which kind of processes lead to chemical changes and still maintain the Frank-Condon requirement for strong zero-phonon lines (large α-values) has been discussed in previous publications [3.13,5]. Here we would like to highlight the main features and to survey the known photochemical systems. First of all, one has to fulfill the Frank-Condon requirements of a comparitively nondisplacive excited state. Two schemes which fulfill the requirement are shown in Figs.3.6b,c. Figure 3.6b shows a one-photon reaction where the excited state is stable for a certain time interval (on the order of the excited state lifetime T_1). The photochemistry occurs via level crossing (as shown in the figure) or via intersystem crossing which can be looked upon as a slightly modified version of the scheme shown in Fig.3.6b. Quite a few molecular systems show one-photon photochemistry (also referred to as "slow photochemistry" [3.5] due to the bottleneck caused by level crossing or intersystem crossing processes). The photochemistry is, in most cases, a proton transfer reaction.

The best-known systems are free-base phthalocyanine and porphine [3.7,14,15] which show tautomeric photoreactions of the inner protons. Other proton transfer systems of lower symmetry are chlorin [3.16] and quinizarin [3.17]. A third class of systems, in which protons are likely to be involved in the photochemistry, are resorufin [3.18] and biological macromolecules like phycocyanobilin and phycoerythrobilin [5.19,20]. The feature common to all of the above photochemical sytems is the proton transfer process, which most probably occurs in the excited triplet state and which leads to a photochromic shift where the photoproduct absorption (frequency ω_x in Fig.3.2) is well separated from the laser frequency ω_o. The frequency shift is usually on the order of 100 to several 100 cm^{-1}.

Due to the sizeable photochromic shift one refers to the above hole-burning systems as *photochemical hole-burning* systems (PHB systems). This nomenclature was used because a proton transfer reaction can be viewed as a reaction producing molecular photochemical changes.

There is, however, a second class of reactions, where there are no molecular changes involved. Molecules like tetracene and perylene [3.6] which are assumed to be photochemically stable, show hole-burning phenomena in appropriate matrices like ethanol glasses [3.6,21] or amorphous hydrocarbon matrices [3.22]. Here one assumes that there is a light-induced rearrangement of the host-guest cluster which leads to a change in site energy, i.e., to a change of the absorption energy of a molecule in a specific local environment. *Hayes* and coworkers [3.21] have shown,that the "photoproduct" of such a reaction absorbs at energies which are close to the laser frequency ω_0 (Fig.3.2) and have therefore utilized the name NPHB (*non-photochemical hole-burning*, see Chap.5). The similarities between PHB and NPHB systems are rather striking and it seems possible that sometimes both mechanisms are involved simultaneously. The latter may occur if one investigates molecules which can undergo photochemical changes in a matrix environment which is appropriate for non-photochemical hole burning [3.23].

A very different class of PHB reactions are the two-photon reactions. Here the first photon produces a stable excited state, as is shown in Fig. 3.6c. This first excited state can be long lived and, hence, can show an absorption line which is narrow and which is dominated by a zero-phonon transition. The second photon subsequently leads to the photochemical reaction. The overall photochemical two-step reaction shows high quantum yields only if the intermediate excited state is long lived [3.24]. Therefore Fig.3.6c shows a second intermediate excited state which acts as an accumulating level. The best-known system exibiting two-photon photochemistry is DMST (dimenthyl-s-tetrazine). It undergoes irreversible photodecomposition; it was also the first system for which the high resolution features of hole-burning were experimentally proven [3.25].

Recently *Moerner* and coworkers have found a two-step photoionization reaction of aromatic molecules in boric acid glass [3.26,27]. They have evidence that the intermediate metastable level (accumulation state) is a long-lived triplet state of the aromatic molecule. Such a reaction scheme for aromatic solutes in boric acid glass had been proposed earlier by *Joussot-Dubien* and coworkers [3.28,29]. The scheme has advantages in the area of applications of hole-burning for data storage (Chap.7).

So far we have summarized the principal examples of organic molecules showing photochemical hole-burning in the visible range of the spectrum (Fig.3.7). Inorganic hole-burning systems like color centers [3.30], mixed organic-inorganic hole burning systems [3.31] and hole-burning systems in the infrared spectral range [3.32,33] will be discussed in Chaps.4 and 6.

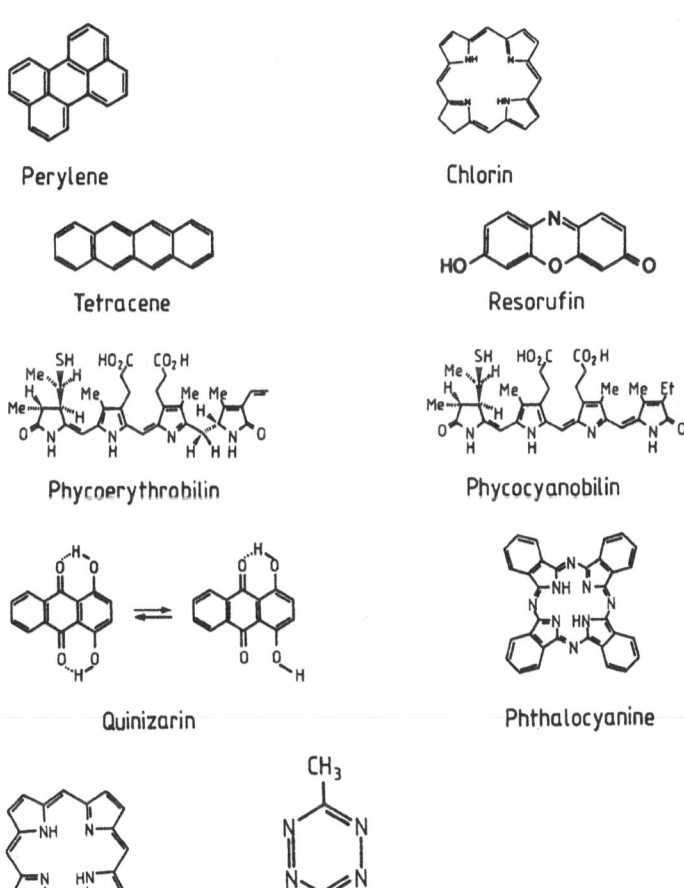

Perylene   Chlorin

Tetracene   Resorufin

Phycoerythrobilin   Phycocyanobilin

Quinizarin   Phthalocyanine

Porphin   Dimethyl – s – tetrazine

Fig.3.7. Various dye molecules which show photophysical and photochemical hole-burning

3.2 Spectroscopic Analysis of Hole-Burning Experiments

3.2.1 General Remarks

In the previous section, the lineshape of a photochemical hole was described using an empirical lineshape function. As shown in Fig.3.4, the line is composed of a Lorentzian zero-phonon origin and a broad phonon sideband, whose envelope is given by a Poisson distribution (see for instance [3.34]). If the electron-phonon coupling is small ($\alpha \simeq 1$), the area under the phonon sideband is small; if the electron-phonon coupling is large ($\alpha < 1$), the phonon sideband dominates the spectrum more and more and is, for small α-values, characterized by a Gaussian envelope function [3.35,36].

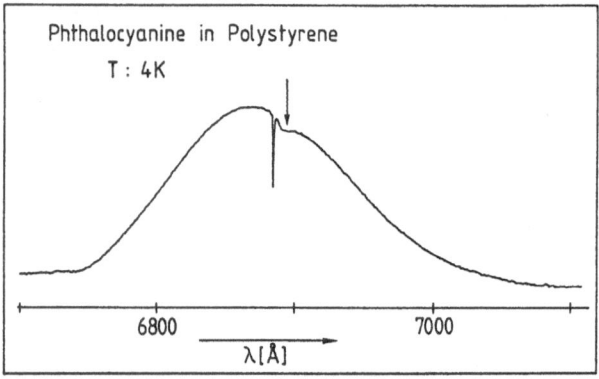

Fig.3.8. Experimental hole spectrum of phthalocyanine in polystyrene at 4 K. The pseudo-phonon wing is marked by the wavy arrow

The experimental hole-burning spectra, as shown in Fig.3.8, resemble the idealized "site spectrum" of Fig.3.4; however, there is one main difference: For large α values (the usual case), the phonon sideband (wavy arrow) appears at the low energy side of the zero phonon line. This is not the case for a "real" phonon sideband, as shown in Fig.3.4, since in order to excite the phonon sideband one needs the the electronic transition energy plus the energy of a one- or multiphonon excitation. The reason for the appearance of the "pseudo-phonon sideband" (wavy arrow) has been discussed extensively in [3.5]. In Fig.3.8, the intensity of the real phonon sideband is very small and thus the band does not show up in the spectrum. Nevertheless, most of the phonon information can be obtained from the pseudo-phonon sideband.

In this section we will deal with the experimental information which can be obtained from evaluating hole-burning spectra. From an analysis of the zero-phonon line, one can obtain information about

92

i)   fast relaxation processes (typically 10^{-13}-10^{-8} s), which may be categorized as non-adiabatic processes, and

ii) slow structural relaxation processes (typically 10^{-1} s extending to 10^{12} s), which may be categorized as adiabatic processes.

The information on both fast and slow relaxation phenomena that can be derived from PHB studies verifies that PHB can be a valuable tool for studying dynamical processes in glasses. Figure 3.9 shows how hole-burning techniques and other known optical methods like NMR and ESR experiments cover a dynamic range of close to 30 orders of magnitude. It is our opinion that, especially for the very slow processes, the potential of the hole-burning technique has not yet been fully realized.

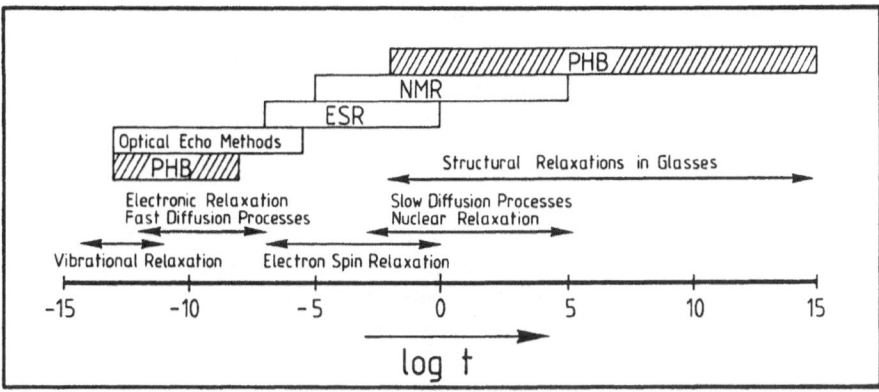

Fig.3.9. Logarithmic scale of relaxation times and of methods of measuring characteristic time constants. Hole-burning is especially well suited for very short and very long relaxation times (see text)

From the intensity and width of the zero-phonon line as a function of light intensity and time, one can obtain information on molecular parameters characterizing the hole-burning system. Such parameters are:

iii) Debye-Waller factors, Debye temperatures, photochemical quantum yields, etc.

Finally, the phonon sideband of the hole-burning spectra can be analyzed. From these data one can obtain information on phonon parameters and their distribution functions. Typical parameters are:

iv) Characteristic phonon frequencies (local mode frequencies), distribution of Debye-Waller factors and distribution functions of the electron-phonon coupling strength [3.5].

3.2.2 Fast Relaxation Processes and Excited State Dephasing

a) *Lineshape Analysis*

The total absorption of a photochemical hole is composed of the absorption of the various molecular sites (spin packets). Each site can be described by a general lineshape function, as given in (3.17).

A hole-burning experiment can be viewed as a "tandem two-photon experiment", i.e. an experiment in which photon number one initiates a photochemical change and a second photon monitors the change via an absoption or emission experiment. Therefore the lineshape of a hole reflects both processes, the burning as well as the monitoring process. This is evident if one writes down the lineshape function L of a photochemical hole (see for instance [3.5,37-39,49]).

$$ L^{\tau}(\omega) = A_0 - \frac{\sigma}{4\pi} \int_0^{2\pi} d\phi \, \sin\theta \int_{-\infty}^{+\infty} d\omega' \, N^{\tau}(f_B(\theta), \omega') \, f_{sc}(\theta,\phi) g(\omega-\omega') . \quad (3.20) $$

In (3.20) A_0 is the absorption intensity before the hole-burning, where it is assumed that the absorption band is flat over the range of the hole-burning spectrum; τ is the burning time during which the sample is irradiated at high intensity to initiate the PHB photochemistry; σ is the total absorption cross section integrated over frequency and N is the total number of dopant molecules in the sample volume.

Since most hole-burning experiments are performed with polarized laser light, one has to take into account the light beam polarization and the sample geometry. We assume that the laser light has a polar angle θ with respect to the molecular transition dipole moment; ϕ is the azimuthal angle. Therefore we can define a differential absorption cross section as follwos

$$ \sigma_{diff} = \frac{\sigma(\theta,\phi)}{4\pi} \sin\theta \, d\theta \, d\phi . \quad (3.21) $$

We can also define the various polarization factors for burning, f_B, and for scanning, f_{sc}, as follows

$$ f_B(\theta) = \cos^2\theta , \quad (3.22a) $$

$$ f_{sc} \parallel = \cos^2\theta , \quad (3.22b) $$

$$ f_{sc} \perp = \sin^2\theta \, \sin^2\phi , \quad (3.22c) $$

assuming parallel and perpendicular scanning experiments (with respect to the primary photochemical experiment).

94

As we can see by inspection, (3.20) factorizes into two parts. The first part represents the photochemical changes $N^{\tau}(f_B(\theta), \omega')$ with

$$N^{\tau}(f_B(\theta), \omega') = N_0(\omega') \exp\left[-\frac{\sigma I}{\hbar\omega_L}\eta\tau\, f_B(\theta)\, g(\omega_L - \omega')\right], \qquad (3.23)$$

where I is the laser intensity, η is the quantum yield of the reaction, $N_0(\omega')$ is the inhomogeneous site distribution before laser irradiation and ω_L is the laser frequency.

The second part of (3.20), namely $f_{sc}(\theta, \phi)\, g(\omega - \omega')$ is the part representing the "scanning photon"; it contains the dynamical part of the linewidth information, i.e. it contains the information on the homogeneous linewidth as given by (3.13). It should be pointed out clearly that extraction of the dynamical homogeneous linewidth from a hole-burning experiment automatically makes the assumption that the site distribution function N^{τ}, as given in (3.23), is static, i.e., by the assumption that there is not spectral diffusion smearing out the N^{τ} curve (see below). If one assumes at this point that spectral diffusion processes can be neglected, then one can indeed extract the $g(\omega)$ lineshape if certain experimental conditions are met (short burning time limit). Under these simplifying conditions and assuming Lorentzian lineshapes the measured hole spectrum L has a zero-phonon part which is roughly twice the homogeneous linewidth given in (3.18).

A more careful analysis of the zero-phonon part of the hole spectrum [3.39-41] yields for the lineshape of the zero-phonon hole $zz(\omega)$

$$zz(\omega) = \sigma\alpha\,\frac{N_0(\omega_L)}{N}\,\frac{1}{4\pi}\int_0^{2\pi} d\phi \int_0^{\pi} \sin\theta\, d\theta \int_{-\infty}^{+\infty} d\omega'$$

$$\cdot\left\{1 - \exp\left[-\sigma\,\frac{I}{\hbar\omega_L}\,\eta\tau\, f_B(\theta)\,\alpha\, z(\omega_L - \omega')\right]\right\} f_{sc}(\theta, \phi)\, z(\omega - \omega') . \quad (3.24)$$

The above expression is lengthy but simple in its structure. It contains the site distribution function nz of the hole to be discussed later, where

$$nz^{\tau}(f_B(\theta), \omega') = \sigma\, N_0\,\frac{\omega_L}{N}$$

$$\cdot\left\{1 - \exp\left[-\frac{\sigma I}{\hbar\omega_L}\eta\tau\cdot f_B(\theta)\cdot\alpha\, z(\omega_L - \omega')\right]\right\}. \qquad (3.25)$$

This site distribution is convoluted with the homogeneous lineshape function $z(\omega - \omega')$ of (3.18).

In this part of the chapter we will assume that the hole-burning experiments were performed in a way which allows extraction of the function $z(\omega-\omega')$ (for the limits of this approach, see below). We also assume that this linewidth function has Lorentzian character. With these assumptions we can say that a hole-burning experiment yields a linewidth which is twice the "homogeneous linewidth" $\Delta\omega_h$. Under these assumptions a fluorescence line narrowing (FLN) experiment will also yield a linewidth of 2 $\Delta\omega_h$.

b) *Temperature Dependence of the "Homogeneous" Linewidth*

Crystalline Systems. The best investigated hole-burning systems in crystalline materials are free-base phthalocyanine [3.7] and porphine [3.15]. However, as far as the temperature dependence of the homogeneous PHB linewidths is concerned, the information is rather scarce. Some data exist for the system free-base phthalocyanine (H_2Pc) in n-octane [3.42] and the related system tetra-tert-butyl-H_2-phthalocyanine in tetradecane [3.43]. A second set of data exists for the system free-base prophine in n-octane [3.44,45]. All of the above-mentioned data on dephasing of organic guesthost systems have to be discussed in the light of earlier experiments and theoretical models dealing with the dephasing of inorganic host-guest systems.

As early as 1963 *McCumber* and *Sturge* [3.46] discussed the temperature dependence of the homogeneous linewidth of the R-lines in ruby. They described the homogeneous linewidth in terms of Raman scattering of phonons by the impurity centers. This Raman two-phonon process involves all thermally accessible phonons and hence does not hinge on specific phonon frequencies. The calculation yields a T^7 linewidth dependence in the low temperature approximation. In the calculation the phonons are approximated by a Debye density of states function and thus the calculation yields a T^7 dependence for the linewidth and a T^4 dependence for the lineshift.

The experimental data on organic systems [3.38,43,45] are scarce. *Rebane* et al. [3.38,43] have compared their linewidth data with a two-phonon Debye model (i.e., A T^7 dependence) and with a model involving a single-mode two-phonon process, in which the linewidth follows the relation

$$\Delta\omega_h = b\bar{n}(\hat{\omega})\,[\bar{n}(\hat{\omega}) + 1] \qquad \text{with}$$

(3.26)

$$\bar{n}(\hat{\omega}) = \left[\exp\left(\frac{\hbar\hat{\omega}}{kT}\right) - 1\right]^{-1}$$

where $\bar{n}$ is the population and $\hat{\omega}$ is the phonon frequency of the local mode. The data [3.38] seem to fit the single-mode picture better than the Debye model yielding a local mode frequency of about 10 cm^{-1}.

In the porphine system [3.44], for which both frequency shifts and linewidths were measured as a function of temperature, an exchange model was used. Here the formal treatment is analogous to the treatment of the NMR linewidths of a hopping spin with site-dependent Larmor frequency [3.47]. In this case the optical excitation is coupled to one specific phonon mode. This coupling causes the transition frequency ω_0 to be shifted to $\omega_0 + \delta\omega$ if the phonon state is populated. If one draws an analogy between the optical frequency ω_0 of the hole-burning experiment and the Larmor frequency of an NMR experiment one can utilize the well-known exchange equations to deduce

$$\Delta\omega(T) - \Delta\omega_0 = \frac{\tau(\delta\omega)^2}{1 + (\delta\omega)^2\tau^2}) \; \exp(-\hbar\hat{\omega}/kT) \; . \tag{3.27}$$

Here $\delta\omega$ is the difference in "Larmor frequencies", τ is the correlation time (i.e., the lifetime in the excited phonon state) and $\hbar\hat{\omega}$ is the energy splitting between the purely electronic state and the phonon state.

Equation (3.27) yields for fast exchange [$(\delta\omega)\tau < 1$] a linewidth proportional to the correlation time τ i.e. *narrow lines* for *fast processes* or fast exchange, a phenomenon which is quite common in the field of NMR and is still rather speculative in the field of optical transitions [3.50].

The slow exchange limit [$(\delta\omega)\tau > 1$] yields a $1/\tau$ dependence. Note that in the case of slow exchange the $\delta\omega$ term in (3.27) cancels out, i.e. there is no explicit dependence on the coupling strength. The data on porphine can be interpreted within the exchange model if one assumes intermediate exchange with $(\delta\omega)\tau$ on the order of unity. From the dependence of the linewidth on the temperature, which is exponential, one obtains local mode frequencies $\hat{\omega}$ of 15 ± 4 cm^{-1}.

One should point out explicitly that the temperature dependence of the exchange model [3.44] and that of the two-phonon scattering process yield identical results for the low temperature regime $\hbar\hat{\omega} \gg kT$. In both cases a plot of $\log(\Delta\omega)$ versus $1/T$ yields a straight line, as is reproduced by the available experimental data.

In conclusion, one can say that the few experimental data for crystalline hosts point towards a dephasing model in which local modes are involved. The frequency of the local modes seems to be on the order of 10-20 cm^{-1}. Optical or IR experiments to correlate the linewidth data with observable IR or phonon sideband transitions do, to our knowledge, not exist.

Amorphous Host-Guest Systems. The first experiments on the homogeneous linewidths in glasses were performed with rare earth ions in silicate glasses [3.51-55]. Some of the transitions investigated were: Eu^{3+}, $^5D_0 \rightleftarrows$ $^7F_0$; Pr^{3+}, $^3P_0 \rightleftarrows$ $^3H_4$ and Yb^{3+}, $^2F_{5/2} \rightleftarrows$ $^2F_{7/2}$. The experiments were performed in a temperature range between 1.6 and 300 K. Two observed phenomena were particularly surprising. First, the linewidth in glasses is typically one to two orders of magnitude larger than in comparable crystalline systems. Second, the temperature variation of $\Delta\omega_h$ is proportional to T^α with α values between 1 and 2 (as compared to $\alpha = 7$; see above).

The first experiments with organic glasses to investigate the lowest singlet transition $S_1 \leftarrow S_0$ of organic molecules yielded similar results [3.56,57], namely comparatively large homogeneous linewidths and algebraic temperature dependencies with $\Delta\omega_h \simeq T^\alpha$.

The experimental studies of homogeneous linewidths in glasses were stimulated by the theoretical interest in the subject of low-lying excitations in glasses. These excitations have a constant density of states which, at low energies, is larger than the Debye phonon density of states (which increases as $\omega^2 d\omega$). The low-energy excitations were introduced to interpret the anomalous specific heat data of glasses [3.58]. The model of low lying two-level systems (TLS) was originally proposed by *Anderson* et al. [3.59] and *Phillips* et al. [3.60]; it was subsequently adopted to interpret the anomalous homogeneous linewidths observed in rare earth ion systems [3.51].

The origin of the dephasing of molecules or ions in amorphous media stimulated many theoretical papers dealing with the general theme of dephasing in amorphous hosts [3.56,57,61-69]. All these theoretical approaches utilize the concept of TLS with a density of states which is approximately constant (on the energy scale). The theoretical models vary with respect to the coupling of the TLS to other degrees of freedom like phonons, fractons or librational excitations. The difficulty at the present time is, in our opinion, the fact that a multitude of experimental data is available yielding T^α dependencies with $1 < \alpha < 2$ and at least three to four different theoretical approaches can be used to interpret the optical data.

In recent years quite a few review articles have appeared on the subject of linewidths in amorphous glasses [3.37,65,70-72]. In order to highlight the main features of the subject rather than discussing the details, which the reader will find in the above review articles, we will select some aspects of current research and also discuss some of the controversies.

One of the most frequently discussed issues is the question of whether the algebraic exponent α allows one to draw conclusions about the dephasing mechanisms of optical centers in amorphous media. Several groups have rencently found an α-value of 1.3; the group of *Völker* et al. was able to measure a $T^{1.3}$ law for practically all PHB experiments using free base porphin (H_2P) [3.73-78]; the group found the same α-law for other molecules such as dimethyl-s-tetrazine (DMST), resorufin and cresyl violet. The pertinent temperature range was $0.3 < T < 20$ K. *Völker* et al. promote the idea tht the exponent of 1.3 reflects a fundamental property of the amorphous state. The T^1 dependence that *Völker* et al. have found for small linewidths (< 50 MHz) complicates the picture somewhat. *Lyo* et al. [3.69] speculated that this may indicate a transition from a quadrupolar process to a dipolar process. They found for a dipole, TLS-quadrupole interaction a theoretical $T^{4/3}$ dependency as experimentally measured by the

Völker-group. The Lyo-Orbach exponent is given by

$$\Delta\omega_h \simeq T^{1+\bar{d}/4} \tag{3.28}$$

where $\bar{d}$ is the spectral dimensionality, as defined by *Alexander* and *Orbach* [3.79]. The latter model hinges on the fact that the density of states of fractals is different from the density of states of phonon-like excitations. Without the elegant conjecture of *Orbach* et al. [3.79], one can also explain a $T^{1.3}$ dependency by assuming a dipolar interaction and a density of states function which has the following form:

$$\rho(E) = \rho_0 E^\mu . \tag{3.29}$$

For this case *Lyo* [3.70] obtained a linewidth of

$$\Delta\omega_h \simeq T^{1+\mu} . \tag{3.30}$$

Inserting a μ value of 0.3 one obtains the above 1.3 dependency. The issue of temperature exponents is further complicated by the fact that other very plausible theories, like the one of *Jackson* and *Silbey* [3.68] can also yield α exponents of 1.3-1.5 by assuming a dephasing process which is initiated by both TLS processes as well as by localized librational modes. This model seems especially realistic, since *Jankowiak* et al. [3.80] have recently found both a $T^{1.4}$ dependence and evidence for a librational mode of low frequency (7-11 cm^{-1}). Their photophysical hole-burning system was 9, 10-diphenylanthracene in 2,3-dimethylanthracene.

To make the situation regarding the $T^{1.3}$ dependence even more puzzling, the Bell-Laboratory/Wisconsin groups [3.65,81,82] have found a $T^{1.3}$ dependence for Nd^{3+}-doped silica (as detected by echo experiments). Their data interpretation, however, is based on a model of spectral diffusion as formulated by *Black* and *Halperin* [3.82] based on previous theories for spin sytems of *Klauder* and *Anderson* [3.83], and *Mims* [3.84]. It should be noted that a spectral diffusion model describes the frequency change of the optical system (A-spins) due to a spin flip of a more abundant species characterizing the amorphous lattice (B-spins). Hence, the transition will "wander away from ω_0", i.e. from the resonance frequency of the site-selected transition. This process of diffusional drift of the resonance frequency is different from a simple T_2 relaxation or dephasing process. It can occur on time scales larger than T_2 and can carry the frequency over an interval which is larger than that defined by the width of a conventional "spin packet". The $\alpha = 1.3$ exponent observed for the Nd^{3+} system, was attibuted to a density of states of the TLS system following an $E^{0.3}$ dependence. If one were to assume a constant density of states in energy space, the *Black* and *Halperin* model would yield a linear temperature dependence.

An independent interpretation of the $T^{1.3}$ dependence based on a spectral-diffusion picture was presented by *Hunklinger* et al. [3.85]. The researchers pointed out that the time and temperature dependence of the linewidth as derived by *Black* and *Halperin* [3.82] is of the form

$$\Delta\omega(t,T) = C_1 T[C_2 + \ln(C_3 T^3)] \tag{3.31}$$

if one considers direct processes. Here the constants C_1, C_2 and C_3 are determined by the sound velocity, the deformation potential and the TLS density of states [3.82]. The authors show numerically that the above dependence can be numerically approximated by a $T^{1.3}$ law assuming reasonable values for the various potentials and density of states. It should also be noted that (3.31) contains a time dependence which is, for the long time limit, logarithmic. This had already been pointed out by *Reinecke* [3.61] who also discussed the short-time limit, $t \ll T_1$, where the linewidth increases linearly with time. In the next section we will discuss the phenomenon of slow spectral diffusion processes from a different point of view.

So far we have focused on the $T^{1.3}$ class of hole-burning systems. There are, however, a significant number of experiments that do not fit into this picture of a generalized exponent. In inorganic glasses the exponents seem to vary between 1 and 2.2 [3.52-55,86-88]. In organic glasses, too, exponents between 1 and 2 are being reported [3.89-95] and thus contribute to a colorful picture. It is our firm belief that the question of temperature exponents will be of further importance; however, the question of spectral diffusion deserves serious discussion. Only if the optical data can be linked to other relevant materials parameters such as the specific heat, the thermal expansion, and dielectric losses will it be possible to finally estabish the physical processes that are responsible for the dephasing of excited optical states in glasses.

3.2.3 Spectral Diffusion in Glasses

a) *TLS Parameters and Tunnelling Rates*

Whereas crystals are characterized by periodic potentials with well-defined and identical potential wells (Fig.3.10a), glasses are characterized by potentials with variable minimum energies and with variable barrier heights. This randomness, as shown in Fig.3.10b, is due to the local disorder of the amorphous state, and reflects the non-periodic nature of the potential curve.

Following the model of *Anderson* et al. [3.59] and *Phillips* [3.60], instead of working with a complex coordinate, as shown in Fig.3.10b, one can choose a double-well potential, as shown in Fig.3.10c. In this simplified representation, one considers only two minima of the multivalley potential of Fig.3.10b. The amorphous nature of the material is taken into account by allowing the main parameters Δ and V_0 of the two-level systems to have broad distributions rather than well-defined values. This

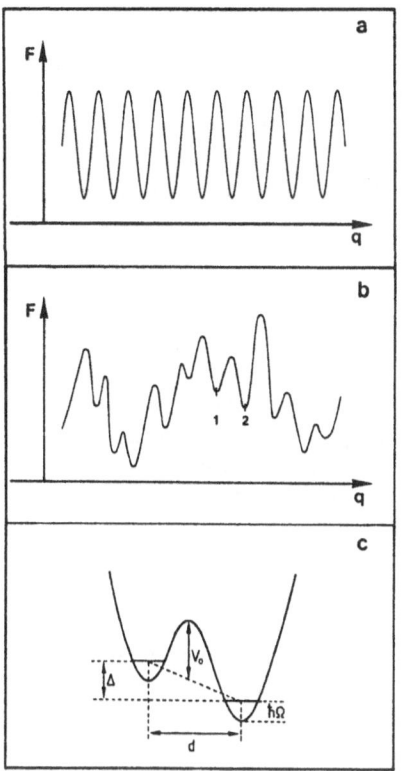

Fig.3.10. (a) Periodic potential of a crystal-line solid. (b) Potential along a fictitious coor-dinate q of a glass: Note the lack of period-icity. (c) Idealized two level scheme with the two parameters Δ and V_0 (see text). The oscillators in the two wells are separated by the distance d and have a zero-point fre-quency on the order of Ω

assumption has far reaching consequences: The glassy system will be char-acterized by dynamical parameters whose distribution is so wide that the slow components characterizing the ensemble of TLS will, within finite times, not reach their equilibrium configurations or energies. In other words, the glassy state is a non-equilibrium state and, if we are able to make our optical experiments sensitive enough and observe the system over long enough periods of time, we should be able to observe changes which are indicative of the non-equilibrium nature of the system.

For the formal description of the glass one uses a 2x2 Hamiltonian matrix, which we write down in the site representation as (see for instance [3.96]):

$$H_{TLS} = \frac{1}{2} \begin{vmatrix} \Delta & \Delta_0 \\ \Delta_0 & -\Delta \end{vmatrix} . \tag{3.32}$$

In the above matrix the Δ parameter characterizes the potential asymme-try and Δ_0 is a tunnelling parameter of the form

$$\Delta_0 = \hbar\omega \exp\left[-d\sqrt{2mV_0/\hbar^2}\right] = \hbar\omega \exp(-\lambda) \tag{3.33}$$

where d is the separation of the two potential wells along the coordinate q, m is the particle mass and V_0 is the barrier height as defined in Fig.3.10c. The energy $\hbar\omega$ is on the order of the quantum energy of a harmonic oscillator. The exponent of (3.33) is frequently written in terms of a tunnelling parameter λ, which contains the various microscopic parameters. The two energy eigenvalues of the TLS system are well known and are given by the following expression

$$E^2 = (\Delta^2 + \Delta_0{}^2) . \tag{3.34}$$

The main assumption of the TLS model, as spelled out in [3.59], is the fact that the distribution of the pertinent tunnelling parameters Δ and λ is assumed to be flat in the range of relevance for the low-temperature specific heat experiments and also in the range that characterizes our low-temperature optical experiments dealing with spectral diffusion problems. Therefore one can write for the distribution function P

$$P(\Delta,\lambda) = \overline{P} = \text{const} . \tag{3.35}$$

Equation (3.35) contains conceptually the most relevant assumption about the distribtion of TLS parameters from which the specific heat anomaly of glasses follows (the specific heat falls off linearly with temperature rather than with a T^3 law, as predicted by the Debye model). The fact that the energy asymmetry parameter Δ has a flat distribution is readily understandable: more difficult, conceptually, is the assumption that the distribution of the tunnelling parameter λ is flat rather than that of the tunnelling matrix element Δ_0. Also λ contains as miscroscopic parameters both the distance between the potential wells as well as the square root of the barrier height, V_0 and can therefore not be visualized as easily as the asymmetry parameter Δ.

It is important to know that a flat distribution in the parameters Δ and λ leads to a distribution which is close to being flat in energy space. A straightforward calculation shows that the density of states as a function of energy can be written as [3.97]

$$\hat{P}(E) \, dE \rightarrow P(\Delta,\lambda) \, \ln(2E/\Delta_{0,min}) \, dE . \tag{3.36}$$

The above energy dependence is only logarithmic and can, for systems with high barriers and with reasonable energy values, be considered as quasi-flat.

One of the interesting questions concerning a system with a distribution of barrier heights and tunnelling distances is the distribution of relaxation rates. This question is intriguing, because it is likely that hole-burning experiments or other experiments will allow the measurement of relaxation rates over wide ranges. In 1972, *Jäckle* [3.98] calculated the distribution of rates P(R)dR in amorphous systems characterized by a constant Δ and λ distribution:

$$P(R)dR = \frac{1}{2}P(\Delta,\lambda) \frac{dR}{R(1 - R/R_{max})^{1/2}} . \tag{3.37}$$

The above equation is taken at a constant energy. It shows that the distribution of rates has two poles; one pole at $R \rightarrow 0$, which is non-integrable and one pole at $R \rightarrow R_{max}$, which is integrable. The rate distribution is depicted in Fig.3.11. Obviously one needs a cutoff value for low values of R at R_{min} in order to provide an integrable expression. Also, a maximum rate is defined at R_{max}. Due to the $1/R$ dependence of the curve for $R \rightarrow 0$, the largest contribution of the observed dynamical processes will stem from the "small-R" region of the curve; this will be the origin of the observed logarithmic decay laws at long times (see below).

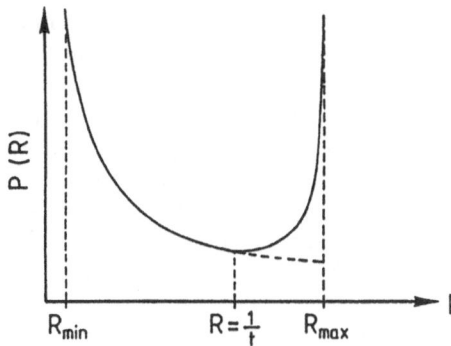

Fig.3.11. Rate distribution as calculated by *Jäckle* [3.98]. For the definition of the two limiting rates R_{min} and R_{max} see text

The absolute value of the rates can also be calculated [3.98], yielding

$$R = \left(\frac{1}{c_L{}^5} + \frac{2}{c_t{}^5} \right) \left(\frac{\partial \Delta}{\partial \varsigma} \right)^2 \frac{\Delta_0{}^2 E}{2\pi\rho\hbar^4} \coth(E/2kT) \tag{3.38}$$

where c_L and c_t are the longitudinal and transverse sound velocities, $d\Delta/d\varsigma$ is the deformation potential parameter, ς is the strain coordinate and ρ is the mass density of the material.

Since the distribution of rates is of major importance for our further considerations we would like to discuss (3.37) by plotting the dependence of the tunnelling parameter Δ_0 on the variables λ and Δ. Note, that the rates R are, under the assumption of constant energy, proportional to $\Delta_0{}^2$ (3.38).

Figure 3.12a shows that Δ_0 falls exponentially with the λ parameter according to

$$\Delta_0 = const \cdot exp(-\lambda) \tag{3.39a}$$

and thus

$$\frac{d\Delta_0}{d\lambda} = - \Delta_0 . \tag{3.39b}$$

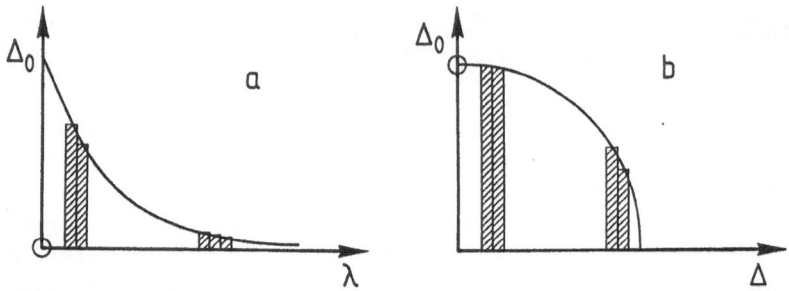

Fig.3.12. Functional dependence of the tunnelling matrix element Δ_0 on the TLS-parameters λ (Fig.3.12a) and Δ (Fig.3.12b) (see text)

In the model description of the amorphous state, one assumes a constant density of states in the λ domain. This is analogous to our usual assumption in solid state physics where one assumes for the calculation of phonon properties, a constant density of states in the wave-vector domain k; here the derivatives dE/dk are zero in regions of k space where the density of states function has a pole. According to Fig.3.12a, the slope $d\Delta_0/d\lambda$ is zero for $\Delta_0 \rightarrow 0$ and $\lambda \rightarrow \infty$, i.e., the density of states diverges for systems having very high barriers and having rates which approach zero. This is one of the two poles in the rate density which is also evident from Fig.3.11. Since the pole is not integrable, one has to introduce a minimal rate R_{min}.

The dependence of the Δ_0 parameter on Δ is depicted in Fig.3.12b. For constant energy, the Δ_0 values lie on a circle defined by the following equation

$$E^2 = \Delta_0{}^2 + \Delta^2$$

and thus

$$\frac{d\Delta_0}{d\Delta} = (\pm) \frac{\Delta}{(E^2 - \Delta^2)^{1/2}} \ . \tag{3.40}$$

Here the pole in the density of states is again at the point with zero slope (see circle in the figure) where Δ is zero, i.e. where we have symmetric tunnelling potentials and maximum rates R_{max}. This pole is the second pole shown in Fig.3.11.

b) *Spectroscopic Parameters*

Changes in the Hole Depths. So far we have only discussed the microscopic parameters entering the TLS description of the glassy state. Now we have to deduce the effect of the above rate concept on the spectroscopic observables. We will do this by focusing on hole-burning linewidths. These linewidths are on the order of $\Delta\omega_h$. The homogeneous linewidth is defined by (3.13); here the relaxation times T_1 and T_2 are the pertinent system parameters. In the absence of spectral diffusion processes, T_1 can be viewed as being the largest intrinsic time constant that enters our simplistic concept of a homogeneous linewidth (at reasonable temperatures, T_1 is larger than T_2). Therefore we define as non-adiabatic processes those which contribute to the homogeneous linewidth in a time t_{na} which is shorter thant T_1 and thus arrive at the following definition [3.99-101]

$$t_{na} \leq T_1 . \tag{3.41}$$

All such fast processes lead to a true homogeneous linewidth.

Other processes which occur on a longer time scale than T_1 we define as adiabatic. For such adiabatic processes occurring on a time scale t_a spectral diffusion models are, in our opinion, appropriate to describe the hole-burning linewidths, where

$$t_a \geq T_1 . \tag{3.42}$$

It is a definite shortcoming of the hole-burning technique that the experimental time scale is usually on the order of seconds or longer. Since typical T_1 times are on the order of 10^{-8} s, there is a large critical time regime T_{crit} for which no information is available from hole-burning techniques. We have:

$$10^{-8} \text{ s} < t_{crit} < 10^0 \text{ s} . \tag{3.43}$$

Nevertheless it is important to realize that there is still a wide experimental time window t_{exp} of

$$10^0 \text{ s} < t_{exp} < 10^6 \text{ s} \tag{3.44}$$

for which the technique of hole-burning is extremely valuable. It should be noted that (3.44) is equivalent to saying that one can measure rates R_{exp} which obey the following equation:

$$R_{exp} < 1/t_{exp} . \tag{3.45}$$

Even though the accessible rates are rather small, they should be of high relevance for amorphous materials, since, according to (3.37), the largest contribution of the TLS comes from the non-integrable pole extending to small rates with a cutoff value R_{min} (Fig.3.11).

It has recently been shown that very slow changes of the integrated area of photochemical holes can be observed for a hydrogen-bonded dye molecule (quinizarin) in alcohol glasses [3.99]. One must assume that the changes are due to a back reaction caused by a reverse tunnelling of the proton from the product state $|p\rangle$ to the educt state $|0\rangle$ (Fig.3.3). The observed change in integrated holewidth can be calculated by integrating the number of "photochemical" TLS or so-called TLS_p which still remain in the photoproduct state after a time t_1 has elapsed. The main assumption, which is currently a conjecture is that the photochemical TLS_p can be treated in analogy to the "normal" microscopic TLS which determine the specific heat of the material, i.e. that the density of states for TLS_p has the same dependence on the various TLS parameters (see above).

One can calculate the number of molecules N_p which are still in the photoproduct state $|p\rangle$ by integrating (3.37) between the limits R_{exp} and R_{min}. Since one performs the experiment at long times $1/t_{exp} \ll R_{max}$, one can omit the integrable pole at $R = R_{max}$ to obtain

$$N_p(\hat{t}) = \frac{1}{2}p(\Delta,\lambda) \int_{R_{min}}^{R_{exp}} \frac{dR}{R} = \frac{1}{2} \, p(\Delta,\lambda) \, \ln\left(\frac{R_{exp}}{R_{min}}\right). \qquad (3.46)$$

Note that the time $\hat{t}$ corresponds to the quantity $1/R_{exp}$.

A relatively straightforward calculation shows that the hole area $A(t)$ decays logarithmically as a function of observation time

$$\frac{A(t)}{A_1} = \frac{N(t)}{N_1} = 1 - \frac{\ln(R_1 t)}{\ln(R_1/R_{min})} . \qquad (3.47)$$

Here A_1 is the hole area at the shortest experimental time $t_1 = 1/R_1$. The above logarithmic decay law has, in fact, recently been observed [3.99] over 3 to 4 decades (Fig.3.13). The logarithmic decay law is caused by the $1/R$ dependence of the TLS rates for small rates. The most remarkable feature of the experiment was the large isotope effect seen upon comparing protonated hydrogen-bonded systems and deuterated systems. The

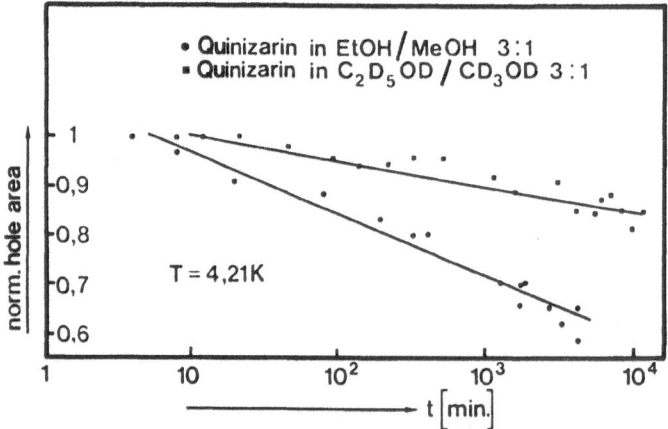

Fig.3.13. Time decay of the area of photochemical holes of quinizarin in alcohol glass for deuterated (*squares*) and protonated systems (*dots*)

deuteration effect is caused by the dependency of the tunnelling rate on the mass of the tunnelling system as defined by (3.33) where we have to insert m_p and m_D for the proton mass and the deuteron mass respectively. The ratio $R_{min,H}/R_{min,D}$, as obtained by the slope of Fig.3.13, is on the order of 1000 cm^{-1}. We have to add, however, that these barriers are characteristic of the slow photochemical back reaction processes of the protons and deuterons, respectively. These processes cause an extremely slow time evolution of the system: The protonated system reaches the half-value of the area of the photochemical hole in 60 days; the deuterated system decays in 20000 years if we extrapolate the logarithmic law to longer times.

Recently *Fearey* et al. have measured spontaneous and photoinduced hole-filling [3.102,103]. The former experiment resembles the quinizarin data described in detail above. It also shows a slow hole-filling process for the systems rhodamine 640, Nd^{3+} and Pr^{3+} in polyvinyl alcohol (the reaction is assumed to be a NPHB reaction). The duration of the NPHB experiment ranges from 10^2 to $3 \cdot 10^3$ s. Within this time range the spontaneous hole-filling can be described, within the accuracy of the experiment, either by a simple exponential decay law or by a "dispersive fit". A dispersive decay comes closer to our above model of distributed rates with a rate density following a 1/R behavior. Interestingly, hole-growth kinetics for color centers have also shown logarithmic time dependencies [3.104], possibly indicative of the underlying distribution of tunnelling barriers. Dispersive decay processes have recently been used quite frequently to describe dynamical processes in amorphous solids which do not follow simple exponential decay laws of the form [3.105-108]

$$K_t = K_t^0 \exp(-\tilde{\lambda}) . \qquad (3.48)$$

Instead of using a fixed $\tilde{\lambda}$ value one can assume a Gaussian distribution function of decay rates $g(\tilde{\lambda})$ with width σ. Such an assumption does not seem unreasonable, because Gaussian distributions of various kinds can, depending on their width σ, have more or less dispersive character. *Bässler* et al. [3.107] have shown that the decay of a normalized concentration of a species M can be described as

$$M(t) = \frac{1}{\sqrt{2\pi\sigma}} \int_{-\infty}^{+\infty} \exp\left\{ -\frac{(\Delta\tilde{\lambda})^2}{2\sigma^2} - k_t^0 \, t \, \exp\left[-\left(\tilde{\lambda}_0 + \Delta\tilde{\lambda} \right) \right] \right\} d(\Delta\tilde{\lambda}) . \qquad (3.49)$$

In the above equation $\tilde{\lambda}$ is the center point of the Gaussian $g(\tilde{\lambda})$-distribution function. In [3.107] it was pointed out that a model based on a Gaussian rate distribution can also describe the data of hole decay presented in Fig.3.13 as well as the process of hole formation [3.109].

In summary, most of the data can be fit with various decay-rate distributions [3.110-112]. To discriminate between the various decay laws to decide which mathematical model is most appropriate would require data taken over longer time ranges (see also [3.113 and 97]). We have presented our own model in most detail, because it follows the well-known TLS model without further assumptions other than that photochemical systems TLS_p can be treated in analogy to normal "TLS" systems.

Changes of Hole Width Caused by Spectral Diffusion. The changes of the hole depth, as discussed above, can be considered as being due to spontaneous dark reactions from $|p\rangle$ to $|0\rangle$ (Fig.3.3). In some systems such as quinizarin in alcohol glasses, there is, however, a concomitant change of the hole width; the latter can, at long times, be on the order of 60-70% of the width extrapolated to short times [3.101] (Fig.3.14). A change in linewidth occurring on a slow time scale (slower than T_1) is an adiabatic change of site frequencies (3.42) which falls under the category of "spectral diffusion". The most interesting aspects of these experiments are the following:

i) Linewidth changes can occur on very slow time scales of 10^0-10^6 s.
ii) The changes of width and depth (Figs.3.13,14) occur on comparable time scales.

In this chapter we do not want to dwell on the details of the spectral diffusion model to explain the linewidth [3.97,101]. It is based on a model of spin diffusion as formulated by *Klauder* and *Anderson* [3.83]. The point which we would like to make is, however, that detectable changes in linewidths are evident at least for some of the hole-burning systems if the experiments focus on the long-time behavior. This means that after a long

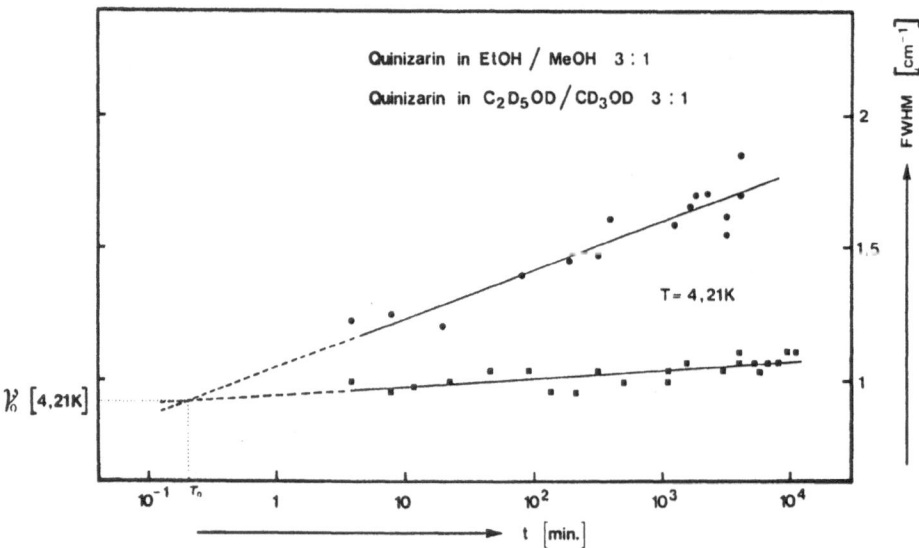

Fig.3.14. Line broadening of the photochemical holes of quinizarin in alcohol glass for deuterated (*squares*) and protonated systems (*dots*)

108

time interval the PHB or NPHB hole does not reflect the true "homogeneous" linewidth but rather a sum of the homogeneous linewidth and a width due to spectral diffusion [3.114]. It is thus dangerous to interpret hole-burning widths as true homogeneous linewidths, without discussing the phenomenon of spectral diffusion. This problem is even more relevant, because the time window between $10^{-8} < t < 10^0$ [s] is not accessible for straightforward hole-burning experiments.

If we draw an analogy between time-dependent phenomena in the specific heat ([3.115] and references therein), we would expect measurable spectral diffusion processes to occur on the μs to ms time scale. There, a clear picture will only emerge if optical echo experiments [3.116,89,95] are combined with hole-burning methods. In this context it is interesting to note that especially the data at very low temperature [3.95] seem to be indicative of rather large contributions to the overall linewidth from spectral diffusion processes [3.117]. In very recent experiments [3.118,119] careful echo measurements indeed show homogeneous linewidths that are clearly narrower than that deduced from NPHB data. This exciting area will continue to grow in the future as evidence accumulates that hole-burning can measure more than just homogeneous dephasing processes.

Changes of Hole Width Caused by Photochemical Depletion. In addition to processes such as spectral diffusion which are dominated by the nature of the glassy state, there are "trivial processes" which contribute to the width of a photochemical hole and which have to be corrected for in order to obtain what can be called a "quasi-homogeneous" linewidth. These "trivial" effects are mostly due to photochemical bleaching. This has also been pointed out by the group of Völker [3.78].

In the following we will formulate the photochemical broadening mechanism mathematically following [3.40]. We have seen in (3.24) that a photochemical hole is a convolution of a site-population function $nz(\omega')$ with a true homogeneous lineshape function $z(\omega-\omega')$.

Fortunately the width of the site distribution function, which we also call the "site memory function", can be calculated exactly. The nz-function can be written as

$$nz(1,\omega) = \sigma \, \frac{N_0(\omega_L)}{N} \left[1 - \exp\left(-\gamma^2 \, B \, \frac{1}{(\omega-\omega_L)^2 + \gamma^2/4} \right) \right]. \qquad (3.50)$$

Here we have, for simplicity, set the polarization factor of (3.24) to unity. We have also used a dimensionless B parameter for the integrated time-intensity product characterizing the extent of the hole-burning experiment with

$$B = \frac{\sigma}{2\pi\gamma} \frac{I}{\hbar\omega_L} \eta\alpha\tau \qquad (3.51)$$

[for the definitions of the parameters, see (3.23-25)].

If we use the above notation, we can calculate a half-width Γ_D of the population hole, normalized to the quasi-homogeneous linewidth γ and deduce [3.40]

$$\frac{\Gamma_D}{2\gamma} = \frac{1}{2}\sqrt{\frac{4B}{\ln[2/(1+e^{-4B})]} - 1} \ . \tag{3.52}$$

For $B \to 0$ this function approaches the value of 1/2, i.e. the site population function contributes 50% of the hole width.

The depth of the site population function at the laser frequency ω_L can be calculated to be

$$nz(1,\omega=\omega_L) = \sigma \ \frac{N_0(\omega_L)}{N} (1-e^{-4B}) \ . \tag{3.53}$$

Figure 3.15 shows the relative depth of the site population function compared to the numerically calculated hole depths for parallel (p) and perpendicular (s for the German word 'senkrecht') scanning [3.40]. It is interesting to note that the site distribution function approaches the bottom of the absorption band rather quickly, quite in contrast to the observed hole. The latter reaches the bottom of the band only if the Debye-Waller factor α is unity. The limiting value of the hole width is given by

$$\lim zz(\omega=\omega_L) = \sigma\alpha \ \frac{N(\omega_L)}{3N} \ . \tag{3.54}$$

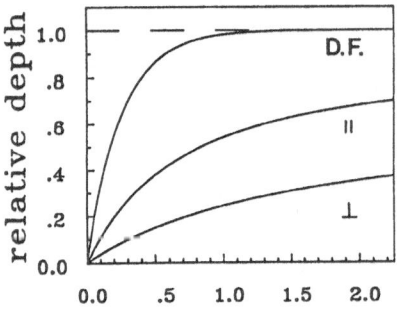

Fig.3.15. Dependence of the relative depth of a photochemical hole on the irradiated light dose B. The upper curve shows the maximally bleached distribution function with a polarization factor of unity. The two curves with parallel and perpendicular polarization are also shown [3.40]

The factor of 1/3 in the above equation stems from the fact that statistically, one-third of the molecules are oriented along each of the three spatial directions.

If α is small, there will still be a remaining absorption due to a summation over all phonon sidebands [note that (3.54) is derived under the assumption that the width of the phonon sideband is small compared to the imhomogeneous linewidth].

So far we have only given the exact equations for the width of the site distribution function and some limiting values for the hole depths. The width of the photochemical hole itself cannot be calculated exactly, but has to be computed by evaluating a power series of the pertinent analytical expressions. Following [3.40] and using the same parameter B for describing the photochemical "maturity" of a photochemical hole, we get the following expressions for the full widths at half maximum (parallel and perpendicular scanning geometry, respectively)

$$\frac{\Gamma_p}{2\gamma} = 1 + \frac{5}{14}B + \frac{5}{168}B^2 - \frac{425}{11319}B^3 , \qquad (3.55a)$$

$$\frac{\Gamma_s}{2\gamma} = 1 + \frac{3}{14}B - \frac{1}{56}B^2 - \frac{809}{135828}B^3 . \qquad (3.55b)$$

Taking the above analytically derived equations we can calculate that for B = 1.0, the linear corrections to the hole width contribute 26% to the total linewidth; the quadratic corrections contribute about 2% (for the parallel case). Therefore we can linearize (3.55) for hole depths less than or equal to 50% with excellent accuracy.

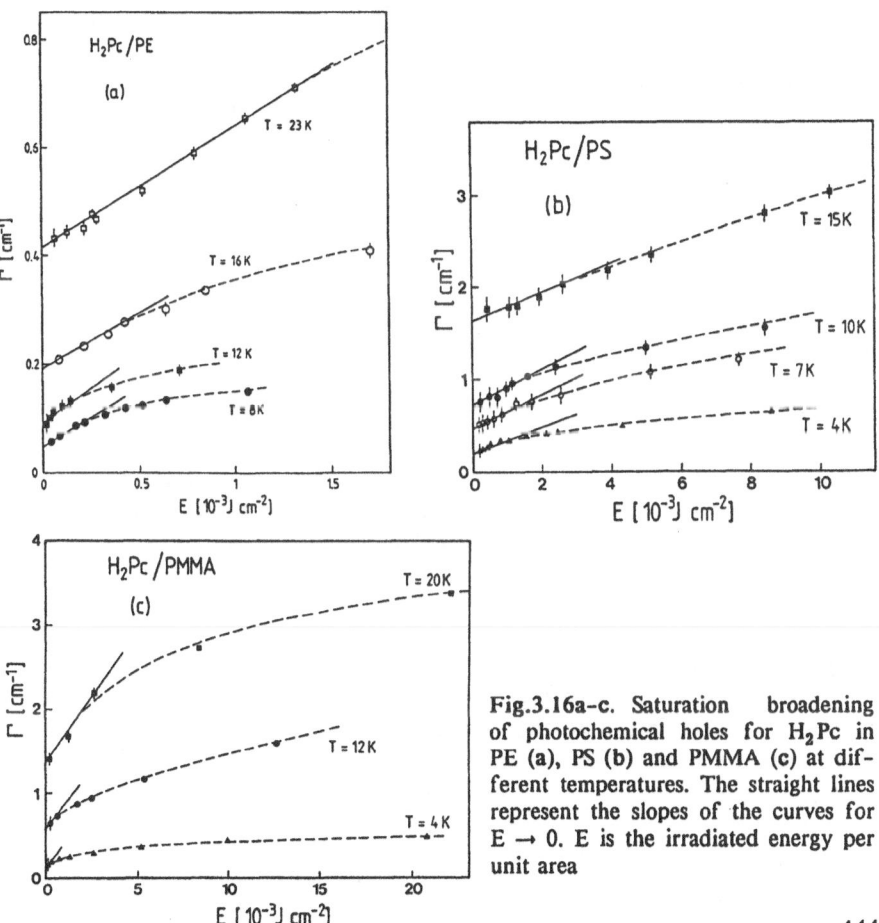

Fig.3.16a-c. Saturation broadening of photochemical holes for H_2Pc in PE (a), PS (b) and PMMA (c) at different temperatures. The straight lines represent the slopes of the curves for E → 0. E is the irradiated energy per unit area

Photochemical hole burning experiments on free base phthalocyanine (H_2Pc) in various polymer matrices (PE,PS,PMMA) show experimentally the predicted linear increase in linewidth (Fig.3.16). It is worth mentioning that the measured slopes of the linewidth versus integrated exposure are independent of the temperature. If one knows the oscillator strength f_{12} of the optical transition of the guest dye molecule, then the initial slope s of the hole-burning curve is only dependent on the photochemical quantum yield η of the reaction. For an experiment with perpendicular scanning polarization one obtains [3.40]

$$\eta = 28 \frac{\epsilon_0 mc\hbar\omega_L}{e^2 f_{12}\alpha} \frac{n}{(n^2 + 2)^2} s \; . \tag{3.56}$$

In the above equation n is the refractive index of the material, α is the Debye-Waller factor and the remaining quantities are natural constants.

In conclusion, one can say that the line broadening of a photochemical hole can be calculated with high accuracy, yielding materials parameters like the photochemical quantum yield. Since the method is based on an extrapolation to B → 0, the photochemcial quantum yield, which is an important hole-burning parameter [3.120], can be obtained to good accuracy.

Since some of the equations of this section follow from a straightforward but tedious calculation, we want to point out that for holes < 50% and for large α values, all equations of this section can be linearized in B without seriously affecting the accuracy of the experimental evaluation.

3.3 Field Effects in Hole–Burning Spectroscopy

3.3.1 Introduction: The Site Memory Function

As has been pointed out in Sect.3.2.2, see (3.20), the lineshape function of a photochemical hole can be viewed as a convolution of a site distribution function $N^r(\omega')$ and a homogeneous lineshape function $g(\omega-\omega')$. In order to keep the equations transparent, we neglect polarization factors for the time being and integrate over the angular coordinates to get

$$\frac{1}{4\pi} \sigma(\theta,\phi) \sin\theta \; d\theta \; d\phi = \sigma_{\text{diff}} \tag{3.57}$$

where σ is the absorption cross section integrated over the angular and wavelength variables. As a second approximation we assume that most of the line intensity is due to the zero-phonon transition and hence we set α = 1. This produces a "site memory function" of the following form, see (3.25),

$$nz^r(\omega') = \sigma \frac{N_0(\omega_L)}{N} \left[1 - \exp \frac{\sigma I}{\hbar\omega_L} \eta\tau z(\omega_L-\omega')\right] . \tag{3.58}$$

Since all of the above lineshape functions are dependent on temperature, we can write the zero-phonon part of the lineshape function as

$$L^\tau(\omega,T) = \int_{-\infty}^{+\infty} nz^\tau(\omega',T)\, z(\omega-\omega',T)\, d\omega'. \tag{3.59}$$

At this point, we want to discuss a subtle aspect of hole-burning spectroscopy whose importance is frequently underestimated and whose implications are rarely discussed in the literature: The site memory function $nz^\tau(\omega',T)$ has a quasi-static character. Once the photochemical changes have been performed via laser chemistry, the hole acts like a photochemical "materials fingerprint" whose shape is only subject to slow changes caused by spectral diffusion and by field effects (see below). The hole also reflects the site spectrum characterized by the temperature T_b, at which the burning (photochemistry) was performed.

The function $z(\omega-\omega')$ on the other hand, has dynamic character and reflects the fast dephasing processes T_1 and T_2 of the system, see (3.13). What has to be realized is that the relevant temperature for the $z(\omega-\omega')$ function is the temperature T, at which the spectroscopy is being performed. This temperature is not necessarily equal to the burning temperature T_b and hence a correct version of (3.59) has to include two temperatures and should therefore be written as

$$L^\tau(\omega,T,T_b) = \int_{-\infty}^{+\infty} nz^\tau(\omega',T_b)\, z(\omega-\omega',T)\, d\omega'\ . \tag{3.60}$$

Most hole-burning experiments which are reported in the literature are performed with $T = T_b$. Therefore the complexity of an experiment in which one performs temperature cycles with a hole-burning spectrum [3.50] is not apparent.

Even if the subtle aspects of the long time memory function of $nz(\omega')$ and the dynamical character of $z(\omega-\omega')$ can be neglected for experiments with $T = T_b$, they cannot be neglected for investigating time effects (t) of slow spectral diffusion and external field effects. The field effects change the site memory function $nz(\omega')$, but, to a good approximation, they do not influence the intrinsic homogeneous linewidth. Characterizing time and field effects with the variables t and q, one has to rewrite (3.60) in a more complete fashion as

$$L^\tau(\omega,T,T_b,t,q) = \int_{-\infty}^{+\infty} nz^\tau(\omega',T_b,t,q)\, z[(\omega-\omega'),T]\, d\omega'\ . \tag{3.61}$$

This rather complex notation for the hole-burning lineshape expresses the origin of the "memory character" of a PHB experiment [3.114]. The experiment monitors all matrix changes occurring after a time interval on the order of $t > T_1, T_2$. These matrix changes can be time and field induced. It is our opinion that this memory nature of the experiment, which stores information about the time and field evolution of the system, is one of the most exciting and challenging aspects of hole-burning spectroscopy. A review of external field effects on persistent spectral holes has recently appeared [3.121].

3.3.2 Electric-Field Effects

In the following we want to treat the field effects in hole-burning spectroscopy in some detail. To do so, one has to realize that the field variable introduced above, can be different for the burning and for the scanning experiments (quite in analogy to the temperature). Therefore one has to rewrite (3.61) in a more general form introducing the field shift $\Delta\omega(q)$.

$$L^r(\omega,T,T_b,t,q,\hat{q}) = \int_{-\infty}^{+\infty} nz^r\{[\omega-\Delta\omega(\hat{q})],T_b,t\}\ z[\{\omega - [\omega'-\Delta\omega(q)]\},T]\ d\omega' \ . \tag{3.62}$$

In the above notation $\Delta\omega(\hat{q})$ characterizes the field shift during the burning experiment and $\Delta\omega(q)$ characterizes the field shift during the scanning experiment. We will see that this will allow us to perform hole-burning experiments with unusual external parameters such as "negative pressure". The latter can be achieved by burning the hole at a higher value of the external field parameter (pressure) and by recording the spectrum at a lower value and, hence, measuring the shift which is due to a system expansion or "negative pressure". Only the "tandem two-photon" character of the hole-burning experiment, as expressed by the above notation, allows the experimental verification of such negative-field effects. Chapter 7 may be consulted for a summary of possible applications of these electric field phenomena. Stark experiments combined with hole-burning spectroscopy were reported early in the history of hole-burning [3.18] and initially analyzed for the PHB case by the Personov group [3.122]. The researchers considered first- and second-order Stark shifts and have derived the following spectroscopic term characterizing the shift caused by the electric field E

$$\Delta\omega(q) \equiv \Delta\omega(E) = \frac{1}{\hbar}(f\Delta\mu E + f^2 E\Delta\hat{\alpha}E/2) \ . \tag{3.63}$$

In the above equation f is a Lorentz local-field correction factor given by

$$f = (\epsilon + 2)/3 \ . \tag{3.64}$$

Here μ and $\Delta\hat{\alpha}$ denote the changes in dipole moment (vector) and polarizability (tensor) which accompany the change in the electronic wavefunction as the molecule is excited from the ground state to the excited state.

If one introduces the field-shift values of (3.63) into the lineshape function (3.62). One has to consider two field values E and $\hat{E}$ for the scanning and the hole-burning parts of the experiment, respectively.

In the recent Stark experiments using hole-burning spectroscopy [3.122,90,123,124], molecules with and without inversion symmetry were investigated. The surprising finding was that both types of experiments were dominated by the linear Stark effect, irrespective of the molecular symmetry. This was a rather unexpected result and we therefore want to discuss the data on molecules with different symmetries separately.

a) Stark Effect for Molecules with Inversion Symmetry

As has been pointed out above, the PHB Stark experiments have yielded a linear effect for molecules with inversion symmetry. This finding leads to the conclusion that the change in dipole moment $\Delta\mu$ is correlated with the molecule-matrix configuration rather than with the molecule itself, because the latter would not exhibit a linear effect due to its inversion symmetry. Since the dipole moment is matrix induced, one can, to a good approximation, assume that the transition dipole μ_0 and the change in dipole moment $\Delta\mu$ are not correlated. In this case it can readily be shown [3.124] that the number of molecules per Stark-shift interval can be represented by a step function for a given $\Delta\mu$ value. This step function extends between the two limits $-(\Delta\mu\cdot E/\hbar)$ and $+(\Delta\mu\cdot E)/\hbar)$, as shown in Fig.3.17. For this case one expects a symmetric line broadening reflecting the step function of the field effect. This is reproduced by the available numerical [3.122] and analytical data [3.123-125]. Figure 3.18a shows the expected hole shapes for different field strengths for a constant $\Delta\mu$ value and Fig.3.18b shows the expected hole shape for a Gaussian distribution of $\Delta\mu$ values (for details see [3.124]). It should be pointed out that the width of the Gaussian distribution and its maximum value $\Delta\mu$ are correlated.

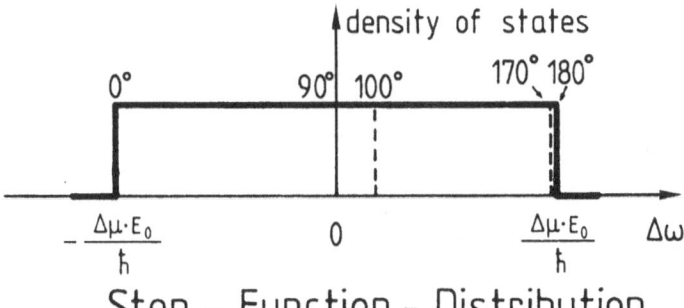

Fig.3.17. Distribution function of the absorption frequencies of guest molecules in an electric field E_0. All molecules absorb at the 0-frequency for E = 0; they have identical dipole moment changes $\Delta\mu$ but are oriented randomly with respect to the external E-field. The numbers indicate the orientation angles for some representative frequency shifts

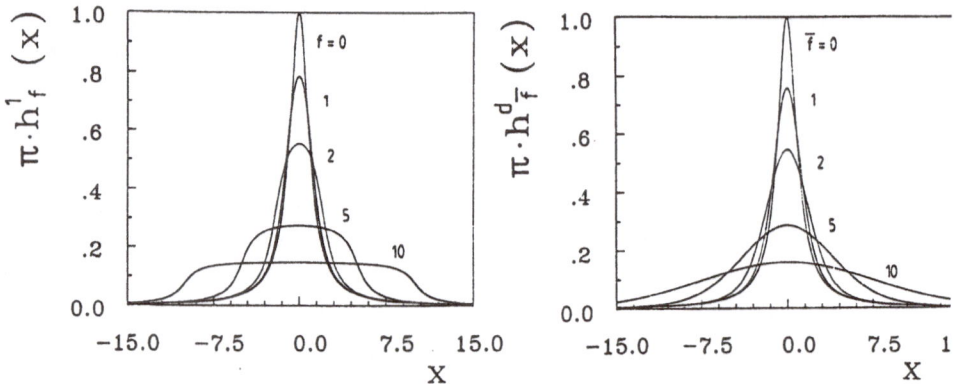

Fig.3.18. (a) Theoretical lineshape of a spectral hole for different external field strengths f. All absorbers have the same $\Delta\mu$ value. For a field value of $f = 1$, the Stark broadening equals the linewidth. (b) The same calculations as a) with a Gaussian distribution of $\Delta\mu$ values

Bogner et al. [3.123] have calculated the hole depth as measured by the intensity I of the fluorescence excitation as a function of the burning field $\hat{E}$ and probing field E, respectively. This calculation can be performed analytically if the probing is performed at the laser frequency ω_L and one obtains

$$I(\hat{E},E) \simeq \left[1 - \frac{r\Delta\omega_{nz}}{\Delta\omega_{nz} + \Delta\omega_z} \pi^{1/2}\hbar \frac{\Delta\omega_{nz} + \Delta\omega_z}{f\,\Delta\mu\,|\hat{E}-E|}\right]$$

$$\cdot \exp\left[\frac{\hbar^2(\Delta\omega_{nz}+\Delta\omega_z)^2}{f^2\,\Delta\mu^2\,|\hat{E}-E|^2}\right] \cdot \left[1 - \mathrm{erf}\left(\frac{\hbar(\Delta\omega_{nz}+\Delta\omega_z)}{f\,\Delta\mu\,|\hat{E}-E|}\right)\right]. \quad (3.65)$$

In the above equation $\Delta\mu$ is the average change in dipole moment and r is a factor characterizing the photochemical bleaching condition of the sample. The values $\Delta\omega_{nz}$ and $\Delta\omega_z$ are the widths of the site memory function and of the spectral lineshape function, respectively. If one assumes short burning time conditions the total linewidth has a Lorentzian envelope function with a width $\Delta\omega_{hole}$ which is the sum of $\Delta\omega_{nz}$ and $\Delta\omega_z$.

The experimental data for two given $\hat{E}$ values are given in Fig.3.19 for the molecule perylene in polyvinylbutyral. The calculated curves (3.65) seem to reproduce the experimental data rather well, see [3.123].

In the literature, $\overline{\Delta M}$ values ranging between 0.1 and 0.6 Debye units [3.122-125] have been reported (Fig.3.20). These experimental data, representing rather large matrix-induced dipole moments are surprising since the $\Delta\mu$ values are not of molecular nature but are induced by matrix effects. So far it is not known why the matrix-induced dipole moments have such high values. To our knowledge the microscopic origin of the

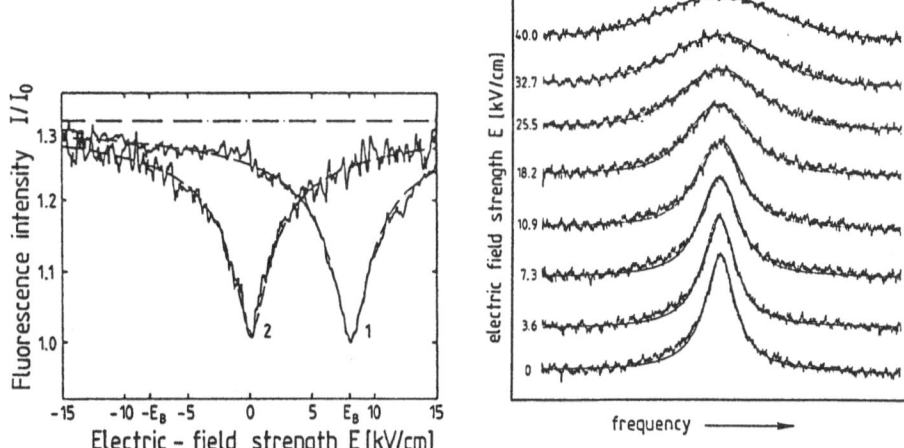

Fig.3.19. Fluorescence excitation spectra of two holes. One spectrum was burned at E = 0 and one was burned at the field E_B [3.123]

Fig.3.20. Measured and calculated hole profiles for the system Zn-TBP $(C_6H_4CH_3)_4$ in polyvinylbutyral (see [3.124]). Here the average change in dipole moment is $\Delta\mu$ = 0.174 Debye units

removal of molecular symmetry in an amorphous matrix is not understood. It is interesting that an amorphous matrix, which is frequently considered as an isotropic environment, seems to have rather low symmetry if one considers length scales typical of molecular dimensions. This does not contradict the isotropic nature of the matrix on a length scale of 100 nm and larger.

A second way of interpreting the measured field shifts is to attribute them to large intrinsic electric fields which induce the measured $\Delta\mu$ values through the molecular polarizability.

b) *Stark Effect for Molecules Without Inversion Symmetry*

In the previous subsection we have assumed that the change in dipole moment $\Delta\mu$ is matrix induced and, hence, it was a plausible assumption to consider the molecular transition dipole moment μ_{01} and the $\Delta\mu$ vector as being uncorrelated. This assumption breaks down if one considers molecules without inversion symmetry. Here one has to assume a correlation between the molecular transition dipole and the molecular $\Delta\mu$ value.

So far, two different cases of correlations have been considered. They are sketched in Figs.3.21a,b. Figure 3.21a shows the case of parallel and perpendicular μ_{01} and $\Delta\mu$ values. It is also assumed that the electric field is parallel to the **k** vector of the incident light. This case is easily realized experimentally by performing the Stark spectroscopy on a sandwich-type

117

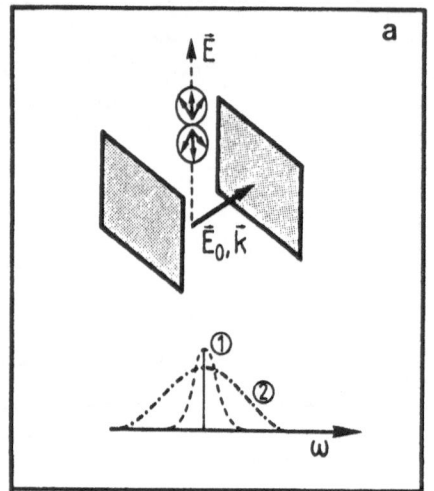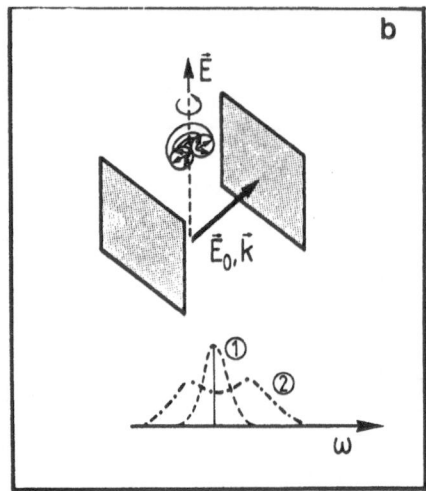

Fig.3.21. Dipolar $\Delta\mu$ configurations for $\Delta\mu$ parallel (a) and perpendicular (b) to the transition dipole moment. E_0 is the electric field strength and k is the wave vector of the incident light

sample having semi-transparent electrodes. In the parallel case, the $\Delta\mu$ ensemble, which couples to the laser light field, is given by a $\cos^2\theta$ distribution. This configuration will yield a symmetric broadening by the Stark field, a conclusion confirmed by numerical calculations [3.122].

The case of a perpendicular arrangement of μ_{01} and $\Delta\mu$ is shown in Fig.3.21b. Here the $\sin^2\theta$ distribution has strong contributions in both "up-field" and "down-field" directions. This gives rise to a double-peaked lineshape, as has also been shown in [3.122]. Fig.3.22 shows experimental data for the molecule chlorin in polyvinylbutyral [3.124].

In summary, we can say that the experimental methods and the analytical tools for investigating $\Delta\mu$ values and $\Delta\hat{\mu}$ distributions are known.

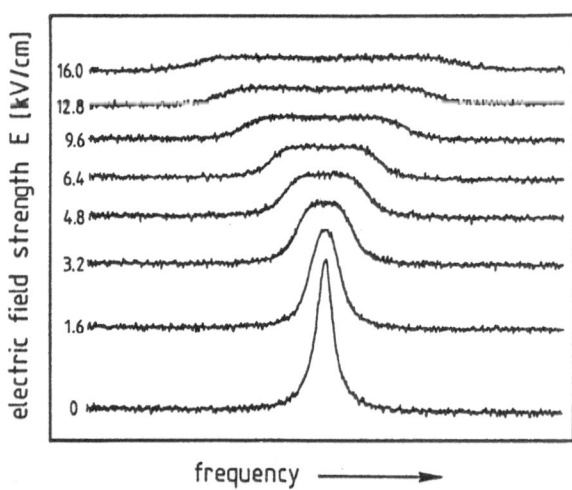

Fig.3.22. Measured and calculated hole profiles for H_2Ch in polyvinylbutyral [3.124]. Here the calculated $\Delta\mu$ value is $\Delta\mu = 0.214$ Debye units

Nevertheless, there are no microscopic models which would, on a molecular level, explain the large measured matrix-induced dipole moments. Due to the large linear field effects, the influence of $\Delta\alpha$ effects also remains unknown. These effects, if accompanied by large enough polarizibilty changes, would yield a quadratic Stark effect. The latter has, to our knowledge, not been reported in the field of hole-burning spectroscopy.

3.3.3 Strain-Field Effects

As the field parameter q one can also use uniaxial or hydrostatic external pressure to achieve a $\Delta\omega(q)$ shift. Again, hole-burning turns out to be an excellent spectroscopic method. The high intrinsic resolution of the experiment gives comparatively large relative lineshifts for small values of uniaxial [3.126] or hydorstatic [3.127] pressure.

Even though there are some similarities in the description of the electric-field and strain-field effects, one has to keep in mind some pronounced differences. For a linear Stark effect, as has been discussed in the previous section, a photochemical hole broadens symmetrically. This is because, statistically, the same number of dipoles $\Delta\mu$ point along and against the field direction. Therefore the center of gravity of the band remains at the center frequency.

For both uniaxial and hydrostatic pressure, the line center shifts to lower or higher frequencies. The most common situation is a shift to lower energies for higher values of the external pressure. Figs.3.23a and b

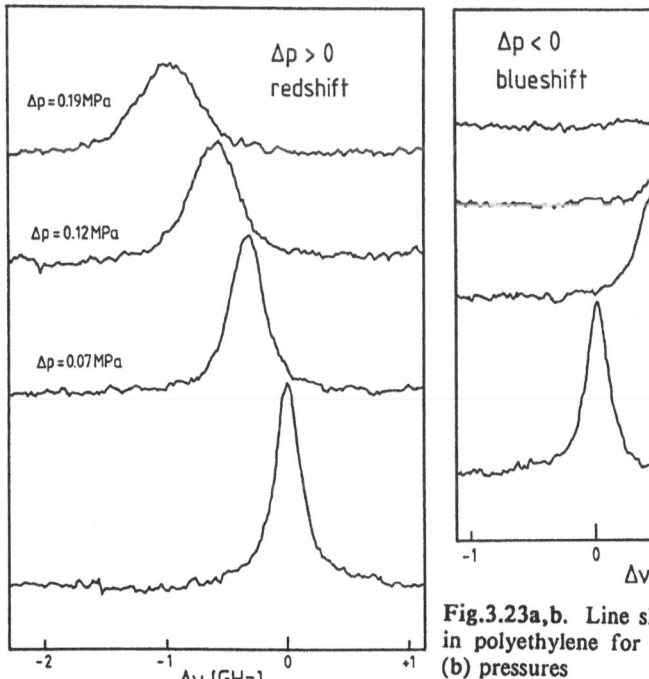

Fig.3.23a,b. Line shift of the molecule H_2Pc in polyethylene for positive (a) and negative (b) pressures

show the line shifts of the molecule free base phthalocyanine (H_2Pc) in polyethylene (PE) for positive and negative pressure variations, respectively [3.127]. It has been pointed out in the introduction that negative pressure variations can be achieved by burning under a high pressure ($\hat{p}$) and recording the spectra under low pressure conditions (p). This is equivalent to measuring a negative pressure effect which can only be achieved experimentally due to the two-photon nature of the hole-burning experiment (see the beginning of this section).

What should be noted is that the observed pressure shifts can be detected with the hole-burning technique at pressures which are 3 to 4 orders of magnitude smaller than those used in straightforward optical spectroscopy under external pressure (see, e.g., [3.128,129]). Therefore the general equation to describe pressure changes [3.130-132].

$$\Delta\omega(p) = Ap + Bp^2 \qquad (3.66)$$

can be linearized. One can show that the line shift can be expressed as

$$\Delta\omega(p) = C_{pr}\kappa p . \qquad (3.67)$$

In this equation κ is the hydrostatic compressibility of the matrix and C_{pr} is a constant characterizing the molecule-matrix configuration.

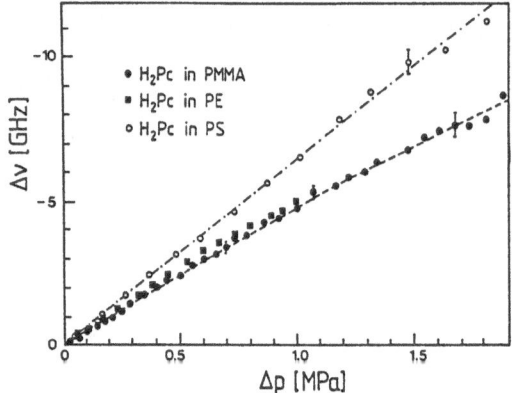

Fig.3.24. Line shifts of H_2Pc in PS, PE and PMMA in the low pressure regime

Figure 3.24 shows that the linearized pressure law, as spelled out in (3.67), holds well for the H_2Pc molecule in PE (polyethylene), PS (polystyrene) and PMMA (polymethylmethacrylate) matrices. It had been shown quite recently [3.127] that the precise determination of the pressure-induced line-shift data and of the molecular matrix shift (solvent shift) parameters allows the optical determinaton of mechanical matrix parameters such as the compressibility.

A second interesting feature of PHB spectroscopy under hydrostatic pressure is the fact that the hole shift and the hole broadening are of

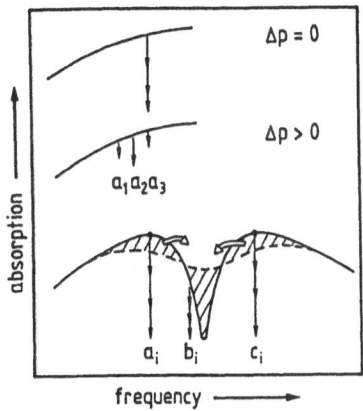

Fig.3.25. Line broadening of sites which had been site-selected at $\Delta p = 0$ and which drift apart at higher pressures $\Delta p \neq 0$. At higher pressures the frequency range of the absorbers extends from a_1 to a_2. The lower part of the figure shows the line broadening for various sets of molecules a_i, b_i and c_i

comparable magnitude (Fig.3.23a,b). To understand this phenomenon we have to assume that various sites of the matrix exhibit different "local pressure" shifts as indicated in Fig.3.25. This means that if we perform the site selection at $\Delta p = 0$ and increase the pressure to a given value of $\Delta p \neq 0$, the various site frequencies ω_i will spread out over a certain frequency range. This spreading out due to the dispersion of local compressibilities is intimately related to the amorphous nature of the material. A perfect single crystal would, to a good approximation, only show pressure-induced lineshifts without appreciable line broadenings. PHB studies under uniaxial pressure produced by ultrasonic waves suggest that this may be the case [3.133,134].

In conclusion, we can say that external pressure effects seem to be a promising tool for the study of amorphous matrices [3.135]. We have demonstrated that one needs only very small pressure values of about 10^{-1} MPa rather than the usual values on the order of 10^3 MPa to achieve measurable line shifts. Under these conditions, the pertinent equations, describing the pressure-induced shifts, can be linearized and can be used to correlate optical data with macroscopic "mechanical" data (for instance compressibilities).

It will be a challenging task to characterize the amorphous nature of the investigated materials by analyzing the pressure-induced line broadening. We expect these experiments to shed light on the "amorphicity", i.e. on the degree of local disorder in the material [3.136,137].

Acknowledgement. The author would like to acknowledge many valuable discussions with J. Friedrich and the enthusiasm of his coworkers in Bayreuth. I also would like to acknowledge support by the Stiftung Volkswagenwerk and support by the U.S. Office of Naval Research and by the IBM Reseach Laboratory (San Jose, California) in the early stages of our work related to hole-burning. I also thank L. Kador for carefully reading the manuscript.

References

3.1   N. Bloembergen, E.M. Purcell, R.V. Pound: Phys. Rev. **71**, 679 (1948)
3.2   K. Blum: *Density Matrix Theory and Applications* (Plenum, New York 1981);
        M. Sargent III, M.O. Scully, W.E. Lamb Jr.: *Laser Physics*, Addison-Wesley,
        Reading, Mass. 1974)
3.3   A. Szabo: Phys. Rev. B11, 4512 (1975)
3.4   A.R. Chraplyvy, W.E. Moerner, A.J. Sievers: Opt. Lett. **6**, 431 (1981)
3.5   J. Friedrich, D. Haarer: Ang. Chemie Int. Ed. **23**, 113 (1984)
3.6   B.M. Kharlamov, R.I. Personov, L.A. Bykovskaya: Opt. Commun. **12**, 191
        (1974)
3.7   A.A. Gorokhovskii, R. Kaarli, L.A. Rebane: JETP Lett. **20**, 216 (1974)
3.8   R.H. Silsbee: In *Optical Properties of Solids*, ed by S. Nudelmann, S.S. Mitra
        (Plenum, New York 1969) p.607
3.9   A.H. Weller: Prog. React. Kinet. **1**, 187 (1961)
3.10   H. Beens, K.H. Grellmann, M. Gurr, A.H. Weller: Disc. Faraday Soc. **39**, 183
        (1985)
3.11   E. Rommel, J. Wirz: Helv. Chem. Acta **60**, 38 (1977)
3.12   F. Graf, D. Haarer: unpublished
3.13   D.M. Burland, D. Haarer: IBM J. Res. Dev. **23**, 543 (1979)
3.14   S. Völker, J.H. Van der Waals: Mol. Phys. **32**, 1703 (1976)
3.15   S. Völker, R.M. Macfarlane, A.Z. Genack, H.P. Trommsdorff, J.H. van der
        Waals: J. Chem. Phys. **67**, 1759 (1977)
3.16   S. Völker, R.M. Macfarlane: J. Chem. Phys. **73**, 4476 (1980)
3.17   F. Graf, H.K. Hong, A. Nazzal, D. Haarer: Chem. Phys. Lett. **59**, 217 (1978)
3.18   A.P. Marchetti, M. Scozzava, R.H. Young: Chem. Phys. Lett **51**, 424 (1977)
3.19   J. Friedrich, H. Scheer, B. Zickendraht-Wendelstadt, D. Haarer: J. Am. Chem.
        Soc. **103**, 1030 (1981)
3.20   J. Friedrich, H. Scheer, B. Zickendraht-Wendelstadt, D. Haarer: J. Chem. Phys.
        74, 2260 (1981)
3.21   J.M. Hayes, G.J. Small: Chem. Phys. **27**, 151 (1978)
3.22   R. Jankowiak, H. Bässler: Chem. Phys. Lett. **95**, 124, 310 (1983)
3.23   E. Cuellar, G. Castro: Chem. Phys. **54**, 217 (1981)
3.24   D.M. Burland, F. Carmona, J. Pacansky: Chem. Phys. Lett. **56** 221 (1978)
3.25   H. de Vries, D.A. Wiersma: Phys. Rev. Lett. **36**, 91 (1976)
3.26   H.W.H. Lee, M. Gehrtz, E.E. Marinero, W.E. Moerner: Chem. Phys. Lett **118**,
        611 (1985)
3.27   W. Lenth, W.E. Moerner: Opt. Commun. **58**, 249 (1986)
3.28   J. Joussot-Dubien, R. Lesceaux: Compt. Rend. Acad. Sci. **258**, 4260 (1964);
        ibidem. **263**, 1177 (1965)
3.29   J. Joussot-Dubien, R. Lesceaux: J. Chem. Phys. **61**, 1631 (1964)
3.30   R.M. Macfarlane, R.M. Shelby: Phys. Rev. Lett. **42**, 788 (1979)
3.31   T. Tani, H. Namikawa, K. Arai, A. Makishima: J. Appl. Phys. **58**, 3559 (1985)
3.32   M. Dubs, H.H. Günthard: Chem. Phys. Lett. **64**, 105 (1979)
3.33   W.E. Moerner, A.R. Chraplyvy, A.J. Sievers, R.H. Silsbee: Phys. Rev. B28,
        7244 (1983)
3.34   T.H. Keil: Phys. Rev. A 140, 601 (1965)
3.35   K. Huang: A. Phys. Proc. R. Soc. A208, 352 (1951)
3.36   J.J. Markham: Rev. Mod. Phys. **31**, 956 (1959)
3.37   G.J. Small: In *Spectroscopy and Excitation Dynamics of Condensed Molecular
        Systems*, ed. by V.M. Agranovich, R.M. Hochstrasser (North Holland,
        Amstardam 1983)
3.38   L.A. Rebane, A.A. Gorokhovskii, J.V. Kikas: Appl. Phys. B29, 235 (1982)
3.39   W. Köhler, W. Breinl, J. Friedrich: J. Chem. Phys. **89**, 3473 (1985)
3.40   L. Kador, G. Schulte, D. Haarer: J. Phys. Chem. **90**, 1264 (1986)

3.41   J. Friedrich, D. Haarer: J. Chem. Phys. **76**, 61 (1982)
3.42   A.A. Gorokhovskii, R. Kaarli, L.A. Rebane: Opt. Commun. **16**, 282 (1976)
3.43   A.A. Gorokhovskii, L.A. Rebane: Opt. Commun. **20**, 144 (1977)
3.44   S. Völker, R.M. Macfarlane, J.H. van der Waals: Chem. Phys. Lett. **53**, 8 (1978)
3.45   S. Völker, R.M. Macfarlane: Chem. Phys. Lett. **61**, 426 (1979)
3.46   D.E. McCumber, M.D. Sturge: J. Appl. Phys. **34**, 1682 (1963)
3.47   A. Abragam: *The Principles of Nuclear Magnetism* (Clarendon, Oxford 1962)
3.48   C.A. Van'T Hof, J. Schmid: Chem. Phys. Lett. **36**, 457 (1975)
3.49   C.D. Harris, R.M. Shelby, P.A. Cornelius: Phys. Rev. Lett. **38**, 1415 (1977)
3.50   A.R. Gutierrez, G. Castro, G. Schulte, D. Haarer: In *Organic Molecular Aggregates*, ed. by P. Reineker, H. Haken, H.C. Wolf, Springer Ser. Solid-State Sci., Vol.49 (Springer, Berlin, Heidelberg 1983)
3.51   P.M. Selzer, D.L. Huber, D.S. Hamilton, W.M. Yen, M.J. Weber: Phys. Rev. Lett **36**, 813 (1976)
3.52   J. Hegarty, W.M. Yen: Phys. Rev. Lett. **43**, 1126 (1979)
3.53   R.T. Brundage, W.M. Yen: Phys. Rev. **B33**, 4436 (1986)
3.54   P. Avouris, A. Campion, M.A. El-Sayed: J. Chem. Phys. **67**, 3397 (1977)
3.55   J.R. Morgan, M.A. El-Sayed: Chem. Phys. Lett. **84**, 215 (1981)
3.56   J.M. Hayes, R.P. Stout, G.J. Small: J. Chem. Phys. **73**, 4129 (1980)
3.57   J.M. Hayes, R.P. Stout, G.J. Small: J. Chem. Phys. **74**, 4266 (1981)
3.58   R.C. Zeller, R.O. Pohl: Phys. Rev. **B4**, 2029 (1971)
3.59   P.W. Anderson, B.J. Halperin, C.M. Varma: Philos. Mag. **25**, 1 (1972)
3.60   W.A. Phillips: J. Low Temp. Phys. **7**, 351 (1972)
3.61   T.L. Reinecke: Solid State Commun. **32**, 1103 (1979)
3.62   S.K. Lyo, R. Orbach: Phys. Rev. **B22**, 4223 (1980)
3.63   S.K. Lyo: Phys. Rev. Lett. **48**, 688 (1982)
3.64   S.K. Lyo: In *Optical Spectroscopy of Glasses*, ed. by I. Zschokke-Gränacher (Reidel, Dordrecht) in press
3.65   D.L. Huber, M.M. Broer, B. Golding: Phys. Rev. Lett. **52**, 2281 (1984)
3.66   P. Reineker, H. Morawitz: Chem. Phys. Lett. **86**, 359 (1982)
3.67   P. Reineker, H. Morawitz, K. Kasser: Phys. Rev. **B29**, 4546 (1983)
3.68   B. Jackson, R. Silbey: Chem. Phys. Lett. **99**, 331 (1983)
3.69   S.K. Lyo, R. Orbach: Phys. Rev. **B 29**, 2300 (1984)
3.70   S.K. Lyo: In *Organic Molecular Aggregates*, ed. by P. Reineker, H. Haken, H.C. Wolf, Springer Ser. Solid-State Sci., Vol.49 (Springer, Berlin, Heidelberg 1983)
3.71   P. Reineker, K. Kassner: In *Optical Spectroscopy of Glasses*, ed. by I. Zschokke-Gränacher (Reidel, Dordrecht) in press
3.72   J.M. Hayes, R. Jankowiak, G.J. Small: see Chap.5 of this book
3.73   H.P.H. Thijssen, A.I.M. Dicker, S. Völker: Chem. Phys. Lett. **92**,7 (1982)
3.74   H.P.H. Thijssen, S. Völker, M. Schmidt, H. Port: Chem. Phys. Lett **94**, 537 (1983)
3.75   H.P.H. Thijssen, R.E. van den Berg, S. Völker: Chem. Phys. Lett. **97**, 295 (1983)
3.76   H.P.H. Thijssen, R.E. van den Berg, S. Völker: Chem. Phys. Lett. **103**, 23 (1983)
3.77   H.P.H. Thijssen, S. Völker: Chem. Phys. Lett. **120**, 496 (1985)
3.78   H.P.H. Thijssen, R.E. van den Berg, S. Völker: Chem. Phys. Lett. **120**, 503 (1985)
3.79   S. Alexander, R. Orbach: J. Physique Lett. **43**, L-625 (1982)
3.80   R. Jankowiak, H. Bässler, R. Silbey: Chem. Phys. Lett. **125**, 139 (1986)
3.81   M.M. Broer, W. Haemmerle, J.R. Simpson, D.L. Hubert: Phys. Rev. **B33**, 4160 (1986)
3.82   J.L. Black, B.J. Halperin: Phys. Rev. **B16**, 2879 (1977)

3.83   J.R. Klauder, P.W. Anderson: Phys. Rev. 125, 912 (1962)
3.84   W.B. Mims: Phys. Rev. 168, 370 (1968)
3.85   S. Hunklinger, M. Schmidt: Z. Phys. B 54, 93 (1984)
3.86   R.W. Macfarlane, R.M. Shelby: Optics Commun. 45, 46 (1983)
3.87   R.M. Shelby, Opt. Lett. 8, 88 (1983)
3.88   H.W.H. Lee, A.L. Huston, M. Gehrtz, W.E. Moerner: Chem. Phys. Lett. 114, 491 (1985)
3.89   L.W. Molenkamp, D.A. Wiersma: J. Chem. Phys. 83, 1 (1985)
3.90   F.A. Burkhalter: G.W. Suter, U.P. Wild, V.D. Samoilenko, N.V. Rasumova, R.I. Personov: Chem. Phys. Lett. 94, 483 (1983)
3.91   R. Jankowiak, H. Bässler: Chem. Phys. Lett. 95, 310 (1983)
3.92   J. Friedrich, H. Wolfrum, D. Haarer: J. Chem. Phys. 77, 2309 (1982)
3.93   T.P. Carter. B.L. Fearey, J.M. Hayes, G.J. Small: Chem. Phys. Lett. 102, 272 (1983)
3.94   T.P. Carter, G.J. Small: Chem. Phys. Lett. 120, 178 (1985)
3.95   A.A. Gorokhovski, V.Rh. Korrovits, V.V. Pal'm, M.A. Trummal: JETP Lett. 42, 307 (1985)
3.96   W.A. Phillips (ed.): Amorphous Solids, Topics Curr. Phys., Vol.24 (Springer, Berlin, Heidelberg 1981)
3.97   J. Friedrich, D. Haarer: In Optical Sepctroscopy in Glasses, ed. by I. Zschokke-Gränacher (Reidel, Dordrecht) in press
3.98   J. Jäckle: Z. Phys. 257, 212 (1972)
3.99   W. Breinl, J. Freidrich, D. Haarer: Chem. Phys. Lett. 106, 487 (1984)
3.100   W. Breinl, J. Friedrich, D. Haarer: J. Chem. Phys. 80, 3496 (1984)
3.101   W. Breinl, J. Friedrich, D. Haarer: J. Chem. Phys. 81, 3915 (1984)
3.102   B.L. Fearey, G.J. Small: Chem. Phys. 101, 269 (1986)
3.103   B.L. Fearey, T.P. Carter, G.J. Small: Chem. Phys. 101, 279 (1986)
3.104   W.E. Moerner, P.Pokrowsky, F.M. Schellenberg, G.C. Bjorklund: Phys. Rev. B 33, 5702 (1986)
3.105   T. Doba, K.U. Ingold, W. Siebrand, T.A. Wieldmann: Chem. Phys. Lett. 115, 51 (1985)
3.106   W. Richert, H. Bässler: Chem. Phys. Lett. 116, 302 (1985)
3.107   R. Jankowiak. R. Richert, H. Bässler: J. Phys. Chem. 89, 4569 (1985)
3.108   C. Aubert, J. Fünfschilling, I. Zschokke-Gränacher, T.A. Wildmann, W. Siebrand: Chem. Phys. Lett. 122, 465 (1985)
3.109   A.A. Gorokhovskii, J.V. Kikas, V.V. Palm, L.A. Rebane: Fiz. Tverd.Tela. 23, 1140 (1981)
3.110   R. Kohlrausch: Ann. Phys. 12, 393 (1847) (Leipzig)
3.111   G. Williams, D. Watts: Trans. Faraday Soc. 66, 80 (1970)
3.112   M. Inokuti, F. Hirayama: J. Chem. Phys. 43, 1978 (1965)
3.113   J. Friedrich, A. Blumen: Phys. Rev. B32, 1434 (1985)
3.114   J. Friedrich, D. Haarer, R. Silbey: Chem. Phys. Lett. 95, 119 (1983)
3.115   R.O. Pohl: In Amorphous Solids, ed. by W.A. Phillips, Topics Curr. Phys., Vol.24 (Springer, Berlin, Heidelberg 1981) Chap.3
3.116   W.H. Hesselink, D.A. Wiersma: J. Chem. Phys. 75, 4192 (1981)
3.117   G. Schulte, W. Grond, D. Haarer, R. Silbey: J. Chem. Phys. in print
3.118   C.A. Walsh, M. Berg, L.R. Narasimhan, M.D. Fayer: Chem. Phys. Lett. 130, 6 (1986)
3.119   C.A. Walsh, M. Berg, L.R. Narasimhan, M.D. Fayer: J. Chem. Phys. 86, 77 (1987)
3.120   W.E. Moerner, M. Gehrtz, A.L. Huston: J. Phys. Chem. 88, 6459 (1984)
3.121   M. Maier: Appl. Phys. B41, 73 (1986)
3.122   V.D. Samoilenko, N.V. Razumova, R.I. Personov: Opt. Sopectrosc. 52, 346 (1982)
3.123   U. Bogner, P. Schätz, M. Maier: Chem. Phys. Lett. 102, 267 (1983)

3.124 L. Kador, D. Haarer, R.I. Personov: J. Chem. Phys. **86**, 5300 (1987)
3.125 U.P. Wild, E. Bucher, F.A. Burkhalter: Appl. Opt. **24**, 1526 (1985)
3.126 W. Richter, G. Schulte, D. Haarer: Opt. Commun. **51**, 412 (1984)
3.127 Th. Sesselmann, W. Richter, D. Haarer: Europhys. Lett. **2**, 947 (1986); Th. Sesselmann: Diplomarbeit, Bayreuth 1985
3.128 P.C. Johnson, H.W. Offen: J. Chem. Phys. **57**, 336 (1972)
3.129 W.W. Robertson: J. Chem. Phys. **33**, 362 (1961)
3.130 S.J. Lin: J. Chem. Phys. **59**, 4458 (1973)
3.131 D. Curie, D.E. Berry, F. Williams: Phys. Rev. **B20**, 2323 (1979)
3.132 C.P. Slichter, H.G. Drickamer: Phys. Rev. **B22**, 4097 (1980)
3.133 A.L. Huston, W.E. Moerner: J. Opt. Soc. Am. **B1**, 349 (1984)
3.134 W.E. Moerner, A.L. Huston: Appl. Phys. Lett. **48**, 1181 (1986)
3.135 Th. Sesselmann, W. Richter, D. Haarer, H. Morawitz: Phys. Rev. B (1987) in print
3.136 L. Kador, R. Personov, W. Richter, Th. Sesselmann, D. Haarer: Polymer J. **19**, 61 (1987)
3.137 Th. Sesselmann, W. Richter, D. Haarer: J. Luminescence **36**, 263 (1987)

4. Persistent Spectral Hole-Burning in Inorganic Materials

R. M. Macfarlane and R. M. Shelby

With 19 Figures

Persistent spectral hole-burning in the electronic transitions of crystalline and amorphous inorganic materials is reviewed. Four classes of materials have been studied so far: color centers, divalent rare earth ions in crystals, transition metal ions in crystals, and rare earth ions in glasses. The mechanism for hole-burning in crystalline systems is usually photoionization of the optical center. Both photon-gated and linear processes occur. In several cases the effects of external electric and magnetic fields on the hole spectrum have been studied and these results are reviewed.

4.1 Introduction

At low temperatures the zero-phonon absorption and emission lines (ZPL's) of most solid-state systems are inhomogeneously broadened due to a distribution of local environments induced by the strains and point defects present in all real materials. Spectral hole-burning is the selective bleaching of such an *inhomogeneously* broadened absorption line (of width Γ_{inh}) by a narrow band irradiation source (Fig. 4.1). The minimum holewidth, W, is just twice the *homogeneous* width (Γ_h) of the optical transition. This is the width experienced by all of the optical centers due to their excited state lifetime, and to dynamical perturbations such as phonons or fluctuating local magnetic fields due to nu-

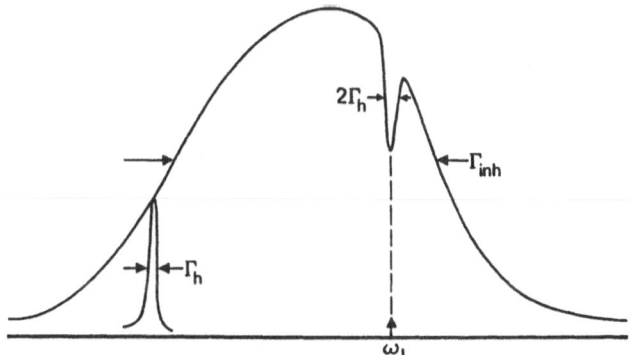

Fig. 4.1. Schematic illustration of an inhomogeneously broadened line of width Γ_{inh} which is the envelope of much narrower homogeneous packets of width Γ_h which can be selectively bleached (for example, at frequency ω_L) leading to spectral hole-burning

clear or electron spins. It can be written

$$W/2 = \Gamma_h = (\pi T_2)^{-1}$$
$$= (2\pi T_1)^{-1} + (\pi T'_2)^{-1}$$

(4.1)

where T_1 is the excited state population decay time and T'_2 the pure dephasing or transverse relaxation time.

The advent of tunable single-frequency lasers has established optical spectral hole-burning as a very useful and increasingly general technique for high resolution spectroscopy of solids at low temperatures. Many mechanisms for hole-burning have been identified and each one has a characteristic time scale for the recovery of absorption. In cases where bleaching occurs by optical pumping of metastable population reservoirs such as optical [4.1] or hyperfine levels [4.2] this recovery is typically ~ 1 s or less. We are concerned here with much longer-lived spectral features, or persistent spectral hole-burning (PSHB) with hole lifetimes of hours or longer. In many cases the lifetimes have not been measured and there is reason to believe that the holes are essentially permanent. A study of the mechanisms of PSHB is in its infancy, and many new mechanisms await discovery or elucidation. Spectral hole-burning has been found to be useful in a number of spectroscopic applications. Among these are the measurement of homogeneous linewidths, as well as the use of narrow (10−100 MHz) spectral holes to measure small frequency shifts or splittings produced by electric [4.3] or magnetic [4.4] fields (or any other frequency shifting perturbation). The latter measurements can provide, for example, precise values of magnetic moments, non-linear Zeeman coefficients and electric dipole moments. In addition, spectral holes can be used to code information in a frequency domain optical storage device [4.5]. This is discussed fully in Chap. 7 of this volume. Measurements of small perturbations and optical storage applications both require long-lived holes; the latter, of course, to a much greater degree.

Since its discovery in organic molecular systems more than a decade ago [4.6, 7] persistent spectral hole-burning has been observed in the electronic absorption bands of many crystalline and amorphous organic systems (see Chaps. 2, 3, and 5 of this volume), in color centers, in rare earth ions in glasses, and in rare earth and transition metal ions in crystals. In addition, persistent hole-burning has been observed in the vibrational bands of molecular impurity systems. Here we review the work which has been done on inorganic materials (excluding the vibrational systems which are covered in Chap. 6 of this volume). An earlier review was given in [4.8]. Following a brief outline of hole-burning mechanisms in Sect. 4.2, the material is organized according to the class of compound: Sect. 4.3 deals with color centers, Sect. 4.4 with rare earth ions in crystals and glasses, and Sect. 4.5 with transition metal ions in crystals. The number of materials exhibiting PSHB continues to increase rapidly and with this expansion will come new scientific interest and new applications.

4.2 Hole-Burning Mechanisms

In contrast to the case of transient hole-burning caused by two-level saturation or optical pumping of hyperfine levels, details of the mechanisms involved in PSHB are often not well known and can be specific to individual materials. However, some general remarks are useful at this stage. In crystalline inorganic materials it is believed that photoionization processes are responsible for all the cases reported so far. The first examples were provided by color centers in alkali halides [4.3, 9] where the burning rate appears to be linear in the photon flux, and the transformation leading to bleaching most likely occurs directly from the selectively excited energy level, or from a metastable level to which it relaxes (Fig. 4.2a). Since the optical transitions involve bound states of the center it was proposed [4.9] that the excited state wave functions were sufficiently extended that tunneling to a nearby trap resulted in photoionization.

(a) Single-photon hole-burning   (b) Photon-gated hole-burning

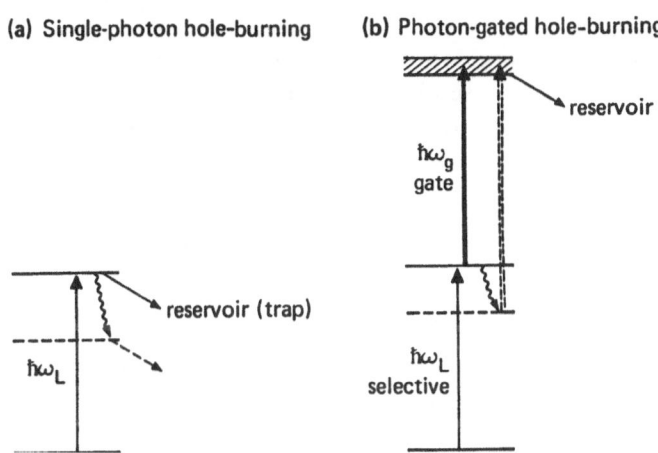

Fig. 4.2. (a) Schematic diagram of single photon hole-burning in which a photoreaction (e.g., ionization to a trap) occurs directly from the frequency-selected excited state or from a level to which the excited state subsequently relaxes. (b) photon-gated hole-burning in which one laser selectively excites population at $\hbar\omega_L$ and a second light source at $\hbar\omega_g$ induces the photoreaction. Again, relaxation to an intermediate state may occur before the gating step

More recently a new class of materials has been identified in which the ionization is a two-step process leading to so-called "photon-gated" hole-burning (Fig. 4.2b) [4.10, 11]. Here one photon from a narrow band laser source ($\hbar\omega_L$) is used to label a frequency-selected subset of the population by exciting it to a metastable level. This can either be the level initially populated by the laser or one to which the system subsequently relaxes. The second step, which may be carried out by a broad band source, takes the excited population produced by the first laser and ionizes some or all of it. Preferably this second photon

$(\hbar\omega_g)$ should be more energetic than the first so that no hole-burning occurs when it is absent, and the background absorption at the gating wavelength should be low enough to prevent broad-band bleaching. It is, of course, possible that "self-gating" occurs, i.e., photoionization results from absorption of two photons at $\hbar\omega_L$ from the frequency-selective or narrow band source. This leads to a greater than linear dependence of burning rate on laser power. Photon-gated materials are attractive for a number of reasons. Their characteristic property is the stability of the holes to reading in the absence of gating light, since under these conditions no further bleaching occurs. This is of crucial importance in applications to frequency domain storage since reading must be accomplished in a short time ($\sim$30 ns) and the photon flux required to do this inevitably leads to erasure of the stored information in single-photon materials [4.12]. Further details of these considerations are given in Chap. 7. For spectroscopic applications the stable holes produced by photon-gating make it possible to measure hole shifts, for example, in external magnetic and electric fields, over long periods of time. In addition it is possible to carry out other measurements, e.g., coherent transients, where persistent bleaching leads to experimental complexity. Because bleaching can be minimized at low power, "self-gated" systems share these advantages to some degree.

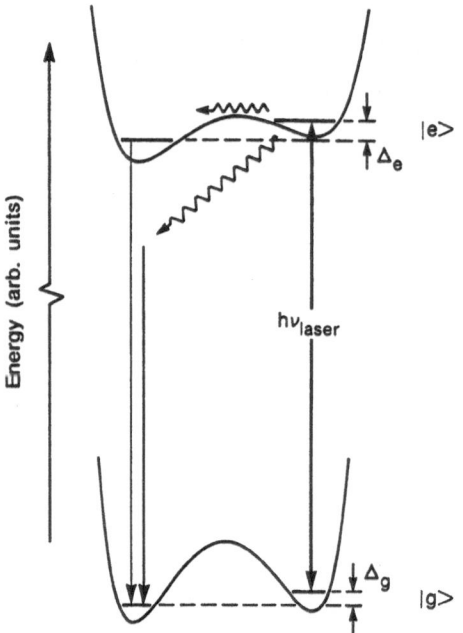

Fig. 4.3. Schematic illustration of the mechanism of hole-burning for a photostable center in a glass. The electronic levels of the defect are coupled to a two-level system (TLS), represented by a double well potential. Tunneling or relaxation produces a transition to the opposite well of the TLS shifting the optical absorption frequency

The second class of hole-burning mechanisms is that where the optical center itself (e.g., the impurity ion) is photostable, but where excitation of the center induces a slight rearrangement of the surrounding environment via phonon coupling. This is the mechanism proposed to explain the observation of PSHB in photochemically stable centers such as trivalent rare earth ions in glasses (Fig. 4.3). A similar mechanism has been proposed and widely studied in organic glasses [4.13]. There are no examples yet of this mechanism in inorganic crystalline systems, although off-center ions might be expected to show PSHB.

4.3 Color Centers

PSHB in inorganic materials was first observed in the F_3^+ color center in NaF [4.9] (Fig. 4.4). This was five years after the first reports of PSHB in organic systems [4.6, 7]. Table 4.1 summarizes the hole-burning measurements that have been made on color center systems. The table shows the holewidths, the homogeneous linewidth of the transition determined by optical coherent transients (which is sometimes different, see below) and the contribution from population decay or T_1 processes [4.8, 14]. Although measurements have been reported only for color centers in NaF, LiF, CaF_2, and diamond, it appears that hole-burning can be observed almost universally in the zero-phonon lines of color centers at low temperatures ($\sim 2K$). Of course, the lines must be inhomogeneously broadened which precludes the broad pho-

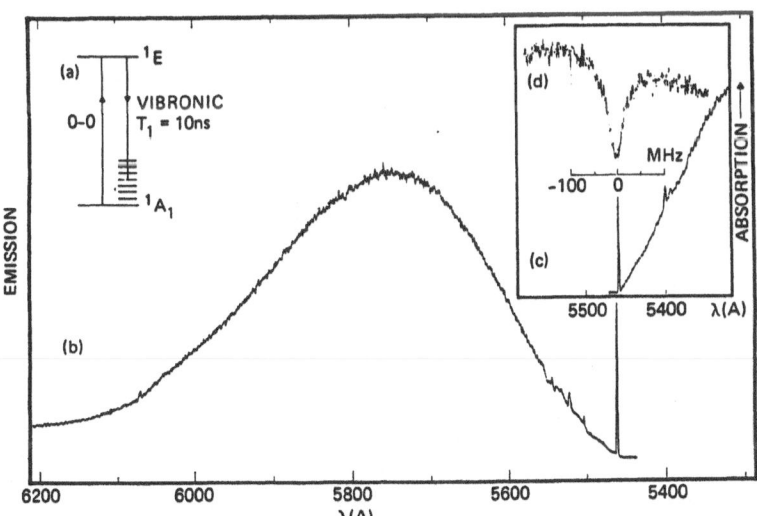

Fig. 4.4. PSHB in the F_3^+ center in NaF. (a) Energy level diagram, (b) emission spectrum showing the zero-phonon line and the vibronic sideband, (c) the fluorescence excitation spectrum, (d) a hole burned in the zero-phonon line at 5456Å

131

Table **4.1.** Hole-burning, coherent transient, and population decay measurements in color center materials

Material	ZPL Wavelength Å	Symmetry/ Assignment	Measured Linewidth			Lifetime	Reference
			Γ_{inh}^a GHz	$W/2^b$ MHz	Γ_h^c MHz	Width[d] MHz	
NaF	5456	C_{3v},F_3^+	32	19	20	16	4.9
	5754		34	~100	>100	20	4.14
	5770	C_s	40	50		20	4.52
	6070	C_s	36	18	16	16	4.3,14,17
NaF:Mn²⁺	6634		80	e			4.8
	6641		50	40			4.8
	6816		100	500			4.8
NaF:(Mn²⁺ or OH⁻)	8892	F_4^-	250	820			4.16
LiF	8330	C_{3v},F_3^-,R'	200	150			4.21
CaF₂	6774	D_{2d},F_3	130	175			4.14
Diamond	4150	C_{3v}, N_3	350	e			4.15
	4960	H_4	800	e			4.15
	6370	C_{3v},N – V	750	25		12	4.15
	7410	T_d,GR1	1000	195		52	4.15

[a] Inhomogeneous linewidth

[b] Half holewidth measured by hole-burning

[c] Homogeneous linewidth measured by coherent transients

[d] Lifetime-limited width = $(2\pi T_1)^{-1}$

[e] Hole-burning demonstrated but width measurement not made

non sidebands which are the dominant feature of color center spectroscopy. This also generally means that only the first excited state of a center, one that typically exhibits fluorescence, will show narrow hole-burning. So far all of the known cases of hole-burning in color centers are associated with fluorescent levels. The mechanisms responsible for bleaching are not understood in any detail, but it is plausible that they involve photoionization by electron tunneling from the excited state to a nearby trap. More studies to identify the nature of the traps and the photoproduct are clearly necessary. In most cases the bleaching can be reversed by irradiation at a different wavelength and the equilibrium hole depth is determined by the relative rates for bleaching and refilling. An additional factor controlling the depth is the availability of suitable traps near the optically excited centers. There are no documented examples of photon-gated hole-burning in color centers although it is expected that photon-gating will occur. Most color centers showing prominent zero-phonon lines, and hence those in which PSHB has been observed, are aggregate centers. It may be that the excited states of such centers are sufficiently delocalized for tunneling processes to be favored, resulting in single-photon hole-burning.

The quantum efficiency (η) of the burning process is typically very low ($\sim 10^{-5}$) and burning times of the order of seconds with laser powers of ~ 1 W/cm^2 are required. Some much more efficient systems with $\eta \sim 10^{-2}$ have recently been discovered, however [4.15, 16]. In all cases, hole-burning is observed on the strong singlet-singlet or doublet-doublet transitions. Most color centers, however, have lower lying excited triplet or quartet states which can store population for milliseconds to seconds. This gives a short-lived component to spectral holes, and it also acts as a bottleneck for the population on its return to the ground state. Because of the low quantum efficiency for burning, many optical cycles are required to burn a readily observable hole. This provides a rather fundamental limitation on the speed of hole-burning in many color center systems. An efficient two-step photon-gated process would overcome this.

The holewidths observed in color centers at low temperatures vary from tens of MHz to approximately 1 GHz. In the simplest case, in the absence of "pure dephasing" terms, the holewidth is just twice the homogeneous width due to population decay, i.e., $W = (\pi T_1)^{-1}$. This was observed, for example, in the F_3^+ (5456Å) [4.9] and 6070Å [4.17] centers in NaF at 1.6K. In these centers the homogeneous width was also measured using coherent transient techniques and was found to agree with the hole-burning result [4.14]. In principle, coherent transient measurements differ in an important way from hole-burning in that the time scale of the measurement is much shorter, i.e., on the scale of T_2 (nanoseconds) rather than seconds. In practice, it is difficult to carry out coherent transient measurements on materials that exhibit permanent hole-burning because the persistent spectral changes have to be avoided, for example, by scanning the laser or reversing the photo-transformation optically. The 5754Å center in NaF shows the same linewidth measured by hole-burning and phase switched coherent transients [4.8]. This may be due to a failure to completely eliminate long-lived hole-burning effects in the phase switching experiment, but the important point to note is that the width at 1.4 K is no longer determined by population decay (see Table 4.1). Similarly, although T_1 measurements are not available for most of the other centers studied, the large widths observed in hole-burning, e.g.,

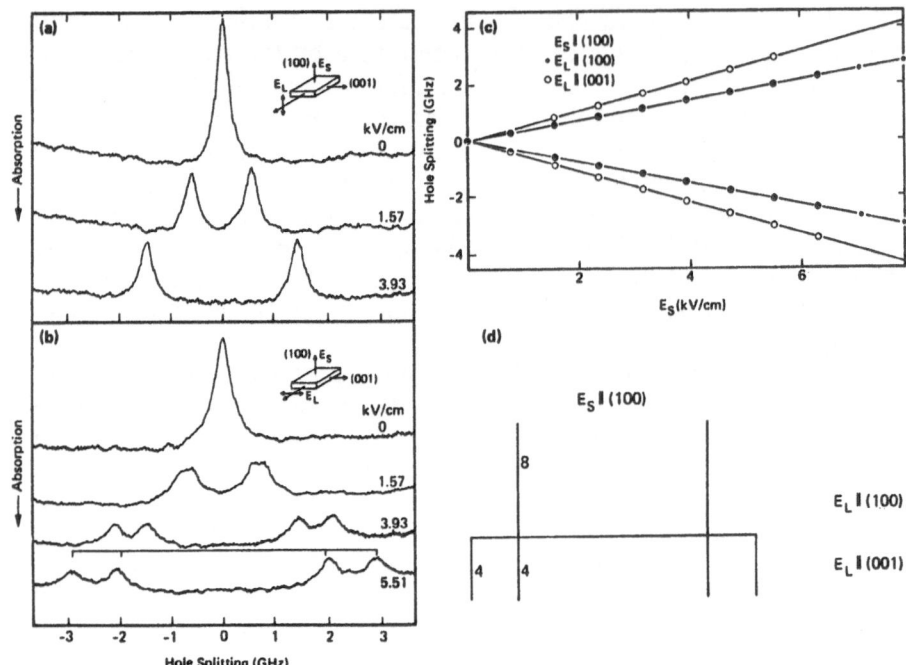

Fig. 4.5. Stark effect on holes burned in the 6070Å center in NaF for an electric field E_s in the (100) direction. **(a)** Pattern of holes for burn and probe laser L polarized with $E_L \| E_s$, **(b)** pattern of holes for $E_L \perp E_s$, **(c)** dependence of the hole splitting on electric field, for two geometries of the applied electric field, **(d)** Stark hole pattern expected in the two geometries

of the F_3 center in CaF_2 and the 6370Å and 7410Å centers in diamond are probably greater than the T_1 limit. This suggests that spectral diffusion due to time varying local fields (e.g., Coulomb fields from trapped electrons) or energy transfer is occurring on the timescale of the hole-burning experiment. A detailed understanding of this point is an important unsolved problem. In spite of this, hole-burning provides the basis for very high resolution Stark [4.3, 9, 18] and Zeeman [4.4, 19] spectroscopy, because level shifts much less than the inhomogeneous linewidth can be resolved with modest external fields of ~100 V/cm, or several hundred gauss. As an example of this, Fig. 4.5 shows the Stark effect of the 6070Å center in NaF. It had been proposed that this center was an N center of D_{2d} symmetry [4.20]. However, by a detailed measurement of the Stark effect in several geometries, two of which are shown in Fig. 4.5, *Harley* and *Macfarlane* [4.3] showed that this is an even electron center of C_s symmetry (i.e., without inversion) but is close to trigonal and could therefore correspond to a perturbed F_3^+ center, for example. The new level of precision available from hole-burning makes the Stark effect a universally useful probe of the symmetry and nature of color centers. This is in contrast to conventional Stark spectroscopy where inhomogeneous broadening almost always precludes the observation of resolved splittings.

In addition to making possible detailed studies of the effect of external perturbations, hole-burning has revealed two cases of fine structure which were not suspected in earlier studies using conventional spectroscopy, because the fine structure was buried in the inhomogeneously broadened line. The first of these was the nitrogen-vacancy (N-V) center in diamond at 6370Å [4.15], and the other the $F_3^-(\equiv R')$ center in LiF [4.21]. In both cases no satisfactory explanation of the structure has yet been given. In the case of the F_3^- center in LiF the structure was somewhat dependent on position in the inhomogeneous line suggesting some role for strain fields. In general, properties probed by hole-burning (and other narrow-band techniques) which show a dependence on position in the line, offer some prospect of progress in understanding the nature of inhomogeneous broadening.

4.4 Rare Earth Compounds

4.4.1 Trivalent Rare Earth Ions in Glasses

The first examples of inorganic impurity ions to show PSHB were trivalent rare earths in glass, in particular Pr^{3+} [4.22, 23] (Fig. 4.6) and Nd^{3+} [4.23] in silicate glasses (Table 4.2). Again a detailed understanding of the hole-burning mechanism is lacking, although the reversibility and burning characteristics suggest that a small change occurs in the local configuration of ions surrounding the rare earth. In glasses, these different configurations can be almost equi-energetic and separated by small energy barriers. A widely used model [4.24, 25] describes these in terms of double well potentials with a distribution of barrier heights and tunneling matrix elements. Tunneling between these wells is modeled at low temperatures in terms of the lowest levels of the two wells: the two level systems (TLS). Non-radiative relaxation following optical excitation produces local vibrations and phonons which assist the tunneling process between the potential wells

Table 4.2. PSHB observed in rare earth doped inorganic glasses

Material	ZPL Wavelength Å	Assignment	Γ_{inh} MHz	$(1/2)W^a$ MHz	$(2\pi T_1)^{-1}$ MHz	Reference
Silicate glass:Pr^{3+}	6060	$^3H_4\leftarrow\rightarrow{}^1D_2$	10^4	275	1.2	4.22,23
Silicate glass:Nd^{3+}	5950	$^4I_{9/2}\leftarrow\rightarrow{}^4G_{5/2}$	7×10^3	2900	b	4.23

aMeasured at 1.6K

bThis level relaxes rapidly by non-radiative decay at a rate which may limit the holewidth.

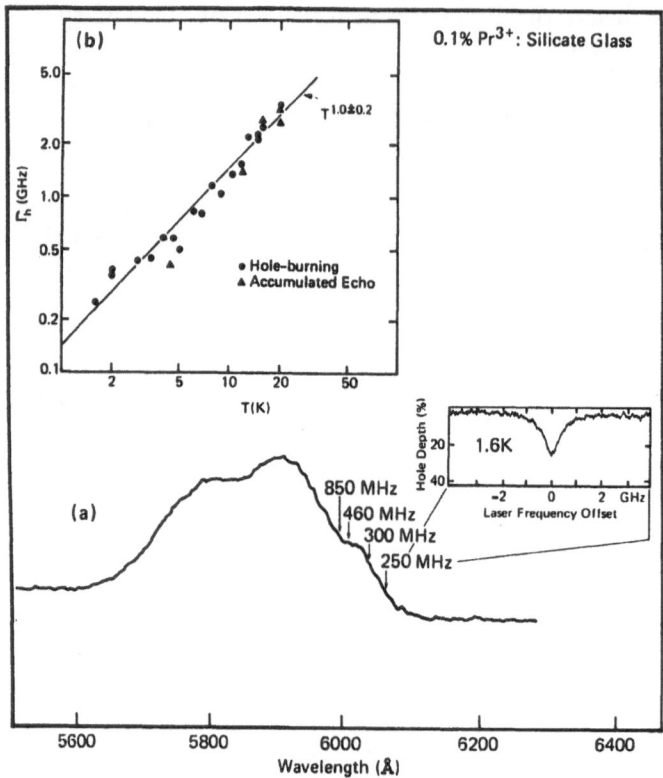

Fig. 4.6. (a) Inhomogeneous $^3H_4 \leftarrow ^1D_2$ absorption of Pr^{3+} doped into a silicate glass and a spectral hole burned on the low energy side at 6050Å, **(b)** temperature dependence of the homogeneous width measured by hole-burning (circles) and measured by the accumulated photon echo technique (triangles)

of the TLS (Fig. 4.3). Because of a coupling between the rare earth (or other) ion and the TLS, the new configuration leads to a slightly different electronic transition frequency (it need only differ by some 100's of MHz) and hence hole formation. This model is also used to describe the hole-burning process for organic molecules doped into organic glasses and polymers [4.13,26].

The holewidths in glasses are generally much larger than in crystals and also much larger than given by the T_1 contribution (e.g., $W/2 = 275$ MHz in Pr^{3+} doped silicate glass (Table 4.2) compared to $(2\pi T_1)^{-1} = 1.2$ kHz for the $^1D_2$ level), and is temperature dependent in the vicinity of 2K. Most of the studies of the temperature dependence of the homogeneous linewidth of rare earth doped glasses have been carried out using fluorescence line narrowing [4.27] which is well suited to higher temperature (≥ 20K) measurements, but generally has insufficient resolution at low temperature. Hole-burning, on the other hand, has good resolution, limited by the laser linewidth of typically ~ 1 MHz, and is useful at low temperatures. However, because hole-burning appears to be more inefficient as temperature increases, only broad, shallow holes are observed at high temperatures. Considerable activity in recent years has centered on the mechanisms and temperature dependence of the homogeneous linewidth of ions and molecules in glasses. This topic is reviewed elsewhere [4.28] but a few comments on the role of PSHB in rare earth doped glasses are in order. The most extensive measurements have been made on the $^3H_4 \leftarrow \rightarrow {}^1D_2$ transition of Pr^{3+} doped silicate glass where a linear temperature dependence of Γ_h was observed between 2K and 20K [4.22] (Fig. 4.6). This was a departure from the almost universal quadratic dependence observed in many Eu^{3+} [4.29] and Nd^{3+} [4.30] doped glasses and closer to the $T^{1.3}$ dependence of organic molecular glasses (see Chap. 5).

There is as yet, no fully satisfactory theoretical analysis of the temperature dependence of Γ_h. All models proposed so far [4.13,28,31] are based on coupling to tunneling states which have a low barrier, so that even at very low temperatures phonon-assisted tunneling occurs producing a rapidly time-varying environment and hence broadening. The hole-burning process, on the other hand, involves TLS's with higher barriers because the hole lifetimes are very long so that once the barrier is surmounted, the system remains in that configuration until it is warmed up to a temperature comparable to the barrier height. These considerations suggest that an important concern in linewidth measurements is the time scale of the measurement process. It is to be expected that at low temperatures there will be motion of the network chains in the glass with a range of time scales so that for a slow process like hole-burning which takes several seconds, spectral diffusion due to low frequency motion could lead to hole broadening. However, it was found that accumulated photon echo measurements on the $^3H_4 \leftarrow \rightarrow {}^1D_2$ transition of Pr^{3+} (in which the characteristic experimental time scale was the $^1D_2$ lifetime of ~ 100 μs) gave the same magnitude and temperature dependence for Γ_h as that obtained from hole-burning. It would be very interesting to probe on a faster time scale, e.g., with two-pulse photon echoes, and see if a narrower linewidth is obtained. Recent measurements on organic glasses show that this can be the case [4.32].

In Nd^{3+} doped glasses the only hole-burning measurement reported so far is that on the $^4I_{9/2} \rightarrow {}^4G_{5/2}$, $^2G_{7/2}$ band at 5800Å in ED-2 glass [4.23]. On the low frequency side of this band the holewidth was 5.8 GHz from which $\Gamma_h = 2.9$ GHz was obtained. This width was ascribed to non-radiative relaxation to a level ~ 1200 cm^{-1} lower in energy, with a population decay time $T_1 = 55$ ps. This is in agreement with accumulated photon echo data [4.33]. As in the case of Pr^{3+}, the holewidths measured at higher frequencies in the absorption band were broader since higher lying electronic states in the $^4G_{5/2}$, $^2G_{7/2}$ manifolds are involved, and these decay rapidly by spontaneous phonon emission.

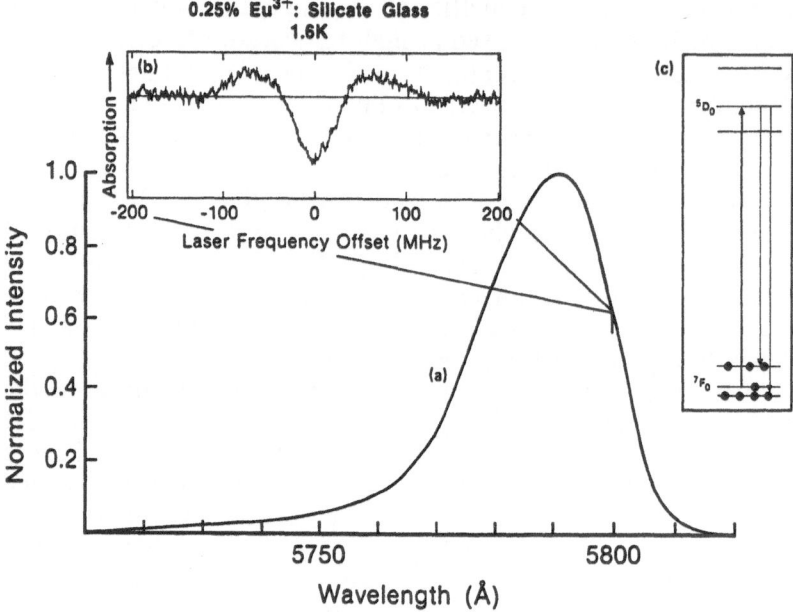

Fig. 4.7. (a) Inhomogeneous line profile of the $^7F_0 \leftarrow \rightarrow ^5D_0$ transition of Eu^{3+} doped silicate glass, (b) hole burned at 5800Å, showing neighboring antiholes. This hole pattern is not persistent and is attributed to optical pumping of nuclear quadrupole levels as shown in (c)

In Eu^{3+} doped glass, the large quadrupole splitting of the $^7F_0$ ground state allows optical pumping of these levels as is observed in crystalline systems [4.34]. This results in antiholes as well as holes (Fig. 4.7), and the hole lifetime is only a few seconds. Many more measurements of hole-burning in inorganic glass systems need to be done. First of all it is necessary to establish how general the phenomenon is and to investigate the time scale issue. In addition the technique is well suited to low temperature measurements and can contribute significantly to unraveling the puzzle of the mechanisms responsible for optical linewidths in glass. This is emphasized by recent measurements [4.35] which suggest that the dephasing of ions in glasses can only be unambiguously assigned to a TLS mechanism at relatively low temperatures ($\leq$5K).

4.4.2 Divalent Rare Earth Ions in Crystals

Most of the rare earth ions – particularly samarium and europium – can exist stably in the divalent state in alkaline earth halide crystals such as CaF_2, SrF_2, $SrClF$, and $SrCl_2$. Since, the trivalent state is more stable it is relatively easy to photoionize divalent ions as was first shown in the work of *Feofilov* and coworkers on CaF_2 co-

Table 4.3. PSHB observed in rare earth doped crystals

Material	ZPL Wavelength Å	Assignment	Γ_{inh} GHz	(1/2)W MHz	$(2\pi T_1)^{-1}$ MHz	Reference
$CaF_2:Sm^{2+}$	6958	$^7F_0A_{1g} \leftarrow \rightarrow 4f^55dA_{1u}$	22	35[a]	0.13	4.37
	6902	$^7F_0A_{1g} \leftarrow \rightarrow 4f^55dT_{1u}$	50	1900	2200	4.37
$SrF_2:Sm^{2+}$	6838	$^7F_0 \leftarrow \rightarrow {}^5D_0$	15	20	1.6×10^{-5}	4.42
$BaClF:Sm^{2+}$	6879.4	$^7F_0 \leftarrow \rightarrow {}^5D_0$	13	25[a]	1.1×10^{-4}	4.10
	6301.6	$^7F_0 \leftarrow \rightarrow {}^5D_1A_2$	11	20[a]	2.0×10^{-4}	4.44
	6298.3	$^7F_0 \leftarrow \rightarrow {}^5D_1E$	12	20[a]	1.1	4.44

[a]Holewidths are somewhat sample dependent and may not be "intrinsic" values.

doped with samarium and europium [4.36]. They observed reversible bleaching of the broad $4f^7 \rightarrow 4f^65d$ bands of Eu^{2+} under irradiation with ultraviolet light. PSHB experiments on crystals doped with divalent rare earth ions are summarized in Table 4.3.

a) $CaF_2:Sm^{2+}$

The existence of PSHB due to photoionization in divalent rare earth systems was first shown by *Macfarlane* and *Shelby* in $CaF_2:Sm^{2+}$ [4.37]. Divalent samarium ions substitute for Ca^{2+} in sites of O_h symmetry. The ground state is $(4f^6)$ $^7F_0$, A_{1g}, with $^7F_1$, T_{1g} at 256 cm^{-1}. The first excited state giving optical absorption belongs to the $4f^55d$ configuration and has A_{1u} symmetry with E_u and T_{1u} levels 57 cm^{-1} and 116 cm^{-1} higher [4.38] (see Fig. 4.8a). The $A_{1g} \rightarrow A_{1u}$ transition at 6958Å is forbidden but can be made weakly allowed by application of a magnetic field [4.39]. Excited state lifetimes are 2 μs for A_{1u} and 150 ps for T_{1u}, the latter being purely non-radiative. Hole-burning was observed in both the $A_{1g} \rightarrow A_{1u}$ (6958Å) and $A_{1g} \rightarrow T_{1u}$ (6902Å) transitions (Fig. 4.9). The latter has a holewidth of 3.8 GHz giving a lifetime of the T_{1u} level of 170 ± 20 ps in good agreement with frequency domain polarization spectroscopy measurements of *Lee* et al. [4.40]. The phonon-induced transition rate between the A_{1u} and E_u levels was measured by *Macfarlane* et al. [4.41] from the temperature dependence of the width of the spectral holes burned in the 6958Å line. From this, the E_u lifetime was determined to be about two orders of magnitude faster than that of T_{1u}, i.e., 2.4 ps, a result predicted by *Akimov* and *Kaplyanskii* [4.38] from a consideration of the symmetry properties of the phonons involved.

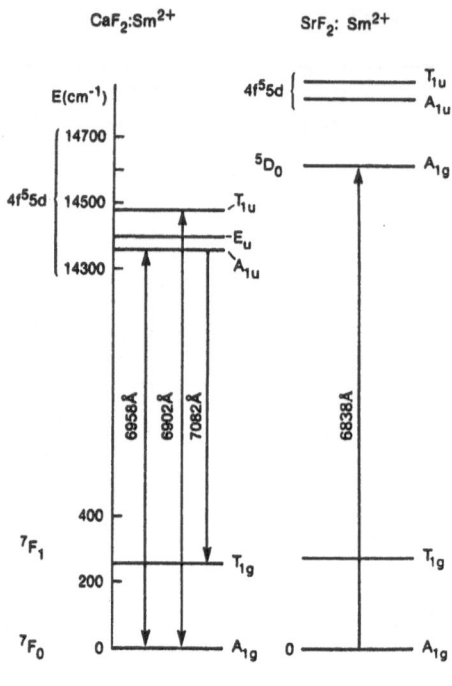

CaF$_2$:Sm^{2+} SrF$_2$: Sm^{2+}

E(cm^{-1})

4f^{5}5d

14700

4f^{5}5d 14500 T_{1u}
                    E_u
           14300 A_{1u}

4f^{5}5d { T_{1u}
                A_{1u}

$^5D_0$ A_{1g}

6958Å 6902Å 7082Å   6838Å

400

$^7F_1$
        200   T_{1g} T_{1g}

$^7F_0$ 0 A_{1g} 0 A_{1g}

◄ **Fig. 4.8.** Energy level diagram for CaF$_2$:Sm^{2+} in which the lowest optical transition is from 4f^6 to the 4f^{5}5d configuration and SrF$_2$:Sm^{2+} where it is within 4f^6

Fig. 4.9. Holes burned by photoionization of Sm^{2+} in two absorption lines of CaF$_2$:Sm^{2+}

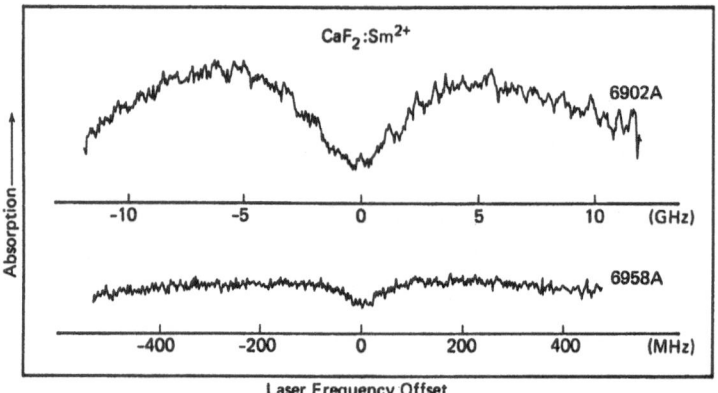

CaF$_2$:Sm^{2+}

6902A

Absorption

-10   -5   0   5   10   (GHz)

6958A

-400   -200   0   200   400   (MHz)

Laser Frequency Offset

b) SrF$_2$:Sm^{2+}

This system differs considerably from the case of CaF$_2$:Sm^{2+} in that the lowest optical transition, $^7F_0 \longleftrightarrow ^5D_0$, is within the 4f^6 configuration (Fig. 4.8b) and hence very weak. It is expected that Sm^{2+} will substitute for Sr^{2+} in sites of cubic symmetry for which the $^7F_0 \longleftrightarrow ^5D_0$ transition is forbidden. Weak emission is, however, seen at 6838Å and assigned to this transition for perturbed Sm^{2+} ions. *Macfarlane* and *Meltzer* [4.42] showed that persistent hole-burning could be observed, and from hole-burning Stark spectroscopy identified the site as having C_{4v} symmetry (Fig. 4.10). The weak linear Stark coefficient

140

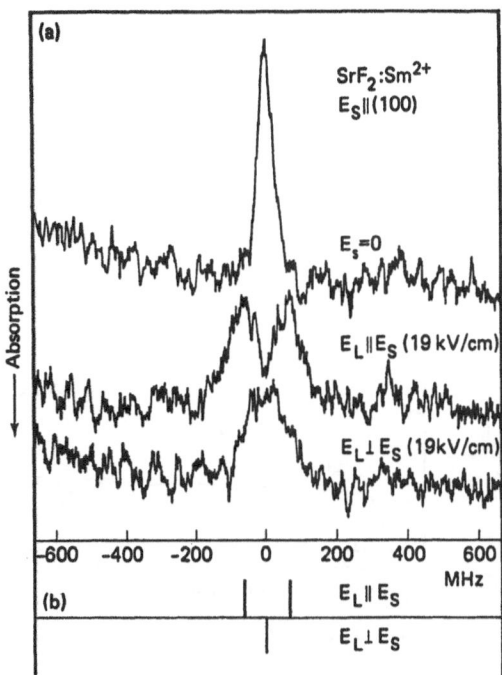

Fig. 4.10. (a) Stark effect on a hole burned in the 6838Å line of SrF_2:Sm^{2+}, (b) the expected Stark pattern for C_{4v} symmetry

(0.006 MHz/V cm^{-1}) and small splitting of $^7F_1$ (~2 cm^{-1}) show that the perturber is not in a nearest neighbor position.

c) BaClF:Sm^{2+}

Perhaps the nicest example of PSHB in inorganic materials is provided by BaClF:Sm^{2+} which shows several novel features. In this system Sm^{2+} replaces Ba^{2+} in sites of C_{4v} symmetry and can be made to be the dominant valence state of samarium. The lowest optical transitions are to $4f^6$ $^5D_0$, $^5D_1$ and $^5D_2$ levels, the $4f^55d$ absorption being above 20,000 cm^{-1} (Fig. 4.11). The oscillator strengths are high for f-f transitions (~10^{-6}) and low concentrations of Sm^{2+} were used (<0.02%).

Hole-burning in the $^7F_0 \leftarrow \rightarrow {}^5D_0$ absorption at 6879Å provided the first example of photon-gated hole-burning [4.10] (see Sect. 4.2). The presence of 5145Å Ar$^+$ laser light enhanced the hole-burning efficiency by more than a factor of 10^4 (Fig. 4.12), and in other experiments this was increased to 10^6. The model proposed to explain this result is illustrated in Fig. 4.11 and involves two-step photoionization from the long-lived ($T_1 = 1.5$ ms) metastable $^5D_0$ level to the conduction band followed by trapping. Although details of the trapping process are not known, two interesting results were obtained. The first is that Sm^{3+} ions appear to be acting as traps since the action spectrum for hole erasure follows the $4f^6 \rightarrow 4f^55d$ absorption profile of Sm^{2+} ions which would be the photoproduct of a Sm^{3+} electron trap:

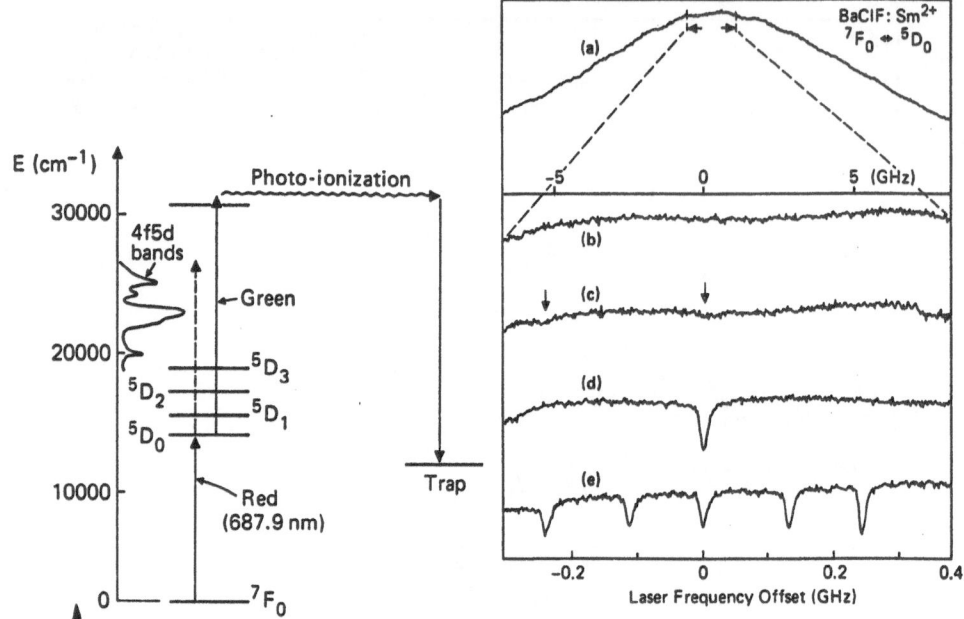

E (cm^{-1})

Photo-ionization

30000

4f5d bands

← Green

20000

$^5D_3$

$^5D_2$

$^5D_1$

$^5D_0$

10000

Red
(687.9 nm)

Trap

$^7F_0$

0

BaClF: Sm^{2+}
$^7F_0 \leftrightarrow {}^5D_0$

(a)

-5   0   5 (GHz)

(b)

(c)

(d)

(e)

-0.2   0   0.2   0.4

Laser Frequency Offset (GHz)

Fig. 4.11. Energy level diagram of BaClF:Sm^{2+} showing the two-step photoionization process

Fig. 4.12: Photon gated hole-burning in the $^7F_0 \leftarrow \rightarrow {}^5D_0$ transition of BaClF:Sm^{2+}. **(a)** The 13 GHz wide inhomogeneous line profile at 6879Å, **(b)** a 500 MHz section of the line before hole-burning. **(c)** after burning at 0 MHz and -220 MHz with 2 W/cm^2 for 2000 s, **(d)** a single hole burned at 0 MHz in 3 s by adding 20 W/cm^2 of 5145Å gating light, **(e)** multiple holes burned 110 MHz apart

$$Sm_A^{2+} + Sm_B^{3+} \underset{\leftarrow erase}{\overset{burn \rightarrow}{}} Sm_A^{3+} + Sm_B^{2+} .$$

The A sites are selectively excited by the narrow-band laser and the B sites are presumably distributed in frequency over the inhomogeneous profile so hole-burning by this mechanism is possible. Some crystals, grown under different conditions, did not show this action spectrum for erasure, highlighting the fact that the nature and availability of suitable traps can be expected to be an important factor in the observation of PSHB by photoionization.

Another novel result was that it was possible to raise the crystal to room temperature following hole-burning, hold it there for several days, and recover the holes at low temperature with very little hole filling or degradation [4.43] (see Fig. 4.13). This is the only case where stability of holes to room temperature cycling is known. At least two conditions are necessary for this. The first is that electron traps are sufficiently deep that they are not thermally emptied at room temperature over long periods of time and the second is that the inhomogeneous crystalline environment does not undergo any significant

142

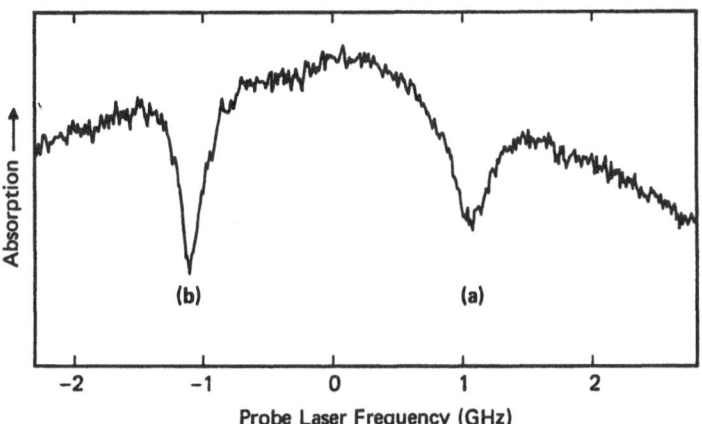

Fig. 4.13. Behavior of spectral holes in the 6879Å line of BaClF:Sm^{2+} under thermal cycling to room temperature. (a) Hole burned at 2 K, held at room temperature for 24 hours and subsequently probed at 2 K, (b) new hole burned at 2 K

annealing under the thermal cycling. Narrow spectral holes can in fact provide a unique and sensitive probe to small displacements of surrounding ions on thermal annealing or other treatment such as ion implantation.

The holewidths observed in BaClF:Sm^{2+} were somewhat sample dependent varying from ~20 MHz to several hundred MHz. At least some of this variation was concentration dependent. Even the narrowest holes are broader than expected from the contributions of laser frequency jitter or fundamental optical dephasing due to thermal and magnetic interactions. This situation is rather common in PSHB experiments and, as mentioned earlier in our discussion of color center results, may be related to the existence of time varying internal fields during photoionization hole-burning.

The stability of photon-gated spectral holes is a very useful feature for spectroscopy, particularly for the measurement of small frequency shifts over long periods of time. It is also of crucial importance for the implementation of frequency domain information storage as discussed in Chap. 7 of this volume. As an example of the former, we cite the Zeeman and Stark effect measurements made on the $^7F_0\leftarrow\rightarrow{}^5D_0$ and $^7F_0\leftarrow\rightarrow{}^5D_1$ transitions of BaClF:Sm^{2+} using photon-gated hole-burning [4.44]. The behavior of the energy levels as a function of magnetic field is shown in Fig. 4.14. Many levels exhibit small quadratic Zeeman shifts due to their interaction with other electronic states both for $H_0\|c$ and $H_0\perp c$. These cannot be measured by conventional spectroscopy but are easily and precisely measured using PSHB. A hole is burned in zero field and serves as a sharp frequency marker to measure the line shift. A typical set of data is shown in Fig. 4.15 where the quadratic nature of the $^5D_0\leftarrow\rightarrow{}^7F_0$ frequency shift is clearly shown. Almost all of this shift comes from the interaction between $^7F_0$ and

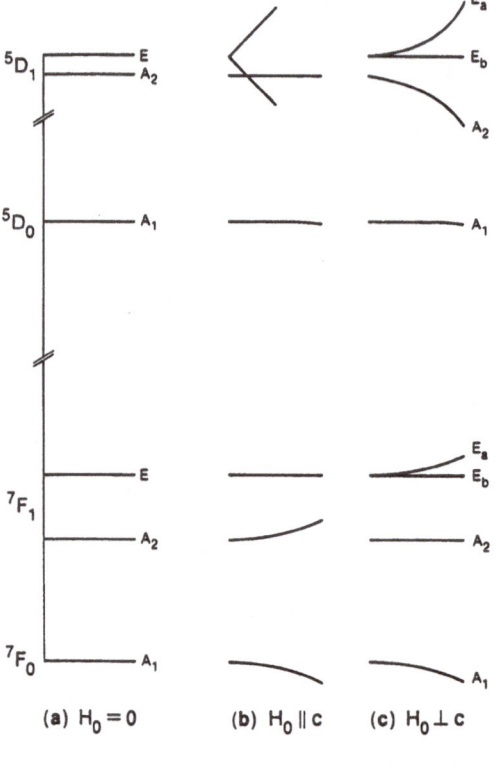

(a) $H_0 = 0$ (b) $H_0 \parallel c$ (c) $H_0 \perp c$

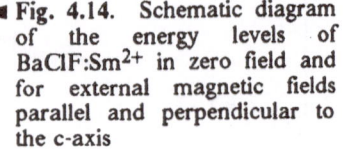

Fig. 4.14. Schematic diagram of the energy levels of BaClF:Sm^{2+} in zero field and for external magnetic fields parallel and perpendicular to the c-axis

Fig. 4.15. Quadratic Zeeman effect of the $^7F_0 \leftarrow \rightarrow {}^5D_0$ transition of BaClF:Sm^{2+} for $H_0 \perp$ c, measured from the shift of the stable photon-gated holes. (a) The hole shift for equal increments of magnetic field, (b) plot showing the quadratic shift for $H_0 \parallel c$ and $H_0 \perp c$

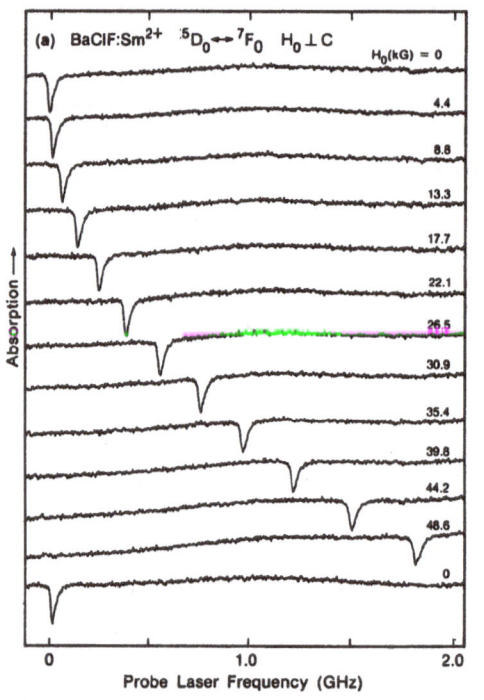

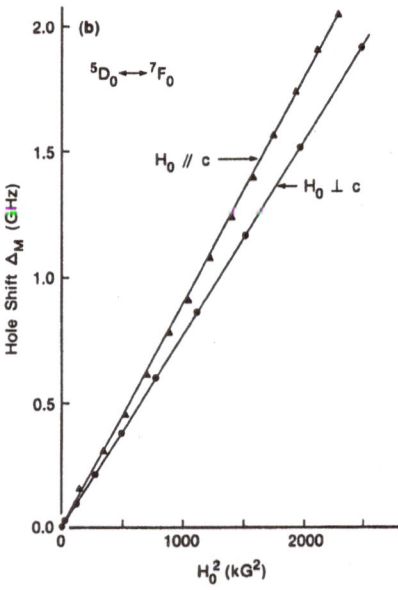

$^7F_1$, the small anisotropy (Fig. 4.14) arising because of the crystal field splitting of the $^7F_1$ level. The non-linear Zeeman coefficients provide a sensitive probe of the nature of the wave functions of the levels involved. It was found that a very good description of the observed shifts could be obtained using free-ion basis states and a perturbation calculation of the magnetic coupling. For higher lying states this may not be the case.

In the C_{4v} site symmetry of BaClF, the effect of an external electric field (E_s) on the levels of Sm^{2+} is to produce a linear pseudo-Stark effect for electric fields parallel to the crystal c-axis. This is due to inequivalent (equal and opposite) shifts of the levels of ions in the unit cell with their electric dipole moments parallel to the field and those anti-parallel to the field. For E_s perpendicular to the c-axis the Stark effect vanishes.

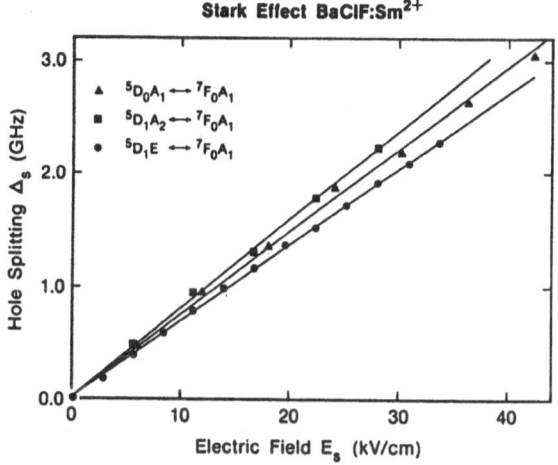

Stark Effect BaClF:Sm^{2+}

$\blacktriangle$ $^5D_0A_1 \longleftrightarrow {}^7F_0A_1$
$\blacksquare$ $^5D_1A_2 \longleftrightarrow {}^7F_0A_1$
$\bullet$ $^5D_1E \longleftrightarrow {}^7F_0A_1$

Hole Splitting Δ_s (GHz)

Electric Field E_s (kV/cm)

Fig. 4.16. Linear pseudo-Stark effect for the transitions to $^5D_0$ and $^5D_1$ in BaClF:Sm^{2+} measured using hole-burning Stark spectroscopy

PSHB was used to measure the linear pseudo-Stark effect for transitions to the $^5D_0A_1$ (0.072 MHz/V cm^{-1}), $^5D_1A_2$ (0.079 MHz/V cm^{-1}) and ^5D_1E (0.068 MHz/V cm^{-1}) levels (Fig. 4.16). In contrast to the magnetic interactions it is very difficult to calculate Stark coefficients since they vanish for pure f-electron states. The fact that there is a Stark effect at all reflects small admixtures of odd parity states by the noncentrosymmetric C_{4v} crystal field. The measured coefficients of $\sim$0.07 MHz/V cm^{-1} are similar for all transitions and comparable to those observed in other f-electron systems where Stark coefficients are known, i.e., CaF$_2$:Ho^{3+} [4.45] and LaF$_3$:Pr^{3+} [4.46].

4.5 Transition Metal Ions

Very little work has been done on persistent hole-burning in transition metal compounds, although they show great promise. In most cases the visible and infra-red absorption is due to transitions within the d^n configuration. Oscillator strengths are generally significantly higher than for rare earth compounds as is the coupling of the d-electrons to the lattice since they form the outer shell of the transition metal ions. The theory of the energy level structure and absorption strengths is covered in the books by *Griffith* [4.47] and *Sugano* [4.48]. Many of the transition metals show several stable valence states, and appear to be good candidates for photoionization hole-burning. This aspect of the solid state photochemistry of transition metal ions is just beginning and it is likely that even the less stable valence states produced by resonant laser excitation can be stabilized in crystals at low temperatures. Recent measurements have demonstrated persistent hole-burning in two transition metal systems $LiGa_5O_8:Co^{2+}$ [4.49] and $Y_5Al_5O_{12}:Ti^{3+}$ [4.50] (see Table 4.4).

Table 4.4. PSHB observed in transition metal doped crystals

Material	ZPL Wavelength Å	Γ_{inh} GHz	(1/2W) MHz	$(2\pi T_1)^{-1}$ MHz	Reference
$LiGa_5O_8:Co^{2+}$	6600	600	500	0.80	4.49
$Y_3Al_5O_{12}:Ti^{3+}$	6502	500	45	0.003	4.50

4.5.1 $LiGa_5O_8:Co^{2+}$

The inverse spinel $LiGa_5O_8$ has both octahedral and tetrahedral Ga^{3+} sites which can be substituted by Co^{2+}. In the tetrahedral sites the oscillator strengths of the spin-allowed transitions of Co^{2+} are typically large ($f\sim10^{-3}$). Hole-burning measurements were made on the zero-phonon line of the $^4A_2 \rightarrow {}^4T_1(P)$ transition at 6600Å (Fig. 4.17). This line has an oscillator strength of $f\sim5 \times 10^{-4}$, which is 10% of the total transition strength and it has a large inhomogeneous width of 20 cm^{-1}.

Persistent hole-burning was observed with a single, resonant, narrow band (~3 MHz) laser for exposures of ~1 s at 1 W/cm^2 (Fig. 4.15). Experiments with two lasers − one resonant and one below resonance − showed that photon-gated or two-step hole-burning was responsible. An enhancement of the burning efficiency by a factor of about 20 was observed with 10 W/cm^2 of nonresonant gating light. The me-

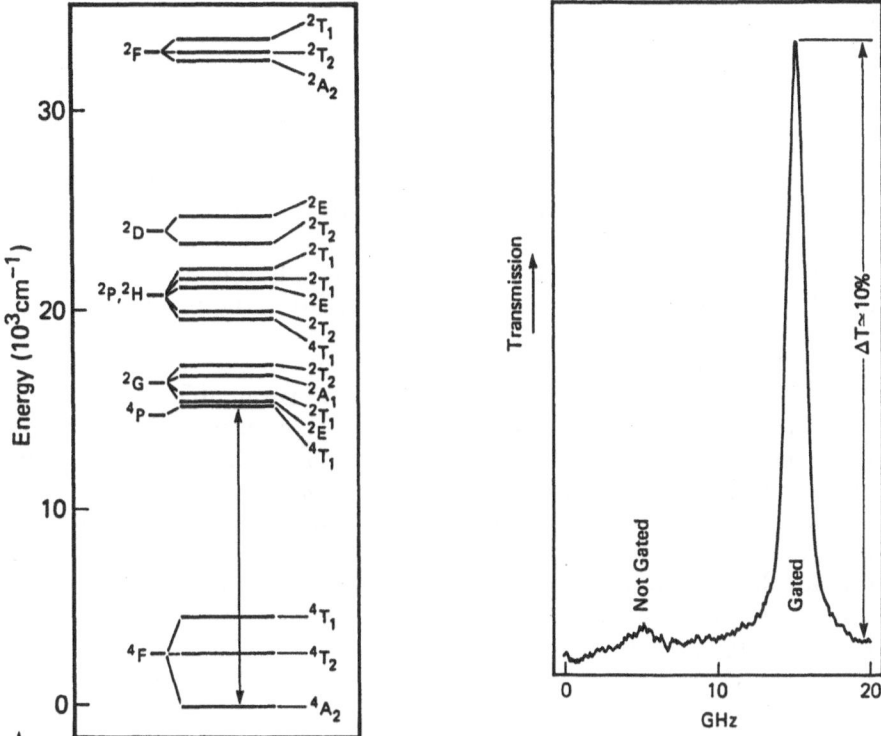

Fig. 4.17. Energy level diagram for Co²⁺ in a tetrahedral crystal environment showing the transition on which photon-gated hole-burning was observed in LiGa₅O₈:Co²⁺

Fig. 4.18. Photon-gated holes in the ⁴A₂←→⁴T₁ absorption of LiGa₅O₈:Co²⁺ at 6600Å. The enhancement of burning efficiency by the gating light is about a factor of 20 in this experiment

chanism of hole burning is presumed to be photoionization of Co²⁺, and this is supported by the hole erasure behavior. The action spectrum for erasure followed the absorption spectrum of Co²⁺. This suggests that one mechanism of erasing is the analog of burning, i.e., photoionization of another Co²⁺ center and transfer of the electron back to the Co³⁺ ions produced by hole-burning. This is similar to the behavior observed in one class of BaClF:Sm²⁺ samples.

The width of shallow, 5% deep holes was 1 GHz and this became larger for deeper holes. The obvious mechanisms for hole broadening such as population decay ($T_1 = 200$ ns,$(2\pi T_1)^{-1} = 0.8$ MHz), magnetic coupling to host nuclei (~1 MHz), or the laser frequency jitter (6 MHz) cannot account for this width. Laser heating was also ruled out. Spectral diffusion may be responsible but no mechanism has yet been identified.

4.5.2 $Y_3Al_5O_{12}$:Ti^{3+}

The Ti^{3+} ion has a particularly simple electronic structure with a single d-electron outside a closed shell. In $Y_3Al_5O_{12}$(YAG) it substitutes for Al^{3+} in sites of C_{3i} symmetry so the zero phonon line of the $^2T_{2g} \rightarrow {}^2E_g$ transition is only magnetic dipole allowed. This transition occurs at 6500Å and has an inhomogeneous linewidth of 15 cm^{-1}. The fluorescence decay time at 4K is 50 μs [4.51]. Persistent hole-burning was observed in this absorption and ascribed to photoionization of Ti^{3+} [4.50]. In contrast to the case of $LiGa_5O_8$:Co^{2+}, the process was not photon-gated but rather the burning efficiency was linear in intensity (Fig. 4.19). This suggests that the excited state of Ti^{3+} lies close to the conduction band of YAG and auto-ionization is possible.

The systematics of persistent spectral hole-burning in transition metal compounds are not yet established. The photon-gated mechanism will probably turn out to be more general since it allows photo-ionization to occur from higher lying energy levels. Very little is known about the position of impurity levels relative to the host conduction band, the cross sections for ionization or in fact the importance of mechanisms other than photoionization for persistent hole-burning in these materials.

4.6 Conclusion

The range of inorganic materials in which persistent spectral hole-burning has been observed continues to increase so that it now in-

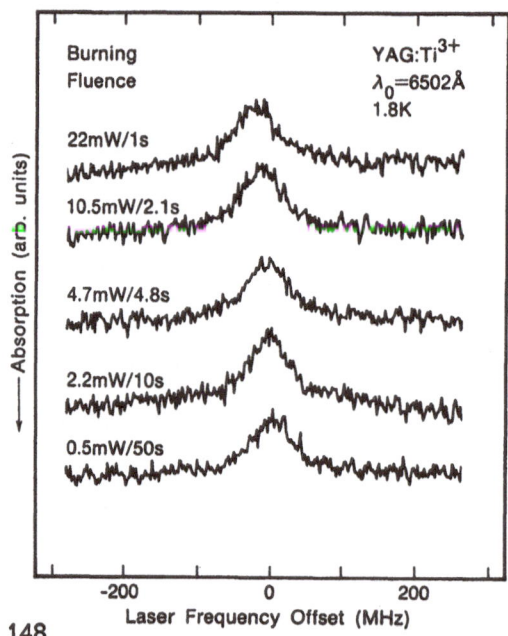

Fig. 4.19. Hole-burning in the 6502Å $^2E_g \leftarrow \rightarrow {}^2T_{2g}$ absorption of YAG:Ti^{3+} for laser powers varied by a factor of 40 but at constant fluence showing the linearity of the burning mechanism

cludes a wide variety of color centers, trivalent rare earth ions in glasses, divalent samarium and two examples of transition metal impurity systems. Thus, while PSHB is by no means a universal phenomenon, there are undoubtedly many new systems and new mechanisms to be discovered. Much more work needs to be done on mechanistic aspects of the problem.

Although it seems clear in the case of color centers and photon-gated materials that the bleaching is due to photoionization of the irradiated centers, the details of these mechanisms have not been elucidated. For example, the nature of the traps involved is not usually known, nor are the energies of the bound states of the impurity ion or of the trap states with respect to the conduction band.

For photon-gated materials, measurement of the absorption spectrum from the selectively excited state as well as the gating action spectrum would allow the gating quantum yield spectrum to be deduced. This would provide valuable information on the energetics of the gating process as well as help to elucidate the nature of the autoionizing states from which the electron is transferred into the conduction band.

In the case of color centers, it is not clear why photon gating has not been observed. One can only suppose that the more extended nature of the excited state wave functions relative to those of ions such as Sm^{2+} or Co^{2+} tends to facilitate electron tunneling directly from the selectively excited state.

Hole-burning has demonstrated some unique capabilities as a technique for high resolution laser spectroscopy. For example, it has provided a way of measuring very small level shifts such as non-linear Zeeman interactions between levels and very small electric field-induced shifts. Non-persistent hole-burning has contributed greatly to our understanding of excited state hyperfine interactions [4.53], and mechanisms can be envisaged which would enable persistent hole-burning to make similar contributions. The spectral resolution of hole-burning is often ~tens of MHz which is significantly better than that of fluorescence line narrowing and so it provides a complementary technique to FLN, particularly at low temperatures. In some cases, homogeneous linewidth information can be unequivocally obtained; in other cases, the source of the holewidth is not fully understood and this remains an important area for further research. A small number of experiments show that the result of a hole-burning measurement and a coherent transient measurement, for example, can measure different dynamical properties of a system because of the different time scales of the measurement. This issue needs to be explored more fully.

The last decade has seen rapid progress in the development of spectral hole-burning as a technique for high resolution laser spectroscopy. Taken together with FLN, optical coherent transients and optically detected magnetic resonance, it can continue to provide new insights into the local structure and dynamics of solids.

Note added in proof: A new mechanism for PSHB in inorganic materials has recently been reported [R.M. Macfarlane, R.J. Reeves, G.D. Jones: Opt. Lett. 12, 660 (1987)]. In CaF_2:Pr^{3+} containing D^- ions, light-induced tunneling of D^- leads to PSHB in the Pr^{3+} absorption.

References

4.1   A. Szabo: Phys. Rev. B 11, 4512 (1975)
4.2   L.E. Erickson: Phys. Rev. B 16, 4731 (1977)
4.3   R.T. Harley, R.M. Macfarlane: J. Phys. C 16, 1507 (1983)
4.4   R.M. Macfarlane, R.M. Shelby: Opt. Lett. 6, 96 (1980)
4.5   A. Szabo: U.S. Patent 3,896,420 (1975); G. Castro, D. Haarer, R.M. Macfarlane, H.P. Trommsdorff: U.S. Patent 4,101,976 (1978)
4.6   V.M. Kharlamov, R.I. Personov, L.A. Bykovskaya: Opt. Commun. 12, 191 (1974)
4.7   A.A. Gorokovskii, R.K. Kaarli, L.A. Rebane: Pis'ma Zh. Eksp. Teor. Fiz. 20, 474 (1974) [JETP Lett. B20, 216 (1974)]
4.8   R.M. Macfarlane, R.M. Shelby: Cryst. Latt. Def. and Amorph. Mat. 12, 417 (1985)
4.9   R.M. Macfarlane, R.M. Shelby: Phys. Rev. Lett. 42, 788 (1979)
4.10   A. Winnacker, R.M. Shelby, R.M. Macfarlane: Opt. Lett. 10, 350 (1985)
4.11   H.W.H. Lee, M. Gehrtz, E. Marinero, W.E. Moerner: Chem. Phys. Lett. 118, 611 (1985)
4.12   W.E. Moerner, M.D. Levenson: J. Opt. Soc. Am. B 2, 915 (1985)
4.13   G.J. Small: Chapter 5 in this book
4.14   R.M. Macfarlane, R.M. Shelby: Radiat. Eff. 72, 1 (1983)
4.15   R.T. Harley, R.M. Macfarlane, M.H. Henderson: J. Phys. Lett. C 17, L233 (1984)
4.16   W.E. Moerner, F.M. Schellenberg, G.C. Bjorklund, P. Kaipa, F. Luty: Phys. Rev. B 32, 1270 (1985)
4.17   M.D. Levenson, R.M. Macfarlane, R.M. Shelby: Phys. Rev. B 22 ,4915 (1980)
4.18   R.T. Harley, R.M. Macfarlane: J. Phys. Lett. C 10, L395 (1983)
4.19   R.M. Macfarlane, R.M. Shelby, A. Winnacker: Phys. Rev. B 33, 4207 (1986)
4.20   G. Baumann: Z. Phys. 203, 464 (1967)
4.21   W.E. Moerner, P. Pokrowsky, F.M. Schellenberg and G.C. Bjorklund: Phys. Rev. B 33, 5702 (1986)
4.22   R.M. Macfarlane, R.M. Shelby: Opt. Commun. 45, 46 (1983)
4.23   R.M. Macfarlane, R. M. Shelby: "Measurement of optical dephasing by spectral hole-burning in rare earth doped inorganic glasses," in Coherence and Energy Transfer in Glasses, P. A. Fleury and B. Golding, eds. (Plenum Press, New York 1984) pp. 189-199
4.24   P.W. Anderson, B.I. Halperin, C.M. Varma: Phil. Mag. 25, 1 (1972)
4.25   W.A. Phillips: J. Low Temp. Phys. 7, 351 (1982)
4.26   G.J. Small: "Persistent nonphotochemical hole-burning and the dephasing of impurity electronic transitions in organic molecular glasses," in Spectroscopy and Excitation Dynamics of Condensed Molecular Systems, V.M. Agranovich and R.M. Hochstrasser, eds. (North Holland, Amsterdam 1983) p. 515
4.27   For a review, see W.M. Yen, R.T. Brundage: J. Lum., 36, 209 (1987)
4.28   R.M. Macfarlane, R.M. Shelby: J. Lum., 36, 179 (1987)
4.29   P.M. Selzer, D.L. Huber, D.S. Hamilton, W.M. Yen, M.J. Weber: Phys. Rev. Lett. 36, 813 (1976); P. Avouris, A. Campion, M.A. El Sayed, J. Chem. Phys. 67, 3397 (1977); G.S. Dixon, R.C. Powell, Xu Gang: Phys. Rev. B 33, 2713 (1986)
4.30   J.M. Pellegrino, W.M. Yen, M.J. Weber: J. Appl. Phys. 51, 6332 (1980); R.M. Shelby: Opt. Lett. 8, 88 (1983)

4.31   R. Silbey, K. Kassner: J. Lum., 36, 283 (1987)
4.32   C.A. Walsh, M. Berg, L.R. Narasimhan, M.D. Fayer: Chem. Phys. Lett. 130, 6 (1986)
4.33   R.M. Shelby: Opt. Lett. 8 88 (1983)
4.34   R.M. Macfarlane, R.M. Shelby, A.Z. Genack, D.A. Weitz: Opt. Lett. 5, 462 (1980)
4.35   D.L. Huber, M.M. Broer, B. Golding: Phys. Rev. Lett. 52, 2281 (1984)
4.36   P.P. Feofilov: Opt. Spektrosk. 12, 531 (1961) [Opt. Spectrosc. 12, 296 (1961)]
4.37   R.M. Macfarlane, R.M. Shelby: Opt. Lett. 9, 533 (1984)
4.38   A.V. Akimov, A.A. Kaplyanskii: Fiz. Tverd. Tela 23, 3226 (1981) [Sov. Phys. Solid State 23, 1932 (1981)]
4.39   M.H. Crozier: Phys. Rev. Lett. 13, 394 (1964)
4.40   J.H. Lee, J.J. Song, M.A.F. Scarparo, M.D. Levenson: Opt. Lett. 5, 196 (1980)
4.41   R.M. Macfarlane, W.S. Brocklesby, P.B. Bloch, R.T. Harley: Opt. Commun. 58, 25 (1986)
4.42   R.M. Macfarlane, R.S. Meltzer: Opt. Commun. 52, 320 (1985)
4.43   A. Winnacker, R.M. Shelby, R.M. Macfarlane: J. de Phys. C 7, 543 (1985)
4.44   R.M. Macfarlane, R.M. Shelby, A. Winnacker: Phys. Rev. B 33, 4207 (1986)
4.45   A.A. Kaplyanskii, V.N. Medvedev, A.P. Skvortsov: Opt. i Spektr. 39, 775 (1975) [Opt. Spectr. 39, 437 (1975)]
4.46   R.M. Shelby, R.M. Macfarlane: Opt. Commun. 27, 399 (1978)
4.47   J.S. Griffith: *The Theory of Transition Metal Ions,* (Cambridge Univ. Press, Cambridge 1961)
4.48   S. Sugano: *Multiplets of Transition-Metal Ions in Crystals* (Academic, New York 1970)
4.49   R.M. Macfarlane, J.C. Vial: Phys. Rev. B 34, 1 (1986)
4.50   R.M. Macfarlane, W. Lenth: Proc. Tunable Laser Conf., Zig-Zag, Oregon, June 1986, A. Budgor, ed., (Springer Verlag, Berlin, Heidelberg 1987)
4.51   P.A. Albers: Ph.D. Thesis, University of Hamburg (1985)
4.52   R.T. Harley, R.M. Macfarlane: J. Phys. Lett. C 16, L395 (1983)
4.53   R.M. Macfarlane, R.M. Shelby: "Coherent Transient and Holeburning Spectroscopy of Rare Earth Ions in Solids," in *Spectroscopy of Solids Containing Rare Earth Ions*, A.A. Kaplyanskii and R.M. Macfarlane, eds. (North Holland, Amsterdam 1987)

151

5. Two-Level-System Relaxation in Amorphous Solids as Probed by Nonphotochemical Hole-Burning in Electronic Transitions

J. M. Hayes, R. Jankowiak, and G. J. Small

With 13 Figures

A survey of the variety of systems in which non-photochemical hole-burning (NPHB) has been observed will be presented in Sect.5.2. In keeping with the spirit of the discussion in Sect.5.1, it should be recognized that in many of the examples cited, the nonphotochemical nature of the holes has not been rigorously proven. Section 5.3 will evaluate and compare the numerous theories which have been advanced to explain dephasing. In Sect.5.4, the density of states for the characteristic two-level system (TLS) will be discussed. A correct expression for this quantity is at the heart of explaining the anomalous optical dephasing of impurities in amorphous solids. Although early work indicated that non-photochemically produced holes were persistent for hours or days if maintained at the burn temperature T_b, as data for more systems have been obtained, a number of hole-filling mechanisms have been identified. In Sect.5.5, the various phenomena, particularly laser-induced hole filling, are examined. Section 5.6 briefly summarizes several very recent developments. A discussion of NPHB mechanisms for vibrational excitations in the infrared is contained in Chap.6.

5.1 Background

In Chaps.3 and 4 various aspects of photochemical hole-burning (PHB) have been discussed. Whereas in PHB holes are produced due to photochemistry involving the absorbing species in the matrix, in non-photochemical hole-burning (NPHB) holes are produced through rearrangements of the matrix itself. Although this distinction between PHB and NPHB appears straightforward, it is better, perhaps, to think in terms of a continuum of hole-burning mechanisms rather than clearly distinct mechanisms. Thus persistent spectral hole-burning (PSHB) mechanisms may be visualized as proceeding from the intramolecular PHB, in which bonds in the absorber may be rearranged; through intermolecular PHB, in which bonds between the absorber and solvent are rearranged; to NPHB in which only matrix bond rearrangement is operative. We will return to the matter of categorizing and distinguishing between hole-burning mechanisms in a moment, but first we will elaborate a bit further on the nature of glasses and the mechanisms of NPHB.

The nature of disordered solids, e.g. glasses and polymers, is fundamental to understanding how non-photochemical hole-burning occurs. That disordered solids are basically different from crystalline media was first noted in measurements of the specific heat and thermal conductivity of glasses at very low temperatures [5.1]. The specific heat was found to contain a linearly temperature-dependent part and the thermal conductivity a quadratic temperature dependence in contrast to the cubic dependence expected for both quantities which is observed in crystals. These observations indicated that while the low-temperature properties of crystals are determined by phonons, a different type of low-energy excitation dominates the low-temperature behavior of disordered media. It was proposed by *Anderson* et al. [5.2] and by *Phillips* [5.3] that glasses are characterized by atoms or groups of atoms, which can occupy nearly isoenergetic configurations, the so-called two-level systems, or TLS.

The exact atomic identification of the TLS has not yet been determined either experimentally or theoretically for any glassy system. However, recent advances in molecular-dynamics computer stimulations of the potential-energy minima of atomic systems promise to be tremendously helpful in understanding the nature of TLS [5.4,5]. These studies, involving 10^2 to 10^3 atoms, follow the classical dynamic trajectories of the atoms as a function of time while, in parallel, identifying the potential minima onto which the instantaneous dynamical configuration would map. Although such studies are as yet not sufficiently refined to distinguish actual TLS configurations and must be carefully interpreted due to the relatively small number of atoms involved, they do illustrate several important features of amorphous materials relevant to hole-burning. First, the complexity of understanding glass structure is dramatized by calculations of the number of possible potential minima of a system, which, for example, has been estimated to be nearly as high as $(10^{10})^{22}$ for a gram of rare gas [5.6]. Second, simulations of amorphous alloys do indicate the presence of "bistable configurations" (Fig.5.1), which can be correlated to

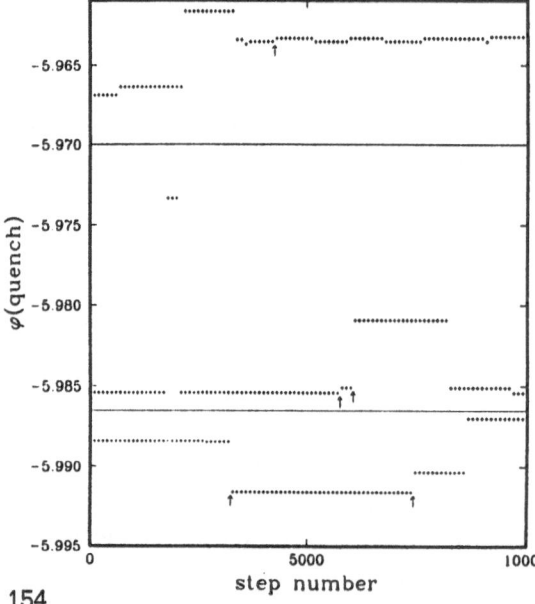

Fig.5.1. Quench potential energy per atom for three different dynamical trajectories. The top panel illustrates the characteristic signature of a TLS in this type of simulation. (From [5.5] by permission of the author)

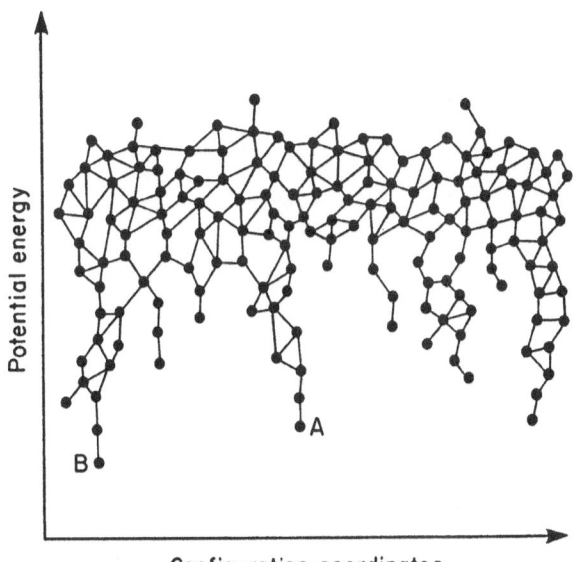

Fig.5.2. Schematic representation of the transition network for amorphous packings. (From [5.4] by permission of the author)

Potential energy

B

A

Configuration coordinates

the groups of atoms responsible for the bistability and to the coordinate along which TLS interconversions occur. Third, these simulations also show that transitions are localized, i.e., interconvertible configurations differ in the arrangement of only a small number of atoms and rearrangement of these atoms does not affect most of the material.

Finally, the simulations clearly illustrate the difficulties attendant upon a glassy system at low temperatures attempting to attain thermal equilibrum (Fig.5.2). This figure presents a transition mapping for an amorphous packing. Each filled circle represents a potential-energy minimum, and lines connecting minima represent feasible localized transitions of the system. The figure shows that there is a sparse and widely separated set of minima at the low-potential limit. The minima labeled A and B are two such minima. As the system is cooled to low temperatures the system will tend to be trapped along one of the downward-hanging tendrils, such as that ending at A. To remain in thermal equilibrum the system must have kinetic access to lower minima such as B. However, such access requires substantial backing up to higher potential configurations. Thus, thermal equilibrum is frustrated by deadends. This explains why NPHB cannot be a driving force for the system to reach thermal equilibrium.

Although the originally proposed two-level-system model was meant to explain phenomena that only occur at temperatures in the vicinity of 1 K, the model has subsequently been extended to a variety of phenomena in which there is an anomaly in either magnitude or temperature dependence in glasses relative to the same quality in crystalline media. Table 5.1 summarizes some of the anomalous glass properties which have been observed. NPHB is one of the basic tools for determining homogeneous optical linewidths in disordered media. An understanding of the observed

Table 5.1. Comparison of glass and crystal properties

PROPERTY	CRYSTAL	GLASS	REL.MAGNITUDE[*]
Specific heat	T^3	$cT + c'T^3$	larger
Therm.conduct.	T^3	$\sim T^2$	smaller
Ultras. atten.		saturates	larger
Sound velocity	T indep.	$\ell n\ T$	10-100
Dielec. const.	T indep.	ℓnT	10-100
Opt. linew.	T^7	$T-T^2$	10-100

[*] "glass value relative to crystal value"

temperature dependence for optical absorption and emission linewidths is one of the more active and challenging areas of NPHB research, *vide infra.*

The absorption spectra of materials disvolved in a disordered matrix are generally very broad due to site inhomogeneous broadening, i.e., the energy of electronic transitions of molecules are influenced by the disordered environment. This results in a distribution of absorption energies rather than a sharp absorption line. For any single absorbing molecule the absorption energy is thus sensitive to the arrangement of solvent molecules around it. If the solvent cage rearranges, then, there is a shift of the absorption energy for that molecule. A driving force for inducing such a shift is the absorption process itself. If, as described above, there is a distribution of low-energy solvent rearrangements and if as is evident from the large absorption bandwidths, there is a coupling between the solute and the solvent, then it is not difficult to imagine that during the excited-state lifetime or during nonradiative decay to the ground state, through this coupling a transition between the minima of a TLS can be induced. The effect of this TLS transition is that the microenvironment around the excited molecule has changed, and following deexcitation, the absorption energy has changed. The coupling mechanism between the TLS and the absorbing species is, of course, the familiar electron-phonon coupling, and as the coupling strength varies, this is manifested in the nature of the hole.

To make the preceding discussion less abstract, Fig.5.3 depicts schematically the NPHB mechanism. TLS_α represents a particular TLS coupled to an impurity molecule in its electronic ground state α. The TLS potential energy is characterized by a barrier height ν, a zero-point energy splitting Δ, and a well separation d. For all of these parameters there is a distribu-

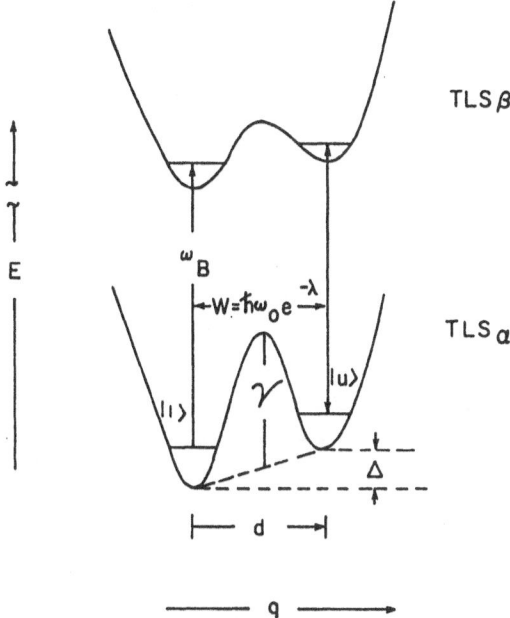

Fig.5.3. Potential energy curves for a TLS coupled to an impurity in its ground state, α, or excited state, β. The tunneling rate between potential wells is given by W

tion within the glass. In addition, there may be a variety of TLS coordinates q. NPHB occurs because, for a subset of the TLS distribution, the rate W of phonon-assisted tunneling while the impurity is in its ground state, is slow on the time scale of the experiment, while for the impurity in its excited state interconversions between tunnel states can occur in a time scale competitive with the excited state lifetimes. In the figure the excited state is represented by TLS_β, and the dramatic rise in the interconversion rate is depicted as being due to a decrease in the potential barrier. In NPHB, an impurity is coupled to a TLS which is trapped in one of the wells of TLS_α but which converts to the other well in TLS_β during electronic excitation and remains in that well during deexcitation. The net result is that the absorption energy of the impurity has been shifted by an amount dependent on the zero-point splittings of TLS_α and TLS_β.

With only slight modification essentially the same figure has also been used to explain PHB. The major difference in the case of a photoreactive species is that it is not necessary to invoke TLS to produce a double-well potential. Rather, the two wells are merely the reactant and product. This may alter the energy scales but serves to unify the mechanisms.

Even more difficult than differentiating between mechanisms conceptually is the problem of discerning the dominant mechanism experimentally. To absolutely identify the hole-burning mechanism as being photochemical requires that the photoproduct be identified and that absorption due to that species (the region of increased absorption, or antihole) be detected following hole-burning. On the other hand, to absolutely identify holes as being due to nonphotochemical processes requires that antiholes be identified and shown not to be due to a photoproduct. In many cases the lack of a photochemical mechanism can be inferred from chemical intuition based upon the stability of the molecular species involved.

A somewhat less rigorous approach to mechanism identification is the hole filling experiments first described by *Hayes* and *Small* [5.7] in the tetracene in alcoholic glass system. In this case it was shown that a major portion of the antihole lay within a few wave numbers of the initially burnt hole. This implied that if photochemistry were responsible for the hole, the absorption spectrum of the photoproduct was essentially identical to that of the starting material. The known, ambient temperature photochemical reactions of tetracene all yield products with substantially shifted absorptions. Rather than postulating a new, low-temperature photochemical reaction it was concluded that the tetracene molecule remained intact but that its absorption energy had been shifted a few wave numbers due to a solvent rearrangement.

5.2 Survey of NPHB Systems

In this section we illustrate some ideas about NPHB by discussing the variety of systems in which the phenomenon has been observed. All of the systems have in common a host material with a disordered nature, which can conceptually be modeled to contain TLS.

5.2.1 Hydrogen-Bonded Crystals

Having explained NPHB in terms of TLS in disordered solids, we discuss first NPHB which occurs in ordered molecular crystals, namely hydrogen bonded crystals of benzoic acid and benzoic acid derivatives. Benzoic acid and related carboxylic acids crystallize as cyclic dimers linked by two hydrogen bonds [5.8,9]. These dimers can rearrange via a simultaneous two-proton transfer into a tautomeric structure. For an isolated dimer the potential diagram describing this tautomer would be a symmetric double well. In the benzoic acid crystal the double well has been shown [5.10,11] to be asymmetric with a barrier height of $\simeq 400$ cm^{-1} and a zero-point splitting of 35 cm^{-1}. Thus, although a crystalline medium is involved, there is a host TLS structure present, which can be coupled to absorbing impurities in the crystal, causing NPHB.

The low-temperature spectroscopy of substitutional impurities in benzoic acid host crystals has been reported for three different inpurities: pentacene [5.12-15], thioindigo [5.16], and tetracene [5.17]. Effects of the host tautomerization are evident in all three systems, but hole-burning is not seen in the last case although it is seen in the first two. However, even for these two cases, there are some disparities in the observations and explanations for those observations.

Pentacene in benzoic acid has been the most studied of the hydrogen-bonded crystal systems. Holes have been burnt in this system by excitation of the pentacene into either the origin band or the first vibronic band of the lowest excited singlet state. The observed hole widths are characteristic of the fluorescence lifetime. Four distinct antiholes, i.e., new absorption features, are observed following burning. These antiholes are spectrally narrow, and exhibit pentacene vibrational structure. The principal antihole, site I, is red-shifted $\simeq 135$ cm^{-1} and contains $\simeq 70\%$ of the molecules which are burnt away. The holes are filled either by spontaneous filling or by laser-induced filling (LIF). The spontaneous filling is site dependent, with a l/e reversion time in the range of 6 minutes (site III) to over 1 hour (site IV). For the principal site, reversion occurs in 45 minutes.

The system also shows interesting deuteration effects [5.14]. Replacing the acidic protons with deuterons causes an order of magnitude increase in the absolute efficiency of hole-burning, although the apparent efficiency drops by two orders of magnitude. This effect is explained as being due to a population bottleneck for the deuterated species which occurs through the lowering of the energy of the second triplet below S_1. There is also a three-fold decrease in the rate of reversion of antiholes.

Finally, it should be mentioned that an alternative, photochemical mechanism for the hole-burning has also been proposed [5.15]. This mechanism, involving hydrogen abstraction from the matrix by the photoexcited pentacene, was proposed because it was not possible to thermally populate the antihole absorptions. This alternative mechanism is not inconsistent with the results of *Clemens* et al. [5.16] obtained for thioindigo in benzoic acid while the proton tunneling rates determined in [5.16] are at odds with the tautomerization mechanism proposed for pentacene/benzoic acid.

For thioindigo in benzoic acid [5.16], the absorption spectrum before hole-burning consists of two systems separated by $\simeq 500$ cm^{-1} and differing in polarization, phonon structure, and relative intensities but with similar vibrational intervals. For each absorption system three distinct origin bands have been assigned. The two absorption systems have been assigned to two distinct orientations of thioindigo within the benzoic acid lattice. For each orientation there are three distinct sites caused by tautomerization of the benzoic acid dimers. Based on variations of relative intensities of these three sites with temperature, it has been concluded that the thioindigo causes a reduction in the zero-point energy splitting of the host double-well potential from 35 to $\simeq 1$ cm^{-1}. Holes have been burnt in all of the sites of thioindigo. The holes spontaneously refill in minutes and no distinct antiholes have been detected. The hole widths are greater than expected from lifetime contributions only.

In contrast to the previous cases, for tetracene in benzoic acid [5.17] no hole-burning is detected. What is observed is a large shift, $\simeq 800$ cm^{-1}, between absorption and emission. This shift is interpreted as being due to a spontaneous reorientation of the tetracene molecule upon excitation. This shift is thought most likely to be a lateral shift of the tetracene within the space of a somewhat larger benzoic acid dimer. The tautomeric forms of the host are thought to manifest themselves as shoulders on the zero-phonon emission line of the tetracene. These shoulders are shorter lived than the bulk of the tetracene emission.

In summary, benzoic acid and other carboxylic acids can exhibit NPHB, with host tautomerization serving as the source of TLS. Although, in principle, these systems are simpler and more well defined than glassy matrices, the three exmples cited show that there is a strong impurity dependence for crystalline systems which is not always as pronounced in the glassy systems.

5.2.2 Molecules in Amorphous Polyacene Films

Another interesting class of host materials in which NPHB has been observed is the linear acenes and their derivatives. Although in spectroscopic applications these materials have most often been used as single-crystal hosts (in which hole-burning is not observed), it has been shown that when deposited rapidly as thin films on a cold substrate an amorphous material results [5.18]. Films of this type have been extensively studied by *Bässler* et al., who have examined electron diffraction [5.18], optical absorption [5.19], and NPHB [5.20-23]. These systems also differ from other NPHB systems in that these glasses can be readily converted from an amorphous state to a non-random (polycrystalline) system by annealing at temperatures well below the melting point. By contrast, alcoholic organic glasses, for example, do not change with regard to NPHB properties if annealed for several hours near the glass transition temperature. The ease of recrystallization has been explained as being due to a high degree of short-range order, derived from crystal-like molecular packing. On the other hand, the amorphous nature characterized by a lack of long-range order is due to local fluctuations of the intermolecular coordinates [5.21].

The impurities which have been studied in these systems have been other acenes so that the coupling between host and impurity tends to be considerably larger than in systems in which there is less guest/host similarity. With regard to hole-burning, the most extensively studied systems have been tetracene in anthracene or in anthracene derivatives. For these systems it has been shown [5.23] that there are two predominant absorption sites, separated by ~200 cm^{-1}, into which the tetracene can be incorporated. The relative population of these two sites can be varied by varying the sample deposition rate. In some hosts, there is also evidence of a third site. Each of these sites is subject to a statistical variation of the intermolecular parameters as evidenced by the large inhomogeneous widths, (200 to 400 cm^{-1}). Site I, the lower-energy site, is the only one for which hole-burning is observed. The holes observed are $\simeq 2$ cm^{-1} wide and the hole-burning quantum efficiency is $\simeq 10^{-5}$. Both of these values, which are larger than observed for tetracene in alcoholic glasses, were interpreted as being due to the stronger coupling between impurity and matrix.

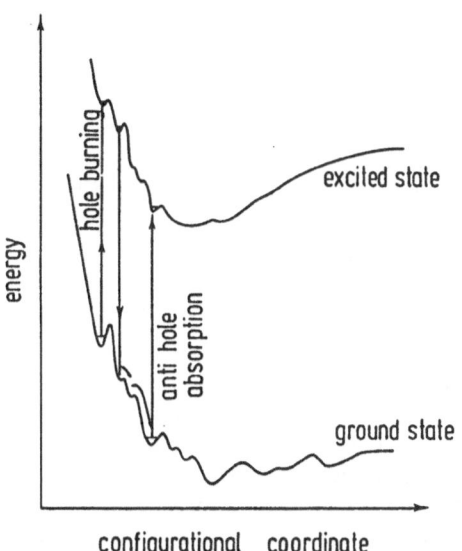

energy

hole burning

anti hole absorption

excited state

ground state

configurational   coordinate

Fig.5.4. Schematic cut of the potential energy surface of a guest molecule in an amorphous matrix along an arbitrarily chosen intermolecular coordinate. (From [5.23] by permission of the author)

Figure 5.4 shows a potential energy curve which has been proposed to explain these observations. Although this potential may seem quite different from that of Fig.5.3, the essential features which cause NPHB to occur are unaltered. In this picture, hole-burning is associated with islands of metastability which are depicted at low values of the configuration coordinate. Molecules in these islands are prevented from relaxing to lower potentials by large barriers; however, there is a barrier reduction in the excited state so that the molecule may tunnel or thermally cross the barrier within the excited-state lifetime. The excited-state barrier does, however, result in the hole-burning efficiency being low so that most exicitations are followed by resonant fluorescence, not NPHB. Up to this point the models depicted by Figs.5.3,4 are identical. In the latter case, though, NPHB may be followed by a cascade through low barrier height local minima to a more stable environment. One consequence of this cascading is that antiholes only occur to higher energy of the burn wavelength. Site II absorption, which cannot be burned, is due to absorption from sites near the absolute minimum of the ground state potential. As depicted in the figure there is a shift between excitation and emission wavelengths for site II.

5.2.3 Molecules in Organic Glasses

In this subsection the glass-forming materials discussed, as distinct from those of the previous subsection, are monomeric, aliphatic, organic substances or mixtures which are also, usually, room-temperature fluids. Materials of this type have been used by spectroscopists for many years because of two properties: first, they are good solvents for a variety of organic materials, and second, they form high optical quality glasses at low temperatures. Tabulations of glass forming materials [5.24-26] and structural explanations of glass forming propensities have been published [5.25,27]. It should be made clear that both the films of the previous subsection and host solids of this subsection are, probably, glasses if one defines a glass as an amorphous (i.e., disordered) solid which exhibits a glass transition [5.28]. However, the two types of materials have been discussed separately because they differ in their NPHB properties, most likely due to differences in electron-phonon coupling strengths.

Although glasses had been shown to be good spectroscopic media with regard to optical quality and solubility, they were not useful for high-resolution studies due to the large inhomogeneous bandwidths of materials dissolved in such media. With the advent of laser sources, however, it was shown that the inhomogeneous broadening could be removed by fluorescence line narrowing [5.29-31], in which a narrow-band emission spectrum is generated by narrow-band excitation. Shortly thereafter hole-burning was reported for several species in ethanol and in ether glasses [5.32-34]. The nonphotochemical nature of the holes and a satisfactory explanation of the mechanism of NPHB were presented in 1978 [5.7,35].

As is the case for other glasses, for the organic glasses also, the exact, atomic nature of the TLS is not known. However, some idea of the nature of modes responsible for NPHB can be deduced from the following facts. First, with only a single possible exception[1], NPHB in the types of glasses of this subsection, has only been observed for impurities dissolved in glasses that exhibit substantial hydrogen bonding. Fluorescence line narrowing has, however, been observed for both hydrogen-bonding and non-hydrogen-bonding glasses. Second, deuteration experiments in which the hydroxyl proton of an ethanol/methanol glass was replaced, showed that the hole-burning efficiency was reduced by a factor of ~5 by deuteration [5.36]. Hole widths, however, did not vary substantially between the deuterated and protonated glasses. These observations tend to indicate that hydrogen-bonded species provide TLS with parameters in the proper range for NPHB to occur. On the other hand, the failure to observe a deuteration dependence of the line width, indicates that there are additional TLS, not involving the hydrogen bond, which influence the dephasing. The independence of hole width on deuteration has also been reported by *Friedrich* et al. [5.39].

5.2.4 Molecules in Polymers

Systems of the type discussed in the previous subsection were the first in which NPHB was studied and those investigations unravelled much about the basic properties of nonphotochemical holes and the mechanism for

---

[1] A weak zero-phonon hole and stronger vibronic holes have been reported for quinizarin in a 2:3:3 :: pentane: 2-methylpentane: methylcyclohexane glass [5.37] although it has also been reported that for the same species hole-burning does not occur in a 3-methylpentane glass [5.38].

their production. There are two major difficulties associated with studies involving those systems. First, the NPHB quantum yields are generally low ($\leq 10^{-5}$). Second, the systems are room-temperature liquids thus necessitating reforming the glass for each new experiment. Since the glass properties are dependent upon formation history, the latter fact results in difficulty in obtaining quantitative reproducibility. These two drawbacks restricted most NPHB research to studies of the phenomena itself rather than to more fundamental applications of the technique, e.g., investigations of disorder.

Organic polymers had been shown to be similar to glasses in their low-temperature thermal properties [5.40,41]. It is not surprising, then, that NPHB should occur in polymers which are a natural choice of matrix to overcome the nonreproducibility associated with the formation of aliphatic glasses. As for the monomeric glasses, NPHB is observed for more impurities when the polymer is capable of hydrogen bonding than in non-hydrogen-bonding polymers. There are, however, more exceptions to this trend for polymers than for monomers. (Bear in mind, however, that the nonphotochemical nature of the hole-burning mechanism is often not rigorously demonstrated). The problem of low efficiency in NPHB studies of disorder has been overcome with the discovery [5.40] that the hole-burning efficiency for ionic dye molecules is one or two orders of magnitude larger than for neutral species, e.g., the linear polyacenes.[2] Ideal systems for studying disorder using NPHB are, thus, hydrogen-bonding polymers, doped with ionic dye molecules. Easily formed hydrogen-bonding polymers in which the ionic dyes are reasonably soluble are polyvinyl alcohol and polyacrylic acid. Most of the increased efficiency of these systems relative to other systems is due to the dyes, although there does seem to be some increase in efficiency relative to the dyes in monomeric alcoholic glasses.

Polymers which do not hydrogen bond but in which hole-burning (presumably non-photochemical) has been detected are polystyrene [5.41], polyvinylcarbazole [5.42] and polymethylmethacrylate [5.43,44] and polyvinylbutyral [5.45]. It is worthy of note that hole-burning has been observed for both chlorophyll and chlorphyll dimers in polystyrene [5.41] as well as for reaction center preparations of both green plants [5.46-48] and photosynthetic bacteria [5.49-52]. Although the interpretation of those observations is beyond the scope of this chapter, they open the possibility of investigating photobiological mechanisms by hole-burning.

A further advantage of polymers as NPHB matrices is that thin films ($\sim 10^{-3}$ mm) may be easily formed. These films are of use for investigating impurities with large absorption cross sections, e.g. the ionic dyes. Nonphotochemical hole-burning has been observed in even thinner materials. In an early paper, *Bogner* [5.53] reported holes for perylene in Cd arachidate Langmuir-Blodgett films and ascribed these holes to a TLS-in-

---

[2] In light of the large number of NPHB studies on tetracene, it should be mentioned that there is nothing special about this molecule, except that it has a reasonably large absorption coefficient and that its absorption spectrum overlaps the wavelengths of argon-ion lasers.

duced structural rearrangement. More recently, NPHB has been reported for species deposited directly on surfaces [5.54]. For optically polished surfaces holes were observed but with poor signal-to-noise ratio. For higher surface area materials, e.g. quartz powder or anodized aluminum, holes were reported with good signal-to-noise ratios for coverages of 0.1 to 0.01 monolayer.

5.2.5 Rare-Earth Ions in Glasses and Polymers

The NPHB examples cited in the previous subsections have all involved polyatomic organic molecules. NPHB has also been observed for rare-earth ions in both inorganic glasses [5.55,56] and organic polymers [5.57-58].

In inorganic silicate glass *MacFarlane* and *Shelby* [5.55] observed hole-burning for Eu^{+3}, Pr^{+3} and Nd^{+3}. For Eu^{+3} hole-burning in the $^7F_0 \leftrightarrows$ $^5D_0$ transition is caused by population redistribution in nuclear quadrupole levels rather than by a glass structural rearrangement. For the Eu^{+3}, it was also noted that the hole width at 1.6 K is consistent with fluorescence line narrowing data that indicates a $T^{1.8}$ thermal dependence for the linewidth [5.59]. Hole-burning was only possible below ~3 K, where the hole widths are less than the quadrupole splittings. The holes refill in ~20 s due to spin-lattice relaxation.

In contrast to Eu^{+3}, for both Pr^{+3} in silicate, phosphate and BeF_2 glasses [5.55] and for Nd^{+3} in silicate glass [5.56] long-lived holes have been reported. Presumably these holes are due to a glass structural rearrangement of the type described in previous subsections. Figure 5.5 shows

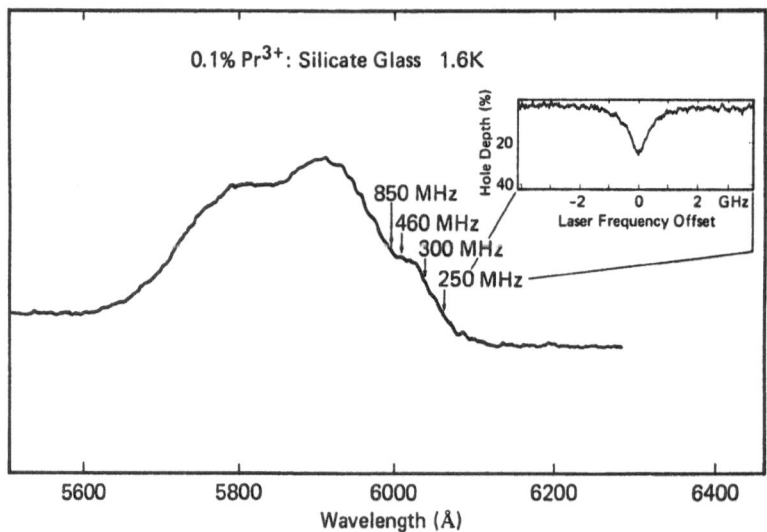

Fig.5.5. $^1D_2 \leftarrow {}^3H_4$ absorption band of Pr^{+3} in silicate glass at 1.6 K. (From [5.48]; by permission of the author)

the absorption spectrum at 1.6 K of the $^3H_4 \leftrightarrows {}^1D_2$ transition of Pr^{+3} in silicate glass, with wavelengths at which holes were burnt indicated by arrows. The widths of the holes at the various wavelengths are included, too. The inset of the figure shows a hole burnt at the longest wavelength. For the temperature range 1.6 to 20 K, the linewidth of holes burnt in the Pr^{+3}/silicate sample exhibits a linear dependence. This is in contrast to the near quadratic dependence seen for dephasing of all other rare-earth-doped inorganic glasses. Since most previous measurements had involved emission, in order to rule out contributions to the linewidth from slow structural relaxation in the glass, *MacFarlane* and *Shelby* verfied the linear temperature dependence by accumulated photon-echo measurements. As can be seen from Fig.5.6, these measurements also yielded a linear dependence.

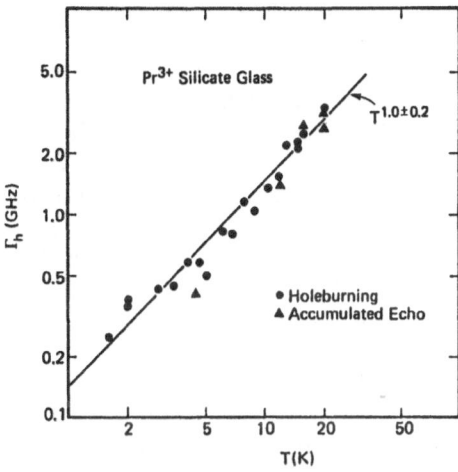

Fig.5.6. Temperature dependence of the homogeneous width of the lowest $^1D_2 \rightleftarrows {}^3H_4$ transition of Pr^{+3} in silicate glass. (From [5.48] by permission of the author)

NPHB has also been observed for both Pr^{+3} and Nd^{+3} in polyvinyl alcohol (PVOH) [5.57,58]. Relative to the inorganic glasses discussed above, the inhomogeneous absorption bandwidths are narrower in the polymer, as evidenced by the resolution of different J levels. Hole-burning is only observed for the lowest energy J-component. On the other hand, the hole widths observed are broader for the polymer although the hole-burning is one to two orders of magnitude more efficient for the polymers. In addition, the strong hole-width dependence on burn wavelength observed for the inorganic glasses is not observed in polymers. Temperature-dependent studies have not been pubished for the polymer system.

5.3 Optical Linewidths and Dephasing in Amorphous Solids

One of the photophysical properties of impurity molecules or ions which is remarkably different in amorphous solids from that in crystalline solids, is optical dephasing. This was first noted for $4f^n$ transitions of rare-earth ions in inorganic glasses from resonant fluorescence line narrowing measurements. The specific transitions studied were $^5D_0 \leftrightarrows {}^7F_0$ of Eu^{3+} [5.59-61] and $^3P_0 \leftrightarrows {}^3H_4(1)$ of Pr^{3+} [5.62]. It was found that the dephasing times (T_2) at low temperatures were 2-3 orders of magnitude shorter than in crystals and that the linewidth followed a near quadratic dependence on T over a very wide temperature range. It was NPHB studies of the $S_1 \leftarrow S_0$ transition of tetracene in an alcoholic glass [5.63,64] which first led to the same conclusion for an organic system. However, at temperatures between ~2 and ~20 K a linear dependence of the hole width on T was observed. Since these early measurements, the optical dephasing of a wide variety of impurities in amorphous solids has been studied and the results of these will be considered later in this section. However, the important conclusions that for the amorphous solids the magnitude of the optical linewidth and dephasing at low T and their T-dependence are markedly different from those in crystals, have not been altered.

Since the above behavior in glasses could not be accounted for by existing dephasing theories developed for mixed crystals, the first theoretical attempts [5.59,63-66] to explain the above results all invoked phonon-assisted tunneling (PAT) of TLS as a basic ingredient. The theories of *Lyo* and *Orbach* [5.66], and *Hayes* et al. [5.63,64] are similar in the sense, for example, that the dephasing is assumed to be due to off-diagonal modulation of the TLS tunnel states due to the impurity-TLS interaction. The approaches of *Reinecke* [5.65] and *Selzer* et al. [5.59] employed the diagonal modulation. The two types of modulation will be defined in the following. Since these initial papers, a number of further interesting theoretical developments have occurred [5.67-72]. These recent theories also assume that it is the TLS which provide the low-energy ($\simeq kT$) excitations responsible for dephasing at low temperatures. It is not surprising, therefore, that the question of how one treats the TLS distribution function is generally crucial to each theory for predictions of the T dependence of $1/T_2$ or line shape. This question will be addressed in this section and the various theories compared. These comparisons raise the question of the relative importance of diagonal versus off-diagonal modulation. The choice between the two appears to hinge on whether one assumes (or can justify) that the dephasing of a given impurity site is dominated, on average, by a single (or small number of) nearby TLS or by a "sea" of more distant and more weakly interacting TLS. With the latter choice, the T dependence of dephasing depends on the nature of the multipolar interaction between the impurity and TLS as well as the TLS density of states (distribution function) [5.67-69].

Table 5.2 provides a summary of the predicted T dependences of dephasing from the various theories. Also included is the result of *Silbey* and *Jackson* [5.73] who have treated the dephasing in organic systems by a combination of PAT of TLS [5.64] and impurity coupling to low-frequency localized phonons (induced by the impurity).

Table 5.2. Theories of optical dephasing in glasses

MODEL USED[a]	TEMP. DEP. OF $\Delta\omega$[b]	REF.
DM[c]; fast modul.	$\simeq\tau$ (TLS relax. time); motional narrowing	5.59
DM	T	5.65
ODM[c]	$\simeq T$ (high–T limit); $\simeq T^2$ (low–T limit)	5.66
ODM	$\simeq T$ (high–T limit); $\simeq T^2$ (low–T limit)	5.64
ODM	$\simeq T$ (high–T limit where $T<<\Theta_D$)	5.71
DM; slow modul.; short-time limit	$\simeq T^{4+\mu-9/s}(\rho(E)\propto E^\mu,$ $V\propto r^{-s}$,W,E correl.)	5.67,68 5.83
DM; slow modul.; long-time limit	$\simeq T^{(1+\mu)s/3}(\rho(E)\propto E^\mu,$ $V\propto r^{-s})$	5.69
DM, slow modul. short-time limit	$\simeq T^{2+\mu-3/s}$ ($\rho(E)\propto E^\mu,$ $V\propto r^{-s}$,no W,E correl.)	5.92
ODM plus exchange coupling via low-frequency impurity induced localized phonons	$\simeq T^n$ with $\simeq1.3$ for high T limit of ODM	5.73
DM with fractons	$\simeq T^{1+\eth/4}$	5.72
DM as *Lyo* [5.67, 68] with new averaging procedure	$\simeq T^{4+\mu-9/s}$ with $0.3<\mu<0.5$ for TLS_{int}[d]	5.97
ODM as *Reineker* et al.[5.71] with new averaging procedure	$\simeq T^{1+\mu_{eff}}$ with $0.2<\mu<0.5$ for TLS_{ext}[d]	5.98

[a]See Sect.5.3.4 for definitions of fast and slow modulation and short and long time limits.

[b]See Sect.5.3.2 for definitions of high and low T limits, Sect.5.3.4 for discussion of DM theories.

[c]DM $\equiv$ diagonal modulation; ODM $\equiv$ off-diagonal modulation. See Sect.5.3.1 for discussion.

[d]For TLS model with averaging, the TLS_{int} were obtained with λ_0 from about 5-7 and σ_2 from 3-4. For TLS_{ext} λ_0 was about 12-16 and σ_2 was 1.5-4 [5.96-99].

In order to facilitate a comparison between the theories we begin by developing the Hamiltonian for a single impurity-single TLS system. Following that the earliest theories which view optical dephasing as due to off-diagonal modulation will be discussed. Then the most recent experimental data which prompted the development of newer theories will be reviewed. These theories and the extent to which they acccount for the data are then discussed and this is followed by a consideration of the relationship between hole widths, spectral diffusion and dephasing. Finally, this section is concluded by a discussion of TLS distribution functions.

5.3.1 Single–Impurity Single–TLS System Hamiltonian

In what follows the symbols and notation of *Reineker* et al. [5.71] will be employed whenever possible. The connection between these symbols and those used by *Lyo* [5.67.68] are given in Table 5.3. The TLS parameters are defined in Fig.5.3 with W the tunneling frequency, and $\Delta = E_u - E_\ell$ the splitting between the local oscillators ($|u\rangle, |\ell\rangle$). The system Hamiltonian is

$$H = H_1 + H_2 + H_3 + H_{12} + H_{23} \qquad \text{where} \qquad (5.1)$$

$$H_1 = \sum_{\rho=\alpha,\beta} E_\rho |\rho\rangle\langle\rho| , \qquad (5.2)$$

$$H_2 = E_\ell |\ell\rangle\langle\ell| + E_u |u\rangle\langle u| + \frac{W}{2}(|\ell\rangle\langle u| + |u\rangle\langle\ell|) , \qquad (5.3)$$

Table 5.3. Symbols for system Hamiltonian

TERM	THIS WORK	*Lyo*
Impurity ground state	α	1
Impurity excited state	β	2
TLS asymmetry parameter	Δ	Δ
TLS tunneling frequency	W	t
TLS tunnel state splitting	E	E
Impurity-TLS coupling energy	V_α, V_β	C_1, C_2
TLS-phonon deformation potential difference	D_{qs}	B_{qs}
Diagonal modulation terms	$V_\alpha \Delta/E$; $V_\beta \Delta/E$	$C_1 \Delta/2E \equiv V_1{}^z/2$; $C_2 \Delta/2E \equiv V_2{}^z/2$
Off-diagonal modulation terms	$V_\alpha W/E$; $V_\beta W/E$	$C_1 t/2E \equiv V_1^\pm$; $C_2 t/2E \equiv V_2^\pm$

and

$$H_3 = \sum_{qs} \omega_{qs} b_{qs}^+ b_{qs} \qquad (5.4)$$

describe the 2-level impurity, TLS and phonon bath, respectively. The ground electronic state of the impurity is $\rho = \alpha$ while q and s in (5.4) represent phonon wave vector and branch labels. The impurity-TLS coupling is written as

$$H_{12} = \sum_{\rho=\alpha,\beta} \left[(V_{\ell\rho}|\ell\rangle\langle\ell| + V_{u\rho}|u\rangle\langle u|) + w_\rho(|\ell\rangle\langle u| + |u\rangle\langle\ell|) \right] |\rho\rangle\langle\rho| .$$

$$(5.5)$$

The w_ρ term has previously not been considered [5.67,68,70,71] and we do so to admit the possibility that the tunneling frequency can depend on the impurity state. The mechanism for NPHB itself indicates that this dependence is strong and so the question of its possible effect on dephasing naturally arises [5.64]. Finally, the TLS-phonon coupling is given in the linear coupling approximation by

$$H_{23} = \sum_{qs} \sum_{j=\ell,u} \frac{1}{\sqrt{N}} h_{qs}^j (b_q + b_{-q}^+) |j\rangle\langle j| . \qquad (5.6)$$

This coupling has generally been viewed as weak so that it can be treated as a perturbation. On the other hand, arbitrary coupling strength for H_{12} has been admitted. It should be noted that H does not allow for TLS-TLS interactions.

The first step involves the diagonalization of the TLS Hamiltonian H_2 to yield the tunnel states. This is straightforward and yields

$$\tilde{H}_2 = \sum_{i=1,2} E_i |\psi_i\rangle\langle\psi_i| , \qquad (5.7)$$

with

$$E_{1,2} = \frac{1}{2} (E_u + E_\ell) \mp \frac{1}{2} E , \qquad (5.8)$$

and

$$E^2 = \Delta^2 + W^2 . \qquad (5.9)$$

The tunnel state wavefunctions have the form

$$\begin{pmatrix} |\psi_1\rangle \\ |\psi_2\rangle \end{pmatrix} = \begin{pmatrix} \sin\phi, & -\cos\phi \\ \cos\phi, & +\sin\phi \end{pmatrix} \begin{pmatrix} |\ell\rangle \\ |u\rangle \end{pmatrix} , \qquad (5.10)$$

where

$$\sin\phi = \sqrt{\frac{E + \Delta}{2E}} \ , \tag{5.11a}$$

$$\cos\phi = \sqrt{\frac{E - \Delta}{2E}} \ . \tag{5.11b}$$

Utilization of (5.10) in (5.5) for the impurity-TLS coupling results in

$$\tilde{H}_{12} = \sum_{\rho=\alpha,\beta} \left[\frac{V_\rho\Delta - w_\rho W}{E} \left(|\psi_1\rangle\langle\psi_1| - |\psi_2\rangle\langle\psi_2| \right) \right.$$
$$\left. + \frac{V_\rho W + w_\rho\Delta}{E} \left(|\psi_1\rangle\langle\psi_2| + |\psi_2\rangle\langle\psi_1| \right) \right] |\rho\rangle\langle\rho| \tag{5.12}$$

where renormalization terms [5.71] which do not affect dephasing have been omitted. The V_ρ term in (5.12) has the following definition

$$V_\rho = \tfrac{1}{2}(V_{\ell\rho} - V_{u\rho}) \ , \quad \rho = \alpha,\beta \ . \tag{5.13}$$

The tunneling frequency correction w_ρ has been defined above in connection with (5.5). At this point it is appropriate to note that the first and second terms of (5.12) are (with $w_\rho=0$) the *diagonal* and *off-diagonal* modulation terms of *Lyo* [5.67,68] (Table 5.3). The former does not couple the $|\psi_1\rangle$ and $|\psi_2\rangle$ tunnel state functions but does modulate the impurity optical transition frequencies. Noting that $\tilde{H}_{12} = \tilde{H}_{12,\alpha}+\tilde{H}_{12,\beta}$, the diagonalization of $\tilde{H}_{12}$ reduces to the diagonalization of two 2x2 matrices yielding

$$\tilde{E}_{1,2;\alpha} = \frac{E_1 + E_2}{2} \mp \frac{\epsilon_\alpha}{2} \tag{5.14}$$

with

$$\epsilon_\alpha = \sqrt{E^2 - 4(V_\alpha\Delta-w_\alpha W) + 4(V_\alpha{}^2+w_\alpha{}^2)} \ . \tag{5.15}$$

The corresponding eigenfunctions are $|\Phi_{1,\alpha}\rangle$ and $|\Phi_{2,\alpha}\rangle$ with

$$\begin{pmatrix} |\Phi_{1,\alpha}\rangle \\ |\Phi_{2,\alpha}\rangle \end{pmatrix} = \begin{pmatrix} \sin\alpha, & -\cos\alpha \\ \cos\alpha, & +\sin\alpha \end{pmatrix} \begin{pmatrix} |\psi_1\rangle \\ |\psi_2\rangle \end{pmatrix} \tag{5.16}$$

and

$$\sin\alpha = \sqrt{\frac{1}{2} + \frac{E - 2(V_\alpha\Delta - w_\alpha W)/E}{2\epsilon_\alpha}} \ ; \tag{5.17}$$

$\cos\alpha$ is obtained from (5.17) by replacing the + sign in (5.17) by a − sign.

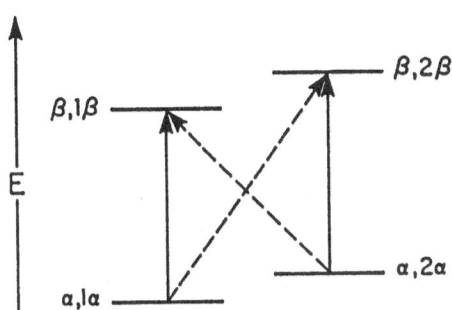

Fig.5.7. Four level optical transition diagram for an impurity and TLS, cf. text for discussion

The equations analogous to (5.14-17) for the excited impurity state ($\rho=\beta$) are obtained by replacing α by β. The resulting 4-level structure is shown in Fig.5.7. With (5.16) and the analogous equation for $\rho=\beta$, the transition moments for the four indicated optical absorption transitions can be expressed as

$$\underset{\sim}{M}_{\beta,1\beta\leftarrow\alpha,1\alpha} = \underset{\sim}{M}_{\beta,2\beta\leftarrow\alpha,2\alpha} = (\sin\alpha\ \sin\beta + \cos\alpha\ \cos\beta)\underset{\sim}{m} \qquad (5.18a)$$

and

$$\underset{\sim}{M}_{\beta,2\beta\leftarrow\alpha,1\alpha} = -\ \underset{\sim}{M}_{\beta,1\beta\leftarrow\alpha,2\alpha} = (\sin\alpha\ \cos\beta - \sin\beta\ \cos\alpha)\underset{\sim}{m} \qquad (5.18b)$$

where $\underset{\sim}{m} = \langle\beta|\underset{\sim}{d}|\alpha\rangle$ is the electronic transition dipole for the impurity. The optical transitions defined by (5.18b) are the dashed arrows in Fig.5.7 which are meant to indicate that in the limit of vanishing impurity-TLS coupling ($V_\rho=w_\rho=0$) they are forbidden. It should be noted at this point that the $\beta,2\beta \leftarrow \alpha,1\alpha$ transition admits the possibility of non-zero dephasing at 0 K. Another such possibility will be considered later.

The final step in the development of H is to rewrite the TLS-phonon interaction H_{23} using (5.10) and the definitions of (5.11). After a little algebraic manipulation one obtains

$$\tilde{H}_{23} = \sum_{qs} \left[\frac{\Delta D_{qs}}{E} (|\psi_1\rangle\langle\psi_1| - |\psi_2\rangle\langle\psi_2|) \right.$$

$$\left. + \frac{W D_{qs}}{E} (|\psi_1\rangle\langle\psi_2| + |\psi_2\rangle\langle\psi_1|) \right] (b_q + b_{-q}^+), \qquad (5.19)$$

where

$$D_{qs} = \frac{h_{qs}^\ell - h_{qs}^u}{2\sqrt{N}}. \qquad (5.20)$$

A renormalization term [5.71] which is unimportant for dephasing has been omitted. By analogy with $\tilde{H}_{12}$ one can refer to the first and second terms of $\tilde{H}_{23}$ as diagonal and off-diagonal but we emphasize that in the literature [5.67,68] diagonal and off-diagonal modulation are referred to $\tilde{H}_{12}$.

In summary, the key equations for a consideration of optical dephasing developed thus far are (5.12,19). The approach taken follows that of others [5.67,68,71] with the symbols and notation used that of *Reineker* et al. [5.71]. It is implicit, therefore, that the treatment assumes that $\tilde{H}_{23}$ is weak. Arbitrary impurity-TLS coupling has been allowed for.

To end this subsection we turn to the last term of $\tilde{H}_{23}$, (5.19), which allows for phonon-assisted tunneling (PAT) of the TLS in the absence of the impurity. For a later discussion of possible motional narrowing effects associated with the 4-level system of Fig.5.7, we provide at this point the expressions for the PAT frequencies associated with a specific TLS. This frequency is denoted by Γ^+ and Γ^- for the tunnel states $|\psi_2\rangle$ and $|\psi_1\rangle$, respectively. Utilization of (5.19) with the Fermi golden rule results in

$$\Gamma^- = \frac{2\pi W^2}{\hbar^2 E^2} \sum_{qs} \langle n_{qs}\rangle_T \, D^2_{qs} \, \delta(\Omega - \omega_{qs}) \tag{5.21}$$

where $\langle n_{qs}\rangle_T$ is the thermal occupation number for the phonon ω_{qs} and $\hbar\Omega = E$, see (5.8). The familiar long-wavelength approximation is made for D_{qs},

$$h^i_{qs} = f^i \sqrt{\frac{\omega_{qs}}{2Mc^2}} \, , \qquad i = \ell, u \tag{5.22}$$

with the deformation term

$$f = h^\ell - h^u \, . \tag{5.23}$$

In (5.22), M is the unit-cell mass and c is an average sound velocity for the three acoustical branches. Utilization of (5.20,22,23) and

$$\sum_{qs} \rightarrow \frac{V}{2\pi^2} \frac{3}{c^3} \int_0^{\omega_D} d\omega \, \omega^2 \tag{5.24}$$

for the Debye density of states result in

$$\Gamma^- = \frac{3W^2 f^2 E}{8\pi\rho c^5 \hbar^5} \langle n_E\rangle_T \, , \tag{5.25}$$

where ρ is the sample density. With $\Gamma^+ = \exp(E/kT)\Gamma^-$ the TLS relaxation time τ is defined by

$$\frac{1}{\tau} \equiv \frac{1}{2}(\Gamma^+ + \Gamma^-) = \frac{1}{2}\Gamma^-(1 + e^{E/kT}) \, . \tag{5.26}$$

5.3.2 Optical Dephasing due to Off-Diagonal Modulation

As previously mentioned, the initial theories by *Lyo* and *Orbach* [5.66] and *Hayes* et al. [5.63,64] considered *pure* dephasing arising from off-diagonal modulation. More recently *Reineker* et al. [5.71] have reexamined the Lyo and Orbach approach using the density-matrix approach for the 4-level system of Fig.5.7. However, because of the complexity of the general expressions their final analysis of the linewidth problem employs a perturbative solution which will now be obtained in a simpler manner. With $\tilde{H}_{12}$ viewed as weak, only the solid-arrow transitions of Fig.5.7 carry significant intensity, i.e. $\sin\alpha,\beta \simeq 1$ and $\cos\alpha,\beta \simeq 0$ in (5.18). With (5.12) one obtains

$$\cos\alpha = \frac{V_\alpha W + w_\alpha \Delta}{E^2} \tag{5.27}$$

and $\cos\beta$ is given by the same expression with α replaced by β. The linewidth from pure dephasing for, say, the $\beta,1\beta \leftarrow \alpha,1\alpha$ transition is given by [5.74]

$$\Delta\omega = \frac{4\pi}{\hbar} \sum_n P_n \sum_m |\; \langle \Phi_{2,\beta} m | \tilde{H}_{23} | \Phi_{1,\beta} n \rangle - \langle \Phi_{2,\alpha} m | \tilde{H}_{23} | \Phi_{1,\alpha} n \rangle \;|^2$$

$$\cdot \rho \, [E - \hbar(\omega_n - \omega_m)] \;. \tag{5.28}$$

In (5.28), n and m are general phonon indices, and the P_n are occupation probabilities for the initial phonon levels. For the evaluation of (5.28), integrals of the type

$$\langle \Phi_{2,\beta}(n_q-1) | \tilde{H}_{23} | \Phi_{1,\beta} n_q \rangle = n_q^{1/2} \left[\frac{2\Delta D_{qs}}{E} \sin\beta \, \cos\beta + \frac{W D_{qs}}{E}(\sin^2\beta - \cos^2\beta) \right]$$

$$\tag{5.29}$$

are required and correspond to phonon absorption. For weak coupling $\sin\beta = 1$ and $\cos\beta$ is given by (5.27). Since pure dephasing depends on the difference of excited- and ground-state matrix elements, only the first term in (5.29) contributes with the neglect of the $\cos^2\beta$ term (we return to this point later). Using the same manipulations which led to (5.25) we find that

$$\Delta\omega = \frac{3(WV + \Delta w)^2 \Delta^2 f^2}{2\pi\rho c^5 \hbar^5 E^3} \operatorname{csch}(E/kT) \tag{5.30}$$

when thermal equilibrium between the two tunnel states is assumed. This result is identical to that of *Reineker* et al. [5.71] when it is noted that they considered only one acoustical branch and assumed that $w = 0$. With the same assumption it is in essential agreement with the result of *Lyo* and *Orbach* [5.66], and *Hayes* [5.64] when it is noted that the latter authors renormalize the TLS for the impurity interaction. The definitions of V and w are

$$V = V_\beta - V_\alpha \qquad (5.31)$$

and

$$w = w_\beta - w_\alpha . \qquad (5.32)$$

The linewidth for the $\beta,2\beta \leftarrow \alpha,2\alpha$ transition is identical to that of (5.30) when thermal equilibration is assumed. Without this assumption this transition will provide non-zero dephasing at 0 K! It should be emphasized that because diagonal modulation in $\tilde{H}_{12}$ has been neglected and the off-diagonal modulation is taken as weak, the two allowed transitions are viewed as degenerate.

Of course, (5.30) is concerned with only a single TLS and the perplexing problem of how to treat the TLS distribution remains. *Lyo* and *Orbach* [5.66], and *Reineker* et al. [5.71] utilized the approach of *Anderson* et al. [5.2] and *Phillips* [5.3], which was developed for inorganic glasses at very low T (Sect.5.4). In essence one assumes that the density of states for the TLS, $\rho(E)$, is constant over the temperature range of the experiment. Details can be found in [5.71]. Suffice it to say here that the TLS distribution parameters for f, V, ω_0 (Fig.5.3), Δ and λ are viewed as uncorrelated except for Δ and λ. *Lyo* and *Orbach* [5.66] concluded that the optical linewidth or $1/T_2$ varies linearly and quadratically with T in the high- and low-temperature limits, respectively. The latter limit is defined by kT less than the maximum TLS splitting E or the Debye frequency $\hbar\omega_D$, whichever is lower. Recently, these limits have been reexamined [5.71]. With TLS parameters which are viewed as appropriate for inorganic glasses, it is found that the quadratic dependence is valid only for temperatures very substantially below T_D. Thus, these theories by themselves could not account for the early observation that for several rare-earth-ion inorganic glass systems, a near quadratic dependence is observed from low to room temperature [5.59-62]. Of course, for higher temperatures the dephasing due to Raman scattering [5.75] cannot be neglected [5.36]. Noting that the high-temperature limit for this scattering yields $1/T_2 \propto T^2$ (independent of the phonon density of states), it seems possible that a combination of the two scattering mechanisms may be able to account for the data. However, this is by no means the complete story on optical dephasing of rare-earth ions in inorganic glasses as has been seen in Sect.5.2.4 (see also Sect.5.3.3).

As already mentioned, *Hayes* et al. [5.63,64] were concerned with the observation of NPHB hole widths of organic systems which exhibited a near-linear dependence on T between $\simeq 2$ and $\simeq 20$ K. A TLS averaging procedure different from the one just considered was utilized. It involves the assumption that the dephasing of each impurity site is dominated, on average, by a single TLS. This TLS is characterized by a prefactor F of

174

the thermal factor $\text{csch}(E/kT)$ lying within some *maximum* range between F_{max} and F_{min}. For each value of F in this range it was further assumed that there is an associated distribution of impurity-TLS, describable by a distribution function $P(\Delta)$ which was taken as a Gaussian centered at 0 cm^{-1}. Configurational averaging of the linewidth showed that the high T limit of the theory where $\Delta\omega \propto T$ obtains when kT is larger than the width of the TLS distribution. To explain the above T dependence, this width for the "dephasing" TLS must be ≤ 1 cm^{-1}. The actual observed line shape is a superposition of Lorentzians having F values in the above range. Conditions under which non-Lorentzian profiles might be observed were defined.

5.3.3 Recent Experiments

The initial experimental studies [5.59-64] and theories [5.59-66] dealing with the anomalous optical linewidths and dephasing of impurities in amorphous solids have led to many new photochemical and nonphotochemical hole-burning, luminescence line narrowing and photon-echo experiments on organic and inorganic systems. Temperature-dependent photochemical hole-burning has been reported for free-base porphin [5.76-78], chlorin [5.79], dimethyl-s-tetrazine [5.77,78], protonated free-base phthalocyanine [5.80], and quinizarin [5.39] in glasses and/or polymers. Temperature-dependent NPHB studies have been reported for cresyl violet perchlorate [5.80,81], resorufin [5.80], chlorophyll *a* [5.82], and pentacene [5.80,81], all in polymers. Pentacene in polymethylmethacrylate has also been studied by the photon-echo technique [5.83]. Tetracene in amorphous anthracene [5.84], 2,3-dimethylanthracene [5.85] and 9,10-diphenylanthracene [5.85] have also been investigated by NPHB. Coronene and 5-bromoacenaphthene in a 1-bromobutane glass have been probed by phosphorescence line narrowing [5.86]. With the exception of quinizarin in ethanol or boric acid glasses [5.39] and protonated phthalocyanine, the power law for $\Delta\omega$ can be said to be $T^{1.3}$ within experimental uncertainties (typically $\pm$ 0.1 - 0.2). Agreement between the NPHB and photon-echo power laws for pentacene in PMMA has recently been achieved [5.80] However, in very recent developments, comparisons between echo and NPHB contain some surprises (Sect.5.6). It should be noted that in a number of systems the $T^{1.3}$ power law has been observed down to 0.3 K [5.77,78,80]. Furthermore, a $T^{1.3}$ extrapolation to lower temperatures has yielded a lifetime-limited value of $\Delta\omega$ for several systems [5.76-78,81,87]. Thus, a T^n power law with n ~ 1.3 appears to govern the optical dephasing in a substantial number of organic systems. More generally, it can be said that $\Delta\omega$ from hole-burning measurements frequently exhibits a $T^{1.3}$ power law (Sect.5.3.6). The effect of laser burn intensity on hole widths is also discussed in Sect.5.3.6.

Most of the T-dependent optical linewidth and dephasing data for rare-earth ions in inorganic glasses have already been discussed (Sect.5.2.4 and the introductory remarks of Sect.5.3). We emphasize again that the

$^3H_4 \leftrightarrows {}^1D_2$ transition of Pr^{3+} in a silicate glass exhibits a T^n power law with $n \simeq 1$, in sharp contrast with many other rare-earth ion transitions for which $n \simeq 2$. It should be clear that the theories discussed thus far cannot account for the Pr^{3+} results. Recent photon-echo studies [5.69,88] on Nd^{3+} in a pure SiO_2 glass have shown that the dephasing of the $^4F_{3/2}(1)$ state follows a $T^{1.3}$ power law between 0.1 and 1.0 K. Thus, near linear T-dependent dephasing may not be all that unusual for rare-earth ions in inorganic glasses. These data for Nd^{3+} are made all the more interesting when it is noted that the same transition exhibits a $\Delta\omega \propto T^2$ dependence for $T > 10$ K [5.89].

Taken as a whole, the above data led naturally to the development of new theories which we now proceed to discuss.

5.3.4 New Theories

In this subsection we will be primarily concerned with the recent theories of *Lyo* [5.67,68] and *Huber* et al. [5.67].[3] Reference [5.90] is a lengthy book chapter which provides many of the theoretical details omitted from earlier work. Thus, it is not necessary to reproduce them here. With Table 5.3 the connection between the equations which follow and those of *Lyo* [5.67,68,90] can be made.

Both of the above theories were prompted by the observation of different T dependences for $\Delta\omega$ or echo decays in "apparently" similar systems but certainly, also, by the fact that many organic and inorganic systems exhibit a near-linear or quadratic behavior over an *extended* T range. The theories are similar in that diagonal modulation is considered to dominate off-diagonal modulation and the impurity is considered to interact with a large number of TLS. As mentioned earlier, this leads to the result that the T dependence depends on the nature of the multipolar interaction between the impurity and the TLS. The major differences between the two theories are the manner in which the spatial distribution of the interacting TLS is handled and the time regimes for which they are valid.

Before discussing the general treatments of *Lyo* [5.67,68] and *Huber* et al. [5.69] which stem from $\tilde{H}_{12}$, (5.12), it is first instructive to expose the essential physics in a simpler manner [5.68]. In order to explain the observed T dependences both theories consider that $\Delta\omega$ is determined by *weak* interactions of the impurity with a "sea" of TLS for which $\hbar\tau^{-1} < V'$ $< E$, $(V' = V\Delta E^{-1})$, see (5.9,12,26,31). These "more distant" TLS are considered to dominate dephasing from the limited number of strongly interacting "nearby" TLS for which V might be greater than E. Because $V' < E$, only the diagonal or solid arrow transitions of Fig.5.7 carry significant intensity. This appears to be reasonable since appreciable intensity for the

---

[3] At the time of writing we were fortunate to have access to very recent unpublished works from both groups [5.90,91]. They have served to clarify certain questions in our minds for which we are thankful.

highest energy transition of Fig.5.7 would lead to asymmetry for the spectral profile at very low T. This asymmetry has not been observed. For a discrete 4-level system the Redfield [5.83] and correlation function theories [5.59] yield

$$\Delta\omega = \frac{(V')^2}{\hbar^2} \operatorname{sech}^2\left(\frac{E}{2kT}\right) \frac{\tau}{1 + (V'\tau/\hbar)^2} \qquad (5.33)$$

for the linewidth with τ the TLS relaxation time in the absence of the impurity. Equation (5.33) is valid for $V' < E$. The fast and slow modulation limits are defined by $\hbar\tau^{-1} > V'$ and $\hbar\tau^{-1} < V'$, respectively. In the former limit, (5.33) leads immediately to the motional narrowing result where $\Delta\omega \propto \tau(V')^2$. That is, the contribution of diagonal modulation is reduced by fast TLS relaxation and the linewidth should decrease with increasing T. Since this has not been observed experimentally we focus on the slow modulation limit in which (5.33) yields

$$\Delta\omega = \frac{1}{\tau} \operatorname{sech}^2\left(\frac{E}{2kT}\right), \qquad (5.34)$$

with $1/\tau$ given by (5.26). Taking into account the T dependence of $1/\tau$, the thermal factor of (5.34) is $\operatorname{csch}(E/kT)$. Equation (5.34) is interesting since the impurity-TLS interaction V has canceled out. The presence of the impurity manifests itself in a more subtle manner! Recall that the condition for applicability of (5.34) is that $V' > \hbar\tau^{-1}$. Thus, in averaging (5.34) over TLS, the cut-off radius r_c is determined from the condition $V' = \hbar\tau^{-1}$. Since V' is governed by the type of multipolar interaction between the impurity and TLS, the nature of the impurity enters into the final linewidth expression, *vide infra*.

The slow and fast modulation (diagonal) results above have also been obtained by a formal Green's function approach [5.68]. Since the equations of motion are many and complex, and given elsewhere [5.90], we do not present the result here. Rather, we consider an approach [5.68,90] which takes into account both the off-diagonal and diagonal modulations from TLS as defined by $\tilde{H}_{12}$ of (5.12). With this Hamiltonian and $\tilde{H}_{23}$, the matrix element of (5.29) was evaluated. Using the definitions of $\sin\beta$ and $\cos\beta$ which follow (5.17) and $E^2 = \Delta^2 + W^2$, the square-bracketed term of (5.29) reduces to

$$[\quad] = \frac{WD_{qs}}{\epsilon_\beta} + \frac{2D_{qs}w_\beta}{\epsilon_\beta}. \qquad (5.35)$$

The corresponding expression for $\rho = \alpha$ is obtained by replacing β by α. In the theories of *Lyo* [5.67,68] and *Huber* et al. [5.69], $w_\alpha, w_\beta = 0$, and for this case the matrix elements for the relaxation processes are independent of the impurity for $\epsilon_\beta \simeq E$. In the reduction to (5.35) a cancellation of terms containing V_β (or V_α) occurs. As discussed above, the major contribution to $\Delta\omega$ is considered to come from more distant weakly interacting TLS for which the neglect of $w_{\alpha,\beta}$ relative to W seems reasonable. In the same vein the dashed transitions of Fig.5.7 can be neglected. The two allowed transitions are broadened by T_1-type TLS relaxation in both the ground and excited electronic states. Determination of $\Delta\omega$ reduces to the evaluation of Γ^- and Γ^+ which has already been done, see (5.25). Thus,

$$\Delta\omega = \frac{2[\Gamma^- + \Gamma^+\exp(-E/kT)]}{\exp(-E/kT) + 1} , \qquad (5.36a)$$

where the occupation probabilities for the tunnel states $|\psi_1\rangle$ and $|\psi_2\rangle$ have been included. In writing (5.36a) it has been assumed that $\sin\alpha \simeq \sin\beta \simeq 1$ in (5.18a). Equation (5.36a) can be written more simply as

$$\Delta\omega = \frac{4\Gamma^-}{\exp(-E/kT) + 1} , \qquad (5.36b)$$

which with (5.26) can be shown to be identical to (5.34) derived by an alternative approach in the slow diagonal modulation limit. Eqs.(5.34,36b) apply to a single TLS and, thus, it is necessary to average over the TLS satisfying the weak coupling criterion and to average over TLS parameters. One need also take into account the *spatial* distribution of TLS. *Lyo* [5.68,90] made the following assumptions: $E \gg W$ with W and E correlated, i.e. $W/E = \kappa$; f and $E(\Delta)$ are uncorrelated; and the spatial distribution is uniform. With (5.34)

$$\langle\langle\Delta\omega\rangle\rangle = \langle\langle\tau^{-1} \operatorname{sech}^2(E/2kT)\rangle\rangle , \qquad (5.37)$$

where the double brackets indicate averaging over TLS parameters and integration over TLS. The spatial density of TLS is denoted by n. The cut-off radius r_c (and $V_c = 4\pi r_c^3/3$) is determined by

$$V' = \frac{V\Delta}{E} = \frac{b\Delta}{Er_c^s} = \frac{\hbar}{\tau} , \qquad (5.38)$$

with τ^{-1} given by (5.26,25). A multipolar impurity-TLS interaction $V = br^{-s}$ has been introduced (s = 3 corresponding to dipole-dipole, s = 4 dipole-quadrapole, etc.). From (5.25)

$$\frac{b\Delta}{Er_c^s} = \frac{\hbar\gamma}{2}\left[\frac{W^2f^2}{E^2}\right] E^3 \operatorname{ctnh}\left(\frac{E}{2kT}\right) \simeq \frac{b}{r_c^s} , \qquad (5.39)$$

from which V_c is determined. The term γ is a collection of constants, see

(5.25). Prior to integration over the TLS for a fixed E, the first term in parenthesis is replaced by an average value. This integration yields $V_c n \rho(E)$ where $\rho(E) = \rho_0 E^\mu$ is the TLS density of states. Only the averaging over E remains; it is straightforward yielding [5.68]

$$\langle\langle\Delta\omega\rangle\rangle = Ab^{3/s} n\rho_0 \langle\frac{W^2 f^2}{E^2}\rangle_{av}^{1-3/s} T^{4+\mu-9/s} \tag{5.40}$$

for $T \ll \theta_D$ (low T limit). The term A is a collection of constants for the system which carries a slight dependence on s. Thus for s = 3, $\langle\langle\Delta\omega\rangle\rangle \propto T^{1+\mu}$ and with $\mu = 0.3$ the theory is in agreement with data on a number of systems. For s = 4, the power law is $T^{1.75+\mu}$. Very recently, *Molenkamp* and *Wiersma* [5.82] have used Redfield's theory, the above assumptions and averaging procedure to rederive (5.40). They argued that the same power law survives for the very weakly interacting TLS which lie outside r_c.

It is interesting to note that if correlation between W and E is not assumed and the above procedure repeated the power law

$$\langle\langle\Delta\omega\rangle\rangle \propto T^{2+\mu-3/s} \tag{5.41}$$

obtains [5.92] when an average value for $W^2 f^2$ is employed. For s = 3 the power law is identical to that of (5.40).

This brings us to the power law

$$\langle\langle\Delta\omega\rangle\rangle \propto T^{(1+\mu)s/3} \tag{5.42}$$

of *Huber* et al. [5.69], which also provides $T^{1+\mu}$ for s = 3. The power laws for s > 3 are different from those in (5.40,41). These researchers relaxed the third assumption used to obtain (5.40) and included the effects of fluctuations from a uniform spatial distribution using an approach based on earlier work [5.93,94]. Very recently, it has been shown that their theory is valid in the long-time limit where $t > \langle\tau\rangle_{av}$ while that of *Lyo* [5.68,69] is valid in the short-time limit $t < \langle\tau\rangle_{av}$ [5.90,91]. Here $\langle\tau\rangle_{av}$ is an average TLS relaxation time. It is probably significant that for s = 3 the two different approaches [as well as (5.41)] yield the same power law.

This completes the presentation of our distillation of the essential features of the linewidth and optical dephasing theories for amorphous solids. We proceed now to further comparisons and discussion of the theories. In addition, the extent to which the theories agree with experiment is considered.

5.3.5 Comparison of Theories and Experimental Data

In Sect.5.3.3, it was seen that $\Delta\omega \propto T^n$ ($1 \leq n \leq 2$) behavior has been observed for a large number of impurity-amorphous host systems. Frequently, a particular power law persists over a substantial T range, several

K to 0.3 K. Moreover, one is struck by the frequent observation of a $T^{1.3}$ power law for organic systems and Nd^{3+} in pure SiO_2 glasses (where the measurements were performed between 0.1 and 1 K). It is appropriate to ask whether it is the off-diagonal (ODM) or the diagonal (DM) modulation theory(ies) which is (are) more in tune with experiment. They stem from (5.30) and (5.34), respectively, and differ considerably. For example, ODM predicts a linear power law in the *high T* limit (see below) which is *independent* of the nature of the impurity-TLS interaction. Diagonal modulation predicts the power laws of (5.40-42) in the *low T* limit. The latter depend on the nature of the interaction (for dipole-dipole coupling $\Delta\omega \, \alpha \, T^{1+\mu}$). For a better comparison it is useful to develop the ODM expression (5.30) further with the same TLS averaging procedure and W, E correlation (W/E = κ) which led to (5.40). Recall also that E $\approx$ Δ. Eq.(5.30) leads to

$$\langle \Delta\omega \rangle = \frac{3 \, \langle f^2 \rangle \rho_0}{2\pi\rho c^5 \hbar^5} \int_0^{E_{max}} \left[\kappa^2 \left(\frac{\langle V^2 \rangle}{E^2} \right) E^3 E^\mu + \langle w^2 \rangle E^{1+\mu} \right] \text{csch}(E/kT) dE \, , \quad (5.43)$$

where $\langle f^2 \rangle$, $\langle V^2 \rangle$ and $\langle w^2 \rangle$ are average values and E_{max} is on the order of $\hbar\omega_D$. The TLS density of states is again taken as $\rho(E) = \rho_0 E^\mu$. The first term is written as it is to underscore the fact that with W, E correlation, $\tau^{-1} \propto E^3$, see (5.26). Eqs.(5.26,34) show that the linewidth due to ODM is proportional to that from DM multiplied by $\langle V^2 \rangle E^{-2}$. The DM is due to a sea of weaky interacting TLS. Even though the $\langle V^2 \rangle$ from these TLS may be considerably smaller than that from a nearby TLS, *Lyo* [5.68,69] argued that the sheer number of more weakly interacting TLS provides a dominance of DM over ODM (at least for dipole-dipole interaction). Before considering this line of argument further, we note that in the high-T limit the first term of (5.43) predicts $\Delta\omega \propto T$. That is, the power law is independent of μ. It is important to note that *Reineker* et al. [5.71] have shown that the linear T regime can extend as low as T $\simeq$ 0.01θ_D (although they do not assume the above W, E correlation). Concerning the second term of (5.43), it also predicts a linear power law in the high-T limit. The low-T limit of (5.43) is $\Delta\omega \propto T^{3+\mu}$.

We may conclude that ODM by itself is not capable of predicting a power law of $T^{1.3}$ over an extended T range when crude averaging procedures are used. This is why *Jackson* and *Silbey* [5.73] invoked both ODM and dephasing due to low frequency impurity librational modes. For several organic systems and T >1 K, good agreement with experimental $T^{1.3}$ laws was obtained for librational frequencies of ~10 cm^{-1}. It is difficult, however, to understand how such librational frequencies can be important for T well below 1 K. Moreover, low frequency localized librations should only be invoked if they can be spectroscopically identified. They are generally conspicuously absent from the optical spectra of impurities in amorphous molecular solids (but see [5.85] for an exception). And, of

180

course, there is the observation of $T^{1.3}$ behavior for Nd^{3+} in SiO_2 for 0.1 $\leq T \leq 1$ K which leads one to suspect that this power law may be fundamentally significant and due to the glassy state itself.

The DM theories are appealing since their low T limits predict $\Delta\omega \propto T^{1.3}$ for $s = 3$ and $\mu = 0.3$. Moreover, they provide a thread which links organic and inorganic systems via the glassy state. However, in considering these theories certain questions arise. For example, the commonly used value for $n\rho_0$ in (5.40) has a value of $\simeq 10^{17}$ TLS per cm^{-1} per cm^3 [5.67,68]. This value is obtained from specific heat measurements which sample TLS relaxation times as long as a second or so. For optical dephasing, the TLS which are important are those whose relaxation times lie in a reasonably narrow range about the radiative lifetime of the impurity, see (5.34). The latter is ~1 ns for organic molecules. Thus, one should consider an effective density of states, $(n\rho_0)_{eff}$, and the fact that its value may be several orders of magnitude less than the aforementioned value. In doing so one immediately runs into a difficulty: the density of states may be too low for the DM theories to work. This question deserves future consideration. Another question which merits more careful exploration pertains to the nature of the dipole-dipole interaction which leads (with $\mu = 0.3$) to the $T^{1.3}$ power law. *Huber* et al. [5.69] have estimated that the elastic dipole interaction is much stronger than the electric dipole coupling for rare-earth ions in inorganic glasses. The former can be viewed as a difference in interaction between a dipole created in the immediate vicinity of the impurity (due to electronic excitation) and the dipole moments associated with the two wells of the TLS. Interestingly, the possibility of a charge (ion)-dipole TLS interaction where $s = 2$ has not been examined. That is, the polarizability of the inner sphere around the ion may produce an effective charge for the ion which is different in the excited state than in the ground state. For many of the organic systems studied, which yield $\Delta\omega \propto T^{1.3}$, the impurity molecule itself is centrosymmetric, albeit sitting at a noncentrosymmetric site. Perylene in polyvinylbutyral provides an example for which the Stark effect in conjunction with NPHB has been studied [5.88]. It is concluded that $S_1 \leftarrow S_0$ excitation produces an "effective" matrix-induced dipole moment change for perylene of $\Delta\mu = 6$ Debye. However, it should be mentioned that there is no precedent for such an enormous dipole moment change associated with centrosymmetric molecules in host crystals. It is possible that the Stark results of [5.95] can be understood in terms of an electrostrictive effect which produces spectral diffusion via TLS transitions. Thus, within the framework of DM and $s = 3$, it is possible that an elastic dipole interaction may be important for organic systems.

A few remarks on the condition of applicability of the slow-modulation DM theories are in order. They follow from the condition $\hbar\tau^{-1} < V' \simeq V < E$. At sufficiently low temperatures where $E \simeq kT < V'$, the two off-diagonal (dashed) transitions of Fig.5.7 can have significant transition dipole moments, see (5.18). The higher energy of the two would exhibit non-zero dephasing at 0 K. We note that persistent hole-burning studies on a number of organic systems which exhibit a $T^{1.3}$ power law down to 0.3 K yield a lifetime limited value for $\lim_{T \to 0}(\Delta\omega)$. However, $T^{1.3}$ extrapolation to 0 K may not be valid. Indeed, given the nature of the glassy state it is unquestionably the case that dephasing due to TLS relaxation must exist at 0 K. The phenomenon of NPHB itself proves this! The question is only whether this dephasing is faster than the excited state lifetime, i.e. observable.

For sufficiently high temperatures one must consider the $\hbar\tau^{-1} > V'$ limit. This corresponds to fast modulation and (5.33) predicts motional narrowing. Such narrowing has not been observed. However, as has already been pointed out, additional line-broadening mechanisms can become operative at higher temperatures and these could obscure the narrowing effect.

Very recently, analytic forms for the distribution functions for the two-level systems (TLS) of amorphous solids have been derived [5.96]. These distribution functions have been used to derive analytic expression for τ^{-1}, $\rho(E)$, and the dephasing frequencies due to ODM and DM [5.96-98] (Table 5.2). With the averaging procedure used in [5.96-98] to obtain the power law $\langle\langle\Delta\omega\rangle\rangle \propto T^{1+\mu}$, no arbitrary assumption about the power law for $\rho(E)$ need be made. With reasonable variations in the distribution-function parameters, it has been shown that the power law $\Delta\omega \propto T^{1+\mu}$ with $0.2 \leq \mu \leq 0.5$ holds for a dipole-dipole interaction (s = 3) [5.97]. However, off-diagonal modulation (ODM) provides the same range in μ (for different set of distribution parameters). This problem of distinguishing between dephasing mechanisms has been discussed in [5.97-99].

In summary, the DM theories with their greater flexibility (via s) would appear to afford better agreement with low-T optical linewidth and dephasing data than ODM theories (Sect.5.4 contains a discussion of TLS distribution functions which pertains to the $\mu = 0.3$ power law for the TLS density of states function $\rho(E)$). However, many questions remain which need to be carefully examined. A number have been raised but we mention also that a combination of DM and ODM (from say the second term of (5.43)) may be necessary for a complete understanding of the dephasing and spectral diffusion due to TLS. Utilization of more realistic density of states functions is highly desirable.

5.3.6 Hole Widths and TLS Relaxation Processes in Organic Systems

The hole-burning and other data of Sect.5.3.3 confirm the early conclusion [5.59-64] that the low-T optical linewidths and dephasing times of molecular impurities in amorphous solids are strikingly different from those in crystalline hosts. The important question of why the power law for optical linewidths and dephasing is nearly linear in T has been tackled by several theoretical approaches. We have seen, for example, that the $T^{1+\mu}$ power law with $\mu \sim 0.3$ can be explained by DM due to TLS relaxation. On the experimental side, it has been established that the laser burn intensities and hole depths above certain values ($\simeq 1$ μW/cm^2, few percent optical density change) lead to hole widths larger than the minimum attainable values [5.87]. Surprisingly, the fact that the power law is not significantly affected has not been commented on [5.87]. Adoption of a narrow viewpoint where this fact and others, *vide infra*, are ignored, amounts to overlooking interesting physics associated with TLS relaxation processes. An important point is that intensity-induced hole broadening in crystalline hosts is not observed at burn intensities 2-3 orders of magnitude higher than $\simeq 1$ μW/cm^2 [5.100]. This observation, and indeed calculations [5.101], prove that the hole broadening in amorphous solids is not due to power broadening or a triplet-state bottleneck associated with the impurity. Clearly, the broadening is due to TLS relaxation processes whose overall effect depends on the laser intensity. An additional clue is the observation that the hole width in amorphous solids appears to plateau at sufficiently high burn intensities [5.87].

An understanding of the intensity/hole-width relationship in amorphous solids is an important goal. It should be remarked that a primary objective of novel approaches to studying TLS relaxation is to go beyond the simple TLS model of *Anderson* et al. [5.2] and *Phillips* [5.3]. We mention a couple of approaches we are investigating which pertain to the above relationship: broadening is due to TLS which fit, with one exception, the mold of the TLS responsible for NPHB itself. It is that their relaxation times, τ, are on the order of the burn time (seconds) when the impurity is in its ground state. But still they are considerably shorter when the impurity is in its excited state. These TLS can provide a "population bottleneck" by themselves; and broadening is due to TLS-TLS connectivity or communication. That is, the more absorption events per unit time the greater the probability of spectral diffusion. In one sense, the very nature of the glassy state demands that the light field - solid interaction be viewed as strong. This approach is suggested by recent laser-induced hole filling experiments (Sect.5.5). This model appears to be consistent with a power law for $\Delta \omega$ which is largely independent of burn intensity [5.92].

5.4 Density of States Functions for TLS

It has been nearly two decades since *Zeller* and *Pohl* [5.1] and *Stephens* [5.102] published their poineering studies which stimulted the investigation of low-temperature properties of amorphous materials. In 1972, *Phillips* [5.3] and *Anderson* et al. [5.2] independently developed the tunneling model (TM) associated with TLS. The model has been remarkably successful in describing the experimental results. A key ingredient is the notion of "uniformity" for the TLS distribution function which leads to a density of states $\rho(E)$, which is approximately constant, i.e. $\rho(E) \simeq \rho_0$. In this section we wish to emphasize that there are a wide variety of distribution functions which have been developed [5.2,3,72,103-115] and not used for the optical linewidth and the dephasing problem. Because of space limitations we will not discuss all of them. Finally, we present a recent development [5.119] which we are currently in the process of applying to a number of low-T properties.

The assumption in the TM of a uniform distribution of the asymmetry Δ and tunneling parameter λ is commonly accepted. The tunnel splitting is given by $W = \hbar\omega_0\exp(-\lambda)$ with $\lambda = d(2mV)^{1/2}/\hbar$. The total energy splitting between the tunnel states E is given by (5.9).

From *Anderson* et al. [5.2]

$$P(\Delta,\lambda) = \begin{cases} \overline{P}(\text{constant}) & \text{for } \lambda_{\text{min}} \leq \lambda \leq \lambda_{\text{max}} \text{ and } \Delta_{\text{min}} \leq \Delta \leq \Delta_{\text{max}} \\ \\ 0 & \text{elsewhere} \end{cases} \tag{5.44}$$

where P is a normalized distribution function, and the λ and Δ parameters are equidistributed. Thus, $\overline{P} = (\Delta_{\text{max}}-\Delta_{\text{min}})^{-1}(\lambda_{\text{max}}-\lambda_{\text{min}})^{-1}$. The resulting density of states (DOS), $\rho(E)$, is approximately constant (equal to ρ_0) and quite good agreement with acoustic and dielectric measurements [5.2,3] is obtained.

Doussineau et al. [5.103] proposed a more complicated function

$$P(\Delta,W) = \begin{cases} A\dfrac{\Delta^{2\mu}E^{2\nu}}{W^{1+2\beta}} & \text{for } W_{\text{min}} \leq W \leq W_{\text{max}} \text{ and } 0 \leq \Delta \leq \Delta_{\text{max}} \\ \\ 0 & \text{elsewhere.} \end{cases} \tag{5.45}$$

They were guided by the fact that special values of the μ, ν and β parameters yield previous distribution functions [5.3,104]. However, (5.45) is simply a new approximation since it has no physical justification. With this equation, $P[E,(W/E)^2]$ can be obtained and was used to show that $\rho(E)$ is an increasing function of E.

Jäckle [5.105], and *Black* and *Halperin* [5.106] have proposed that $P(r) = (\overline{P}/2)r^{-1-q}(1-r)^{-1/2}$ with $r = W^2/E^2$. The parameter q allows the relative weights of symmetric and asymmetric TLS to be varied. It can be shown that when q = 0, λ and Δ are uniformly distributed as in [5.2,3]. The distribution function P(r) was applied by *Golding* et al. [5.104] but discrepancies with the experimental data were found, suggesting that the power law for P(r) is over-simplified.

Jäckle and *Jüngst* [5.107] have shown that

$$\rho(E) = \rho_0 \text{ arctanh} \sqrt{1 - r_0(E)}, \tag{5.46}$$

when a reasonable assumption about the distribution of tunneling frequencies is made. Here $1/r_0(E)$ measures the width of the distribution of relaxation times for a given value of E. For $r_0 \ll 1$, the energy dependence of ρ is logarithmic with E.

Lasjaunias et al. [5.108] invoked a cut-off in $P(\Delta,\lambda)$ to obtain

$$\rho(E) = \begin{cases} \rho_0 \ln\{W_{min}/[E-(E^2-W_{min}^2)^{1/2}]\} & \text{for } E \geq W_{min} \\ \\ 0 & \text{for } E < W_{min}. \end{cases} \quad (5.47)$$

This leads to an obvious gap in the DOS for $E < W_{min}$.

Also, *Geszti* [5.109] examined the DOS problem within the framework of a Frenkel-Kontorowa-type model. He found that $\rho(E) \propto E^{-1}$ which leads to a T-independent specific heat and a diverging DOS.

Very recently *Schilling* [5.110] showed that for chaotic configurations in one dimension, the existence of TLS can be proven theoretically and, moreover, that the TLS form a Cantor set of fractal dimension. Extension of this to three dimensions may show a connection between the $\rho(E) = \rho_0 E^\mu$ DOS and fractal behavior. We should note at this point that the dephasing theories discussed earlier employ a Debye density of states for the phonons. *Lyo* and *Orbach* [5.72] have considered the fractal model within the context of optical dephasing. The DOS for the fractal analogue of phonons, so-called fractons, has a $\omega^{1+\kappa}$ dependence with $\kappa < 1$. Utilization of this DOS in the dephasing theories will obviously affect the predicted temperature vs power laws. But the importance of fractons at very low temperatures seems doubtful because fractons are excitations of short wavelength whereas long-wavelength excitations are important at these temperatures.

Returning to the TLS DOS problem, *Anthony* and *Anderson* [5.111] have explained the T and frequency dependences of ultrasonic velocity and the dielectric constant by explicitly including an energy-dependent DOS and a distribution of relaxation times τ for each E. They postulated $\rho(E) = \rho_0(1+aE^2/k^2)$, where k is the Boltzmann constant. However, the quadratic dependence is only employed to emphasize that the DOS has to be an increasing function of E.

Frossati et al. [5.112], for a comparison of their dielectric absorption and capacitance data, have used a similar DOS function: $\rho(E) = E^q(\tau_{min}/\tau)^{q/2}$ with $q = 0.28$.

An interesting consequence of a distribution for τ is that the observed specific heat should depend on the time scale of the measurement. One expects [5.106]

$$C_V = (\pi^2 k^2 T/12) \, \overline{P} \, \ln(4t/\tau_{min}) \quad (5.48)$$

from the standard TLS tunneling model ($\rho(E) = \rho_0$), i.e. the linear term (in T) of C_V should depend logarithmically on the time t. The time dependence arises because the low-energy excitations have $\tau \geq \tau_{min}$. A few relevant short-time-scale experiments have been reported [5.116,117]. But they showed a logarithmic dependence only for 10 μs < t < 100 μs. For t

≥ 100 μs a much stronger t dependence (i.e., larger coupling of low-E excitations) was observed [5.116]. These results are in disagreement with the standard tunneling model. *Knaak* and *Meissner* [5.118] observed similar behavior for both types of vitreous silica. These facts and the non-linear T dependences of the specific heat and homogeneous optical linewidths (dephasing) suggest that the DOS function $\rho(E)$ may not be constant at temperatures below ~30 K. It may also indicate that the somewhat arbitrary cut-off procedures used for say λ and Δ in the development of DOS functions are not valid. These procedures mean that neither very fast or slow processes which contribute to relaxation occurring during the time scale of the experiment are taken into account.

However, it does appear possible to account for experimental discrepancy with the tunneling model by deriving a more realistic DOS function for TLS. Because of space limitations, only an outline of the derivation will be given, see [5.96] for details.

Recently, *Jankowiak* et al. [5.119] have shown that the kinetics of the spontaneous filling of non-photochemical holes can be described in terms of a dispersive first-order reaction of noninteracting reaction centers. It is assumed that site relaxation is a tunneling process, the tunnel parameter λ being subject to a Gaussian distribution function (GDF). In what follows we employ the idea that a Gaussian is the proper distribution function for any quantity that depends on a multitude of internal parameters in a random fashion. The standard deviation of λ is the convolution of the fluctuations of both tunneling distance d and barrier height V for a fixed tunneling mass.

In our model, the parameters Δ and W ($E^2 = \Delta^2 + W^2$) are taken as stochastically independent. This is reasonable since the strain field causes changes in the environment of TLS, i.e. a modulation $\delta\Delta$ of Δ, but has a negligible effect on W [5.3,103]. A random distribution for $\delta\Delta$ is assumed and the distribution function is taken as symmetric (since neither the right- or left-hand well in Fig.5.3 is to be preferred). In order to get dimensionless quantities $\tilde{\Delta}$, $\tilde{W}$, and $\tilde{E}$; Δ and W have been normalized, e.g. $\tilde{E} = E/\hbar\omega_0$. A GDF for the asymmetry parameter $\tilde{\Delta}$ is postulated, $g_{\tilde{\Delta}}(x)$ centered at $\tilde{\Delta} = \tilde{\Delta}_0$. Similarly, a GDF for the tunneling parameter λ is postulated, from which the distribution function $f_{\tilde{W}}(x)$ can be determined which converges for both very small and large W. With $g_{\tilde{W}}(x)$ and $f_{\tilde{W}}(x)$ it is possible to derive the distribution functions for $\tilde{\Delta}^2$, $\tilde{W}^2$ and $\tilde{E}^2$.

Since $\tilde{\Delta}$ and $\tilde{W}$ are stochastically independent so are $\tilde{\Delta}^2$ and $\tilde{W}^2$ and hence, the distribution function of $\tilde{E}^2$ is the convolution of $f_{\tilde{\Delta}^2}$ and $f_{\tilde{W}^2}$. From the relationship

$$f_{\tilde{E}}(x) = f_{\tilde{E}^2}(x^2)2x \qquad (5.49)$$

the DOS function $\rho(\tilde{E})$, can be exactly determined. We give here only the first approximation:

$$\rho(\tilde{E}) \simeq 2\alpha_1\alpha_2 \; \tilde{E}^{\mu_1} \tilde{E}^{-\mu_2 \ln \tilde{E}} \; B(1/2, \gamma(\tilde{E})) \;, \tag{5.50}$$

where α_1 and α_2 are the dimensionless constants depending on $(\tilde{\Delta}_0, \tilde{\sigma}_1)$ from $g_{\tilde{\Delta}}$ and $(\sigma_2, \tilde{W}_0)$ from $f_{\tilde{W}}$, respectively. The other terms are defined as follows: $\mu_1 = 2\theta$; $\theta = -\lambda_0/2\sigma_2^2$; $\mu_2 = (2\sigma_2^2)^{-1} \ll 1$; $B(1/2, \gamma(\tilde{E}))$ is the Beta function; and $\gamma(\tilde{E}) = \theta - \mu_2 \ln \tilde{E}$, with $\gamma(\tilde{E}) > 0$.

We note that $\rho(\tilde{E})$ given by (5.50) is only valid for $\gamma(\tilde{E}) > 0$. The analytical form of $f_{\tilde{E}^2}$ for $\gamma(\tilde{E}) < 0$ is extremely complicated and integral evaluation is most easily done numerically. The condition $\gamma(\tilde{E}) > 0$ is met for tunnel splittings $\leq 10^{-4}$ eV, comparable to phonon energies below about 4 K but much smaller than $\hbar\omega_0/2$ ($\simeq 10^{-2}$ eV for inorganic glasses [5.3]).

With appropriate choices for the distribution of $\tilde{\Delta}$ and $\tilde{W}$ (which have not been measured for any amorphous solid), (5.50) predicts that $\rho(\tilde{E})$ *is an increasing function of $\tilde{E}$ for $\tilde{E} \ll \hbar\omega_0$*, [5.96] which has been phenomenologically assumed to explain many properties of glasses at low T.

The equation can be written in the form

$$\rho(\tilde{E}) \simeq \rho_0(\tilde{E}) \; \tilde{E}^{\mu(\tilde{E})} \cdot B(1/2, \gamma(\tilde{E})) \tag{5.51}$$

where $\mu(\tilde{E}) = \mu_1 - \mu_2 \ln \tilde{E}$, $\tilde{E} = E/\hbar\omega_0$.

A few brief remarks concerning this DOS function are in order. Because of the $\ln \tilde{E} < -\lambda_0$ restriction, (5.51) can be applied only for some (λ_0, σ_2) values. Nevertheless, it does provide a reasonable approximation over the appropriate E range. For $\hbar\omega_0 = 10^{-2}$ eV, σ_2 values between 4 and 5, and for $\lambda_0 = 4$ (5.50) yields an effective $\tilde{E}$ exponent of $\tilde{\mu}_{eff} \simeq$ 0.3-0.4 over a broad energy range, $10^{-6} < E < 10^{-4}$ eV. However, the extension of the DOS curve to higher energies and to a larger range of (λ_0, σ_2) parameters, (i.e., the $\gamma(\tilde{E}) < 0$ region) can be accomplished by either numerical evaluation of the integral for $f_{\tilde{E}}(x)$ or by a Monte-Carlo-type procedure.

Very recently, *Jankowiak* et al. [5.120] used the TLS distribution functions with the Monte-Carlo simulation technique to determine $\rho(\tilde{E})$ over the entire range of tunnel state splittings. For the intrinsic TLS ($\lambda_0 \simeq 5$-7, $\sigma_2 \simeq 3$-4) it has been shown that $\rho(\tilde{E})$ is a slowly increasing function of the tunnel splitting E($\propto E^\mu$) with $0.2 \leq \mu \leq 0.5$ over a broad energy range, as predicted above and in [5.96]. Thus, the results of many low-temperature experiments (T < 1 K) can be explained: for example, those from specific heat measurements when $C_V \propto T^{1+\mu}$ [5.111,113] or from optical linewidth measurements when $\Delta\omega \propto T^{1+\mu}$ (Sect.5.3). The value of μ is determined principially by the values of λ_0 and σ_2, as predicted earlier [5.96]. The crossover to a negative slope $\rho(\tilde{E})$ at a higher $\tilde{E} = E/(\hbar\omega_0)$ value occurs at $\tilde{E} \simeq \tilde{\sigma}_1$ ($\tilde{\sigma}_1$ is the normalized width of the $\tilde{\Delta}$ distribution).

The distribution functions yield a gap in the DOS which has an onset at very low tunnel state splitting (E_{min}). The value of E_{min} depends on λ_0, σ_2, and σ_1, but the gap has its origin in the distribution function for the tunneling frequency W [5.120]. This is very important in terms of the question raised in Sect.5.3.5, if $T^{1+\mu}$ ($\mu \simeq 0.3$) extrapolation to 0 K, for the dephasing problems is valid.

This prediction that the DOS does *not* approach a constant value as $E \simeq kT \rightarrow 0$ is interesting and provides agreement with the experimental data of *Lasjaunias* et al. [5.113]. They studied the specific heat of two types of vitreous silica down to 25 mK and found that $C_V \propto T^{1+\mu}$ with $\mu = 0.22$ for Suprasil and $\mu = 0.3$ for Suprasil W. Taking into account the intrinsic and extrinsic TLS, this difference can be explained [5.121].

Finally, we remark that our DOS function stems from justifiable or reasonable assumptions. It is our hope that precise measurements between very low and high temperatures will soon be available for its stringent testing.

5.5 Laser-Induced Hole Filling

In addition to optical linewidth data, non-photochemical hole-burning (NPHB) provides other types of data which relate to TLS tunneling processes. They are thermal annealing, spontaneous filling (SPHF) and laser (or light)-induced filling (LIHF) of holes. The first phenomenon will not be discussed here since it has been considered in an earlier review article [5.122]. The SPHF (or spectral diffusion) phenomenon is discussed in Chap.3 and will only be referred to in the context of LIHF which is the primary subject of this section. By LIHF we mean the partial or complete filling or erasure of a hole burnt at ω_{B_1} which results from subsequent laser irradiation at frequencies removed from ω_{B_1}. Interestingly, LIHF is the least studied aspect of hole-burning in amorphous hosts even though it may be the most intriguing one. The first LIHF experiment was perfomed on tetracene in an ethanol/methanol glass in order to argue that the hole-burning mechanism is non-photochemical (photophysical) rather than photochemical in nature [5.7]. It is conceivable that LIHF can also occur for photchemical holes burned in glassy hosts. Undoubtedly, LIHF and hole erasure due to white light irradiation are related phenomena. *Gutierrez* et al. [5.123] have studied the latter for quinizarin in the ethanol/methanol glass. White light erasure has also been observed by us [5.124] for cresyl violet perchlorate (CV) in PVOH. It is important to note that we do not ascribe LIHF or white light erasure to a bulk heating effect. It is easy to identify and, consequently, prevent interference due to thermal erasure since it is generally accompanied by spectral diffusion and hysteresis whose T dependences can be characterized by separate thermal erasure (cycle) experiments [5.65,124].

Returning to the tetracene in ethanol/methanol glass system, it was reported that significant LIHF occurs only when the secondary irradiation frequency lies within $\simeq 2$ cm^{-1} of the primary hole at ω_{B_1} [5.7]. Although a mechanism for LIHF was not given, it might be inferred that the LIHF is due to reversion of originally burnt impurity-TLS sites back to their configurations prior to the burn (presumably due to excitation of anti-hole sites). LIHF due to reversion following excitation of anti-holes has been firmly established for pentacene in benzoic acid crystals [5.13,14].

Following the observation [5.40,74] that LIHF for cresyl violet perchlorate (CV) in PVOH (polyvinylalcohol) is facile for secondary irradiation frequencies (ω_{B_1}) far removed from ω_{B_1}($|\omega_{B_2}-\omega_{B_1}| \sim 100$ cm^{-1}), it was decided to study the phenomenon in far greater detail. Given the large inhomogeneous absorption linewidths of dyes in polymers ($\sim$500-1000 cm^{-1}) and apparent large widths of the anti-holes ($\sim$100 cm^{-1} [5.58]), the above observation for CV/PVOH suggested that reversion resulting from anti-hole excitation may not be the dominant mechanism for LIHF in dye/polymer systems. Other possibilities include a mechanism based on spectral diffusion which depends on connectivity between the different two-level systems (TLS) of the polymer which couple to the impurity. That is, secondary irradiation (or the resulting hole formation) may trigger configurational changes at impurity-TLS sites which are spatially removed from those involved in the primary burn or secondary light absorption process. Another possibility is that LIHF can result from intermolecular energy transfer. To begin to understand the LIHF process it was judged important to first determine relative LIHF efficiencies as a function of the sign and magnitude of $\omega_{B_2}-\omega_{B_1}$ while, at the same time, to ascertain whether LIHF is accompanied by broadening of the primary hole. Some of the results of such studies on R640, Nd^{3+} and Pr^{3+} in PVOH are presented here. Further details can be found in [5.127].

5.5.1 Rhodamine 640 in Poly(vinylalcohol)

Since SPHF occurs during the time scale of the LIHF experiments it was essential to subtract out the contribution of SPHF to the filling. The results of [5.58] allow this to be done in a reliable manner. Two examples of how this was accomplished are shown in Figs.5.8,9. Both figures are derived from a primary burn at ω_{B_1} = 582.5 nm with secondary irradiation frequencies at ω_{B_2} = 576.2 and 588.9 nm, respectively. Burn times and intensities are indicated in the figure captions. It should be noted that the vertical segment of the solid line in each figure indicates the time at which the ω_{B_2} irradiation terminates. Prior to and following the ω_{B_2} irradiation, the SPHF is monitored and the primary hole decay fit to [Ref.5.58, Eq.(1)]. Good agreement for the SPHF parameters with those reported [5.58] was found. Thus, the dashed line extensions, for example, are calculated. The difference between the upper and lower curves of Figs.5.8,9 yield the percent hole filling due only to LIHF. These percen-

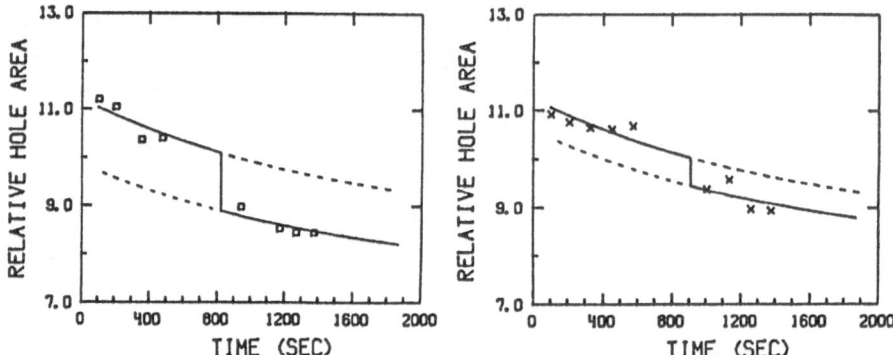

Fig.5.8. Laser-induced hole filling (LIHF) of R640 in PVOH in superfluid liquid helium (T = 1.7 K). Burns were performed with a ring dye laser with a flux of 1 mW/cm² for 75 s at ω_{B_1} = 582.5 nm (initial burn) and for 250 s at ω_{B_2} = 576.2 nm. The data points (...) show the measured holes in time, to which are fit the spontaneous hole filling (SPHF) curves as per [5.51]. The vertical line indicates the end of the fill burn where the percent laser-induced hole filling is 12%

Fig.5.9. LIHF of R640 in PVOH in superfluid helium (T = 1.7 K) with ω_{B_1} = 582.5 nm and ω_{B_1} = 588.9 nm. The data points (x) indicated the measured holes. All else is as stated in Fig.5.8. The percent laser-induced hole filling is 6%

tages are 12 and 6% for Figs.5.8,9, respectively. Thus, LIHF for R640/PVOH is significantly easier for $\omega_{B_2} > \omega_{B_1}$ than for $\omega_{B_2} < \omega_{B_1}$. Nonetheless the observation that LIHF does occur for the latter case is important.

Figure 5.10 summarizes the results of our 1.7 K LIHF studies on R640/PVOH. The arrow from ω_{B_1} locates the primary burn frequency (582.5 nm) and the smaller arrows locate the secondary irradiation frequencies, ω_{B_2}. For each and every ω_{B_2}, a fresh primary hole at ω_{B_1} was burned for 75 s with a burn flux of 1 mW/cm². A flux of 1 mW/cm² was also used for all ω_{B_2} irradiations. Focusing on the small arrow near 594.5 nm, the x, o and + percent hole depth change (filling) data points correspond to ω_{B_2} irradiation times of 75, 150 and 225 s, respectively. We note that for the burn flux and times utilized, the zero-phonon holes were not saturated (i.e., were not burned to maximum depth) and that phonon side band and vibronic satellite holes were not observed. Separate thermal annealing and cycling experiments confirmed that hole filling from bulk heating does not contaminate the data [5.128].

The absence of bulk heating is also consistent with one of the two most striking aspects of the data in Fig.5.10, the "discontinuity" or step in LIHF efficiency which occurs at ω_{B_1}. The second is the apparent insensitivity of the LIHF facility to ω_{B_2} on both the low and high energy sides of ω_{B_1}. Clearly, LIHF facility is not obviously related to optical density. A third important observation is that LIHF of the primary hole *occurs without*

190

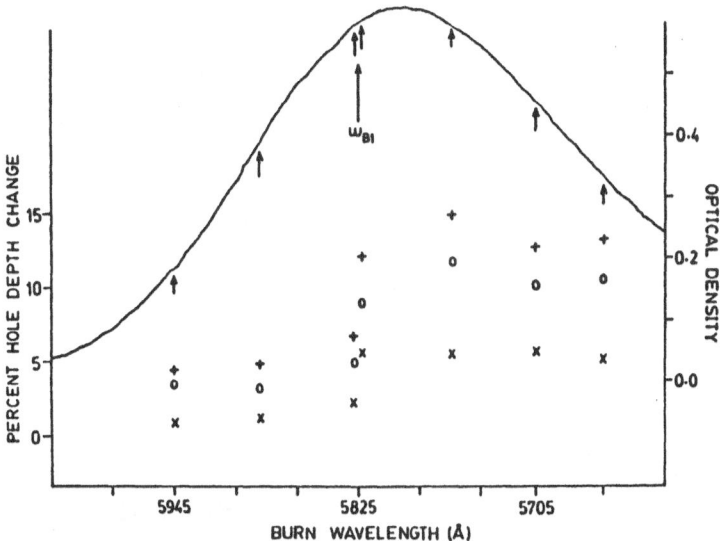

Fig.5.10. LIHF of R640/PVOH as a function of filling frequency. The initial burn at 582.5 nm (ω_{B_1}) was performed by a ring dye laser with 1 mW/cm² for 75 s. The observed percent hole depth change versus secondary burn frequency (marked with small arrows) is shown. The irradiation times used for filling were 75 s (x), 150 s (o) and 225 s (+) with the same flux. Each data set represents a different "identical" sample. T = 1.7 K

hole broadening. Our linewidth measurements are accurate to ~6% and the FWHM of the initial primary holes burned at 582.5 nm was constant at 0.45 cm⁻¹ for T = 1.7 K.

An obvious question is whether the LIHF of the primary hole is the result of the secondary hole formation at ω_{B_2} *itself* (rather than just absorption of ω_{B_2}). Data have been presented in [5.127] which indicate that the LIHF efficiency is not simply related to the depth of the secondary hole. Also, LIHF experiments on a binary dye mixture in PVOH argue against intermolecular energy transfer as the mechanism for LIHF when $\omega_{B_2} > \omega_{B_1}$ [5.127]. Energy transfer cannot be operative for $\omega_{B_2} < \omega_{B_1}$.

5.5.2 Nd³⁺ and Pr³⁺ in Poly(vinylalcohol)

The optical transitions of Nd³⁺ and Pr³⁺ studied were $^2G_{7/2}$, $^4G_{5/2} \leftarrow {}^4I_{9/2}$ and $^1D_2 \leftarrow {}^3H_4$, the same as in [5.58] where their SPHF properties have been discussed. Hole-burning was performed on the lowest energy J-component of each transition. In fact, attempts to burn holes in the higher energy components were not successful, presumably due to their rapid radiationless decay [5.129].

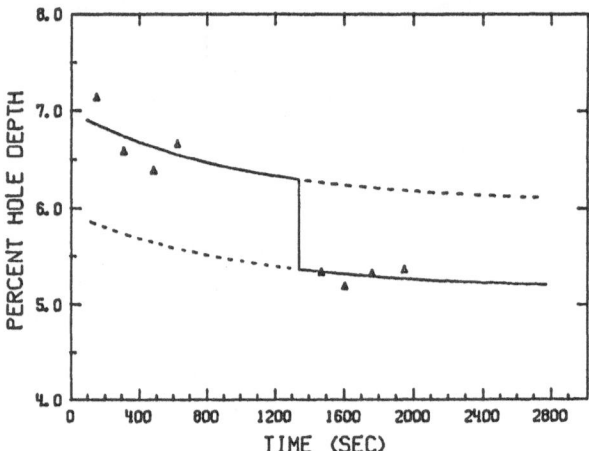

Fig.5.11. Laser-induced hole filling of Nd^{3+} inPVOH with initial burn (ω_{B_1}) = 579.0 nm and the fill burn (ω_{B_1}) = 578.75 nm, where both burns were performed with 100 mW/cm^2 for 10 minutes. The data points (Δ) indicate the measured holes after each burn. The calculated curves are fits for spontaneous hole filling (see text). The vertical line represents the end of the fill burn where percent hole filling was measured to be 15%

The experimental procedure used for LIHF in the rare-earth/PVOH materials was essentially identical to that used for R640/PVOH. Figure 5.11 shows the results of one LIHF run for Nd^{3+} which are of the type shown in Figs.5.8,9 for R640. We note that high RE^{3+} concentrations corresponding to a number density of $\simeq 10^{21}$ cm^{-3} or to an average ionic separation of $\simeq 1$ nm were employed. The LIHF results for Pr^{3+} and Nd^{3+} are shown in Figs.5.12,13. The indicated LIHF (filling) percentages are corrected for SPHF. Numbers in square brackets correspond to the percent optical density change due to hole-burning at the indicated burn frequencies. Numbers without brackets correspond to the percent filling of a hole due to subsequent irradiation at a displaced frequency. With reference to trace b) of Fig.5.12, irradiation at ω_{B_2} produces a 17% hole at ω_{B_2} while filling the ω_{B_1} hole by 4%. Similarly, trace c) shows that irradiation at ω_{B_3} fills the hole at ω_{B_2} by 15%. Trace d) was obtained with secondary irradiation at 594.0 nm into the second J-component of the $^1D_2 \leftarrow {}^3H_4$ transition.

From Fig.5.12 we conclude that for Pr^{3+}/PVOH, LIHF is significantly more efficient when the secondary irradiation frequency is displaced to higher energy of the hole being filled, as is the case for R640/PVOH. Another similarity is that the LIHF does not broaden the hole being filled.

The absence of broadening accompanying LIHF is also observed for Nd^{3+}/PVOH. However, this system behaves in a qualitatively different

192

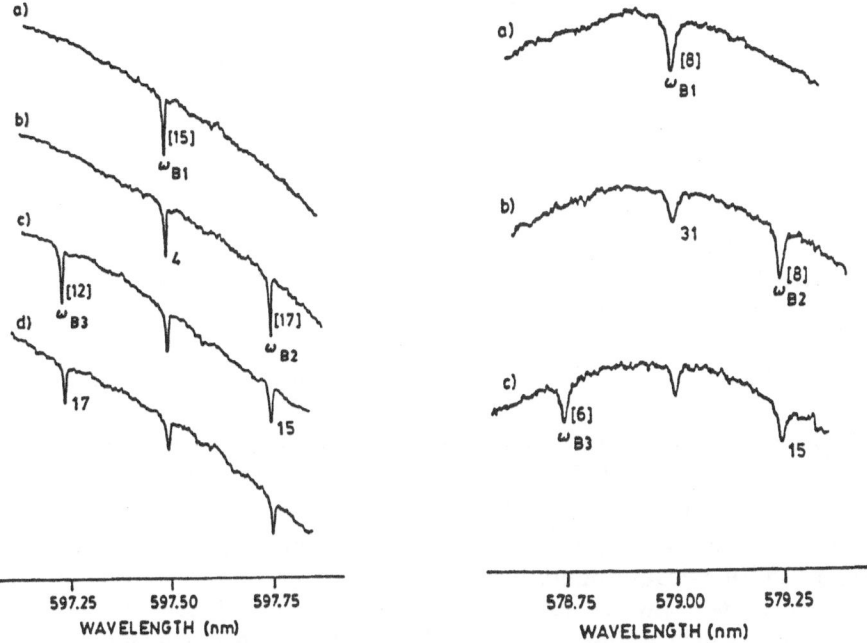

Fig.5.12. Hole burning and laser-induced hole filling data for Pr^{3+} in PVOH. All burns were performed with a ring dye laser with ~100 mW/cm^2 for 10 minutes. For spectra $(a-c)$, all burns were into the lowest energy component of the $^2G_{7/2}, {}^4G_{5/2} \leftarrow {}^4I_{9/2}$ crystal field split transitions as indicated, i.e. $\omega_{B_1} = 597.5$ nm, $\omega_{B_2} = 597.75$ nm and $\omega_{B_3} = 597.25$ nm. For spectrum (d), the burn was into the next higher energy component at 597.0 nm. The numbers in brackets refer to the initial percent hole depths while the others indicate the percent hole filling due to a subsequent burn. All samples were immersed in superfluid helium at 1.7 K

Fig.5.13. Hole-burning and LIHF of Nd^{3+} in PVOH. Bruns were into the low energy component of the crystal-field-split transition of $^1D_2 \leftarrow {}^3H_4$. The sequence of burns is as indicated with the bracketed numbers referring to initial hole depths while the others refer to hole filling. Burns were performed with $\simeq 100$ mW/cm^2 for 10 minutes at 1.7 K

manner than Pr^{3+} or R640 in that the LIHF is more efficient for a secondary irradiation lying to the *red* of the hole being filled. We return to this point in the following section.

5.5.3 A Tentative Model for LIHF

We begin by summarizing the key experimental results:

1) LIHF of a primary hole at ω_{B_1} occurs for both ω_{B_2} (secondary irradiation) $> \omega_{B_1}$ and $< \omega_{B_1}$.

2) Except for Nd^{3+}, the LIHF efficiency for $\omega_{B_2} > \omega_{B_1}$ is significantly higher than for $\omega_{B_2} < \omega_{B_1}$.

193

3) For R640 the LIHF efficiency exhibits a step behavior at ω_{B_1}. For $\omega_{B_2} > \omega_{B_1}$ and $< \omega_{B_1}$, the dependence of efficiency on $|\omega_{B_2} - \omega_{B_1}|$ is weak.

4) LIHF does not produce any perceptible broadening of the primary hole (for hole-width measurements accurate to $\simeq 6\%$ and hole-widths of $\simeq 0.3$ cm^{-1}).

In reference to Item 2, the behavior of R640 and Pr^{3+} has also been observed for chlorophyll a [5.82], chlorophyll b [5.130] and the self-aggregated dimer of chlorophyll a [5.131] in polystyrene films. With regard to Item 1, the observation that LIHF is operative for $\omega_{B_2} < \omega_{B_1}$ establishes that a filling mechanism other than intermolecular energy transfer (ET) exists.

Taken as a whole the above results present an intriguing puzzle. A number of possible LIHF mechanism can be considered not to be significant [5.127]. Included are: direct antihole site excitations (by ω_{B_2}) followed by reversion to original (preburn) configurations; local heating and trivial emission-reabsorption.

We consider now a recently proposed model for LIHF [5.127]. The glassy state which we associate with glasses and polymers depends on how the state was prepared. Prior to a burn there are points in configuration space which are thermally accessible to the initially prepared state. The simple TLS model was designed to model this state of affairs [5.7]. With this in mind NPHB leading to stable holes can be viewed as a process that produces a glassy state which is *thermally inaccessible* to the pre-burn state. In our view, LIHF or thermal erasure of a hole in excitation frequency space should not be considered as a return of the system to its orginal pre-burn state. Rather, both are likely irreversible in nature, given the essential continuum of glassy configurations which exist.

An obvious difficulty with the TLS model as it is applied to most systems is that something as fundamental as the spatial extent of the TLS is not known. Without this knowledge for the host TLS (intrinsic) and those which may be introduced by the impurity (extrinsic), the concept of TLS connectivity is impossible to quantify. Nevertheless, we proceed to explore a model for LIHF which is based on the premise that filling involves communication between impurity sites excited by secondary irradiation (ω_{B_2}) and those which are not and are, therefore, spatially removed from sites in the ω_{B_2} isochromat. In doing so we pursue the idea that we can divide the TLS distribution into two sets, one whose members are weakly coupled to the impurity and the other whose members are strongly coupled [5.132]. For convenience we refer to the former and latter as intrinsic and extrinsic. The extrinsic TLS would include those which, for example, lead to hole formation by phonon-assisted tunneling. Their spatial extent may be considered as localized relative to that of the intrinsic TLS. For polymers, in particular, a large spatial extent for intrinsic TLS is not

194

unreasonable since chain "snaking" along the channel or "tube" formed by its neighbors can presumably occur to some degree even at helium temperatures. *De Gennes* [5.133,134] has referred to such motion as reptation. Now for each and every extrinsic TLS we associate a potential energy curve $V^j_{ex}(q,\varsigma)$ where q is the usual intermolcular double-well coordinate associated with the extrinsic (ex) TLS [5.64]. The superscript j labels the ground or excited electronic state of the impurity. The ς variable denotes the coordinates of the atoms associated with the intrinsic TLS. Within the Born-Oppenheimer approximation one can view V^j_{ex} as depending parametrically on the ς coordinates. Thus, $\langle V^j_{ex}(q,\varsigma)\rangle_{gls}$ represents the extrinsic TLS potential for a particular glassy state (gls). The gls defines the value of ς and is governed by the intrinsic TLS(s) which surround the extrinsic TLS.

LIHF is then viewed as resulting from irreversible gls → gls' glassy state transitions triggered by secondary light absorption by extrinsic impurity-TLS different from those involved in the primary burn. The initial excitation (via secondary irradiation) and coupling between the extrinsic excited impurity-TLS and intrinsic TLS leads to the irreversible evolution of gls'. The time scale may be long with much of the configurational change associated with gls → gls' occurring with the impurity in its ground electronic state. The energies V^j_{ex} respond adiabatically to the glassy state transition (we have used glassy state interchangeably with the state of the intrinsic TLS which surround and influence the extrinsic TLS). Thus, provided the spatial extent of the intrinsic TLS is large, the excitation frequencies of extrinsic impurity-TLS spatially removed from those excited can be altered. The picture which emerges is that excitation of a spectrally narrow isochromat causes excitation frequency diffusion over a broad segment of excitation frequency space.

We proceed now to discuss the results in Items 1-4 (*vide supra*) in terms of this model (referred to hereafter as A). Consider first the first part of result 3 and Fig.5.10. Our initial response to the step behavior at ω_{B_1} was that there are two LIHF mechanisms, only one of which is operative for $\omega_{B_2} < \omega_{B_1}$. At the present time this possibility has not been excluded, although we are not able to come up with a plausible model (B). Thus, we consider here only model A. We suggest that the behavior of Fig.5.10, the results of Figs.5.12,13 and other LIHF results for polystyrene [5.82,127] are explicable by one mechanism, i.e. model A. An important clue is provided by Fig.5.13 which shows that Nd^{3+} is intriguingly different in the sense that LIHF is more efficient for $\omega_{B_2} < \omega_{B_1}$.

This leads us to consider correlation effects between impurity site excitation energies and glassy state (intrinsic TLS) absolute energies. By positive (negative) correlation we mean that increasing excitation energy is associated with increasing (decreasing) *absolute energy* of the intrinsic TLS. Although, the zero-point splittings, barrier heights, etc. of extrinsic impurity-TLS may be dominated by interatomic interactions from *within* their domains, the intrinsic TLS which envelop them may be a major contribution to the determination of excitation frequencies within the inhomogenously broadened absorption profile. It is reasonable to suggest that the absolute energies of intrinsic TLS are quite insensitive to the state of the impurity due to their assumed large spatial extent. In addition to correlation, we introduce the notion that an arbitrary perturbation (like secondary irradiation of LIHF) is, on average, far more disposed to promote gls $\rightarrow$ gls' configurational transformations in which E(gls') < E(gls). This is not unreasonable for transformations at helium temperatures. Thus, for positive correlation and within the framework of model A, secondary irradiation would cause red shifts of a wide distribution of site excitation energies while for negative correlation, blue shifts would occur. Given the gradient in excitation frequency space due to the primary hole, these shifts lead to hole filling.

With this in mind one can qualitatively understand the step behavior for R640 (Fig.5.10) and the results for Pr^{3+} (Fig.5.12). For both there is a significant amount of positive correlation, albeit not perfect since some filling is observed for $\omega_{B_2} < \omega_{B_1}$. On the other hand for Nd^{3+} (Fig.5.13) a significant amount of negative correlation is indicated.

Consider next the result 4 which is that the hole broadening accompanying LIHF is imperceptible. The analogous spectral diffusion problem pertaining to thermally induced spectral diffusion (hole broadening) and hysteresis has been treated theoretically by *Friedrich* et al. [5.135] (Chap.3). The associated problem for spontaneous hole filling (SPHF) is considered in [5.58]. The theory of *Friedrich* et al. [5.135] is, with slight modification, a reasonable starting point for LIHF and model A. If the probability function for an excitation frequency shift $\omega' \rightarrow \omega'+\Delta\omega$ has an effective width σ and is independent of ω', the theory shows that hole broadening would be difficult to observe (at our 6% accuracy and percent fills) for $\sigma \geq 10\gamma$, where γ is the width of hole prior to filling. Thus, failure to observe hole broadening is not necessarily in contradiction with model A.

Thus far, this model has qualitatively accounted for all the results listed at the beginning of this section except the second part of Item 3 (Fig.5.10). Although, one can understand why LIHF can occur for large $|\omega_{B_2}-\omega_{B_1}|$ (with ω_{B_2} whithin the absorption profile), it is difficult to explain why for $\omega_{B_2} > \omega_{B_1}$ or $< \omega_{B_1}$, the percent LIHF is essentially constant. For the same ω_{B_2} flux and irradiation time one would reasonably assume that the percent filling would mimic the variation in optical density across the absorption profile (Fig.5.10). Further experiments are required to understand the above behavior including studies with substantially reduced (<< 1 mW/cm^2) ω_{B_2} fluxes. Given the saturation (bottleneck) behavior for NPHB discussed in [5.58], it is possible that LIHF itself is being affected by a saturation effect.

5.6 Recent Developments

Since this chapter was written, a number of new developments have occurred in the area of NPHB and dephasing in amorphous solids. On the theoretical side, the derivation of the density of states for TLS partially described in Sect.5.4 has appeared [5.96]. Applications of the nonphenomenological distribution function to diagonal [5.97] and off-diagonal [5.98] modulation mechanisms, for dephasing have also been published. A thorough review of dispersive kinetic processes, optical linewidths, and dephasing that extends the concept of intrinsic and extrinsic TLS has also appeared [5.136], along with a new review of experimental dephasing data [5.137].

Regarding experimental advances relating to dephasing, the *Fayer* group has provided a new, careful comparison of optical linewidths as measured by NPHB versus photon-echo techniques [5.138-140]. For the case of resorufin in ethanol glass [5.138,139], they find temperature-dependent NPHB hole widths following a $T^{1.3}$ power law in agreement with [5.78b]. However, two-pulse photon-echo measurements yield homogeneous linewidths that are four times narrower than that that would be deduced from the NPHB hole widths. For resorufin in glycerol glass [5.140], the low-temperature linewidths obtained from NPHB are also substantially wider than the linewidths $(1/\pi T_2)$ obtained from photon-echo measurements. They model the dephasing using TLS as well as librational contributions. These intriguing data provide clear evidence that NPHB experiments can measure more than just the homogeneous linewidth; hole-burning can also probe other processes such as spectral diffusion by virtue of its long time scale.

5.7 Concluding Remarks

Given the limited space available it has not been possible to discuss several important aspects of NPHB or, more generally, persistent hole-burning. These include thermal annealing and hysteresis in the thermal cycling of holes, spontaneous hole filling and the theoretical aspects of power and population bottleneck broadening effects. For this we apologize. The topics considered in this chapter were carefully chosen. First, we felt that the fact that NPHB has been shown to be a quite general phenomenon since 1982 [5.122] is very important. Noteworthy is the class of laser dyes in polymers, acenes in amorphous aromatic hosts and rare earth ions in inorganic glasses and polymers. Second, we felt it timely to give a rather detailed discussion and comparison of the various theories which have been proposed for the optical linewidth and dephasing problem. Such a comparison was not available. It was during the writing of this section that we became intensely interested in the TLS distribution function problem and developed a non-phenomenological density of states function, $\rho(E)$. Thus we are indebted to the Editor, W.E. Moerner, for inviting us to contribute to this book. Finally, we decided to discuss laser-induced hole filling because it is a new phenomenon and one which bears (we think) on the important problem of TLS-TLS connectivity.

Although considerable progress has been made in our understanding of NPHB in amorphous solids, TLS relaxation processes and impurity optical dephasing and linewidths due to the impurity-TLS coupling, we are still at a rather primitive stage. There is a real need for a better understanding of the microscopic structures of bistable configurations or TLS in glasses and one can hope that molecular dynamics simulations will soon provide us with such for simple atomic systems. Such structures would provide a test for the approximations which have led to TLS distribution functions. With regard to the "non-phenomenological" TLS density of states function we have developed, it will be interesting to examine to what extent it can account for the T-dependent (power law) data which now exist. It should not, for example, be difficult to utilize it with existing optical dephasing and spectral diffusion theories. The testing of this density of states function by impurity optical dephasing and linewidth measurements will require measurements over a fairly extended temperature range on well chosen systems with lower uncertainties than have been achieved thus far. Of course, a proven density of states function of the type we propose will not take us beyond the simple TLS model of *Anderson* [5.2] et al. and *Phillips* [5.3]. By simple, we mean the absence of TLS-TLS connectivity or communication. It seems unlikely that optical "pure" dephasing measurements will shed light on this important question. In this regard, experiments of the LIHF type (Sect.5.5) are much more promising.

Acknowledgements

This work was supported by the Division of Materials Researach of the National Science Foundation, Grant No. DMR-8400.

The authors wish to thank D. Huber, K. Lyo, R. Reineker and F. Stillinger for providing us with preprints of their work. We are also grateful to D. Huber, K. Lyo and H. Morawitz for very useful discussions of dephasing theories.

References

5.1 R.C. Zeller, R.O. Pohl: Phys. Rev. B 4, 2029 (1971)

5.2 P.W. Anderson, B.I. Halperin, C.M. Varma: Phil. Mag. 25, 1 (1972)

5.3 W.A. Phillips: J. Low Temp. Phys. 7, 351 (1972)

5.4 F.H. Stillinger: Science 225, 983 (1984)

5.5 T.A. Weber, F.H. Stillinger: Phys. Rev. B32, 5402 (1985)

5.6 F.H. Stillinger, T.A. Weber: Phys. Rev. A28, 2408 (1983)

5.7 J.M. Hayes, G.J. Small: Chem. Phys. 27, 151 (1978)

5.8 G.A. Sim, J.M. Robertson, T.H. Goodwin: Acta Cryst. 8, 157 (1955)

5.9 S. Hayashi, J. Umemura, R. Nakamura: J. Mol. Struct. 69, 123 (1980)

5.10 S. Nagaoka, T. Terao, F. Imashiro, A. Saika, N. Hirota, S. Hayashi: Chem. Phys. Lett. 80, 560 (1981)

5.11 S. Nagaoka, N. Hirota, T. Matushita, K. Nishimoto: Chem. Phys. Lett. 92, 498 (1982)

5.12   F.G. Patterson, H.W.H. Lee, R.W. Olson, M.D. Fayer: Chem. Phys. Lett. 84, 59 (1981)

5.13   R.W. Olson, H.W.H. Lee, F.G. Patterson, M.D. Fayer, R.M. Shelby, D.P. Burum, R.M. Macfarlane: J. Chem. Phys. 77, 2283 (1982)

5.14   C.A. Walsh, M.D. Fayer: J. Lumin. 34, 37 (1985)

5.15   R. Casalegno, H.P. Trommsdorff: *Photochemistry and Photobiology*, Proc. Int'l. Conf., Univ. of Alexandria, Egypt (1983)

5.16   J.M. Clemens, R.M. Hochstrasser, H.P. Trommsdorff: J. Chem. Phys. 80, 1744 (1984)

5.17   H.B. Levinsky, D.A. Wiersma: J. Chem. Phys. 79, 2677 (1983)

5.18   R. Eiermann, G.M. Parkinson, H. Bässler, J.M. Thomas: J. Phys. Chem. 87, 544 (1983)

5.19   R. Jankowiak, K.-D. Rockwitz, H. Bässler: J. Phys. Chem. 87, 552 (1983)

5.20   R. Jankowiak, H. Bässler: Chem. Phys. Lett. 95, 124 (1983)

5.21   R. Jankowiak, H. Bässler: Chem. Phys. Lett. 95, 310 (1983)

5.22   R. Jankowiak, H. Bässler: Chem. Phys. Lett. 101, 274 (1983)

5.23   R. Jankowiak, H. Bässler: J. Mol. Electron. 1 73 (1985)

5.24   J.D. Winefordner, P.A.St. John: Anal. Chem. 35, 2211 (1963)

5.25   A.V. Lesikar: J. Chem. Phys. 66, 4263 (1977)

5.26   C.A. Angell, J.M. Sare, E.J. Sare: J. Phys. Chem. 82, 2622 (1978)

5.27   D. Turnbull: Contemp. Phys. 10, 473 (1969)

5.28   J. Wong, C.A. Angell: *Glass Structure by Spectroscopy* (Marcel Dekker, New York 1976) Chap.1

5.29   A. Szabo: Phys. Rev. Lett. 25, 924 (1970)

5.30   R.I. Personov, E.I. Al'shitz, L.A. Bykovskaya: JETP Lett. 15, 609 (1972); Opt. Commun. 6,169 (1972)

5.31   J.H. Eberly, W.C. McColgin, K. Kawaoka, A.P. Marchetti: Nature 251, 215 (1974)

5.32   B.M. Kharlamov, R.I. Personov, L.A. Bykovskaya: Opt. Commun. 12, 191 (1974)

5.33   B.M. Kharlamov, R.I. Personov, L.A. Bykovskaya: Opt. Spectrosk. 39, 240 (1975)

5.34   B.M. Kharlamov, L.A. Bykovskaya, R.I. Personov: Chem. Phys. Lett. 50, 407 (1977)

5.35   J.M. Hayes, G.J. Small: Chem. Phys. Lett. 54, 435 (1978)

5.36   R.L.Fearey, R.P. Stout, J.M. Hayes, G.J. Small: J. Chem. Phys. 78, 7013 (1983)

5.37   R.P. Stout: *Nonphotochemical Hole Burning and the Nature of Amorphous Solids*, Ph.D. Thesis, Iowa State Univ. (1981)

5.38   J. Friedrich, D. Haarer: J. Chem. Phys. 76, 61 (1982)

5.39   J. Friedrich, H. Wolfrum, D. Haarer: J. Chem. Phys. 77, 2309 (1982)

5.40   B.L. Fearey, T.P. Carter, G.J. Small: J. Phys. Chem. 87, 3590 (1983)

5.41   T.P. Carter, G.J. Small: Chem. Phys. Lett. 120, 178 (1985)

5.42   E. Cuellar, G. Castro: Chem. Phys. 54, 217 (1981)

5.43   A.F. Childs, A.M. Francis: J. Phys. Chem. 89, 466 (1985)

5.44   L.W. Molenkamp, D.A. Wiersma: J. Chem. Phys. 83, 1 (1985)

5.45   U. Bogner, K. Beck, M. Maier: Appl. Phys. Lett. 46, 534 (1985)

5.46   V.G. Maslov, A.S. Chunaev, V.V. Tugarinov: Mol. Biol. (USSR) 15,788 (1981)

5.47   J.K. Gillie, B.L. Fearey, J.M. Hayes, G.J. Small, J.H. Golbeck: Chem. Phys. Lett. 134,316 (1987)

5.48   J.M. Hayes, J.K. Gillie, D. Tang, G.J. Small: Biochem. Biophys. Acta, in press

5.49   S.G. Boxer, D.J. Lockhart, T.R. Middendorf: Chem. Phys. Lett. 123, 476 (1986)

5.50   S.R. Meech, A.J. Hoff, A.A. Wiersma: Chem. Phys. Lett.121, 287 (1985)

5.51   S.G. Boxer, T.R. Middendorf, D.J. Lockhart: FEBS Lett. 200, 237 (1986)

5.52   S.R. Meech, A.J. Hoff, D.A. Wiersma: Proc. Natl. Acad. Sci. USA 83, 9464 (1986)

5.53   U. Bogner: Phys. Rev. Lett. **37**, 909 (1976)
5.54   U. Bogner, P. Schatz, M. Maier: Chem. Phys. Lett. **119**, 335 (1985)
5.55   R.M. Macfarlane, R.M. Shelby: Opt. Commun. **45**, 46 (1983)
5.56   R.M. Macfarlane, R.M. Shelby: *Proc. of the NATO Workshop on Coherence and Energy Transfer in Glasses* (Plenum, New York 1984) p.189
5.57   B.L. Fearey, T.P. Carter, G.J. Small: J. Lumin. **31&32**, 792 (1984)
5.58   B.L. Fearey, G.J. Small: Chem. Phys. **101**, 269 (1986)
5.59   P.M. Selzer, D.L. Huber, D.S. Hamilton, W.M. Yen, M.J. Weber: Phys. Rev. Lett. **36**, 813 (1976)
5.60   P. Avouris, A. Campion, M.A. El-Sayed: J. Chem. Phys. **67**, 3397 (1977)
5.61   J.R. Morgan, M.A. El-Sayed: Chem. Phys. Lett. **84**, 215 (1981)
5.62   J. Hegarty, W.M. Yen: Phys. Rev. Lett. **43**, 1126 (1974)
5.63   J.M. Hayes, R.P. Stout, G.J. Small: J. Chem. Phys. **73**, 4129 (1980)
5.64   J.M. Hayes, R.P. Stout, G.J. Small: J. Chem. Phys. **74**, 4266 (1981)
5.65   T.L. Reinecke: Solid State Commun. **32**, 1103 (1979)
5.66   S.K. Lyo, R. Orbach: Phys. Rev. **B22**, 4223 (1980)
5.67   S.K. Lyo: Phys. Rev. Lett. **48**, 688 (1982)
5.68   S.K. Lyo: In *Electronic Excitations and Interaction Processes in Organic Molecular Aggregates*, ed. by P. Reineker, H. Haken, H.C. Wolf, Springer S. Solid-State Sci., Vol.49 (Springer, Berlin, Heidelberg 1983)
5.69   D.L. Huber, M.M. Broer, B. Golding: Phys. Rev. Lett. **52**, 2281 (1984)
5.70   P. Reineker, H. Morawitz: Chem. Phys. Lett. **86**, 359 (1982)
5.71   P. Reineker, H. Morawitz, K. Kassner: Phys. Rev. **29**, 4546 (1984)
5.72   S.K. Lyo, R. Orbach: Phys. Rev. **B29**, 2300 (1984)
5.73   B. Jackson, R. Silbey: Chem. Phys. Lett. **99**, 331 (1983)
5.74   K.E. Jones, A.H. Zewail: In *Advances in Laser Chemistry*, ed. by A.H. Zewail, Springer Ser. Chem. Phys., Vol.3 (Springer Berlin, Heidelberg 1978)
5.75   D.E. McCumber, M.D. Sturge: J. Appl. Phys. **34**, 1682 (1963)
5.76   H.P.H. Thijssen, A.I.M. Dicker, S. Völker: Chem. Phys. Lett. **92**, 7 (1982)
5.77   H.P.H. Thijssen, S. Völker, M. Schmidt, H. Port: Chem. Phys. Lett. **94**, 537 (1983)
5.78   H.P.H. Thijssen, R. van den Berg, S. Völker: Chem. Phys. Lett. **97**, 295 (1983); Chem. Phys. Lett. **120**, 503 (1985)
5.79   F.A. Burkhalter, G.W. Suter, U.P. Wild, V.D. Samoilenko, N.V. Rasumova, R.I. Personov: Chem. Phys. Lett. **94**, 483 (1983)
5.80   H.W.H. Lee, A.L. Huston, M. Gehrtz, W.E. Moerner: Chem. Phys. Lett. **114**, 491 (1985)
5.81   T.P. Carter, B.L. Fearey, J.M. Hayes, G.J. Small: Chem. Phys. Lett. **102**, 272 (1983)
5.82   T.P. Carter, G.J. Small: Chem. Phys. Lett. **120**, 178 (1985)
5.83   L.W. Molenkamp, D.A. Wiersma: J. Chem. Phys. **83**, 1 (1985)
5.84   R. Jankowiak, H. Bässler: Chem. Phys. Lett. **95**, 124 (1983)
5.85   R. Jankowiak, H. Bässler, R. Silbey: Chem. Phys. Lett. **125**, 139 (1986)
5.86   J. Fünfschilling, I. Zschokke-Gränacher: Chem. Phys. Lett. **110**, 315 (1984)
5.87   H.P.H. Thijssen, S. Völker: Chem. Phys. Lett: **120**, 496 (1985)
5.88   M.M. Broer, B. Golding, W.H. Haemmerle, J.R. Simpson, D.L. Huber: Phys. Rev. B **33**, 4160 (1986)
5.89   J.M. Pellegrino, W.M. Yen, M.J. Weber: J. Appl. Phys. **51**, 6332 (1981)
5.90   S.K. Lyo: In *Optical Spectroscopy of Glasses*, ed. by I. Zschokke-Gränacher (Reidel, Dordrecht 1986)
5.91   D.L. Huber, M.M. Broer, B. Golding: submitted to Phys. Rev. B
5.92   G.J. Small: unpublished results
5.93   J.R. Klauder, P.W. Anderson: Phys. Rev. **B16**, 2879 (1977)
5.94   W.B. Mims: Phys. Rev. **B16**, 2879 (1977)
5.95   U. Bogner, P. Schaetz, R. Seel, M. Maier: Chem. Phys. Lett. **102**, 272 (1983)

5.96   R. Jankowiak, G.J. Small, K.B. Athreya: J. Phys. Chem. **90**, 3896 (1986)
5.97   R. Jankowiak, G.J. Small: J. Phys. Chem. **90**, 5612 (1986)
5.98   R. Jankowiak, G.J. Small: Chem. Phys. Lett. **128**, 377 (1986)
5.99   R. Jankowiak, G.J. Small: Science **237**, 618 (1987)
5.100 S. Völker, R.M. Macfarlane, A.Z. Genack, H.P. Trommsdorff, J.H. van der Waals: J. Chem. Phys. **67**, 1759 (1977)
5.101 H. DeVries, D.A. Wiersma: J. Chem. Phys. **72**, 1851 (1980)
5.102 R.B. Stephens: Phys. Rev. B**8**, 2896 (1973)
5.103 P. Doussineau, C. Frénois, R.G. Leisure, A. Levelut, J.-Y. Prieur: J. Phys. (Paris) **41**, 1193 (1980)
5.104 B. Golding, J.E. Graebner, A.B. Kane, J.L. Black: Phys. Rev. Lett. **41**, 1487 (1978)
5.105 J. Jäckle: Z. Phys. **257**, 212 (1972)
5.106 J.L. Black, B.I. Halperin: Phys. Rev. B**16**, 2879 (1977)
5.107 J. Jäckle, K.-L. Jüngst: Z. Phys. B**30**, 243 (1978)
5.108 J.C. Lasjaunias, R. Maynard, M. Vandorpe: J. Phys. Paris Colloq. C 6, 973 (1978)
5.109 T. Geszti: Phys. Rev. B**30**, 1181 (1984)
5.110 R. Schilling: Phys. Rev. Lett. **53**, 2258 (1984)
5.111 P.J. Anthony, A.C. Anderson: Phys. Rev. B**20**, 763 (1979)
5.112 G. Frossati, J. le G. Gilchrist, J.C. Lasjaunias, W. Meyer: J. Phys. C 10, L515 (1977)
5.113 J.C. Lasjaunias, A. Ravex, M. Vandorpe, S. Hunklinger: Solid State Commun. **17**, 1045 (1975)
5.114 M.W. Klein, B. Fischer, A.C. Anderson, P.J. Anthony: Phys. Rev. B**18**, 5887 (1978)
5.115 R. Rammal, R. Maynard: J. Phys. (Paris) Colloq. C6, 970 (1978)
5.116 M. Loponen, R. Dynes, V. Narayanamutri, J. Garno: Phys. Rev. B**25**, 1161 (1982)
5.117 M. Meissner, K. Spitzmann: Phys. Rev. Lett. **46**, 265 (1981)
5.118 W. Knaak, M. Meissner: In *Photon Scattering in Condensed Matter*, ed. by W. Eisenmenger, K. Lassmann, S. Döttinger, Springer Ser. Solid-State Sci., Vol.51 (Springer, Berlin, Heidelberg 1984) p.416
5.119 R. Jankowiak, R. Richert, H. Bässler: J. Phys. Chem. **89**, 4569 (1985)
5.120 R. Jankowiak, G.J. Small, B. Ries: Chem. Phys., in press
5.121 R. Jankowiak, J.M. Hayes, G.J. Small: unpublished results
5.122 G.J. Small: In *Molecular Spectroscopy*, ed. by V.M. Agranovich, R.M. Hochstrasser (North Holland, Amsterdam 1983)
5.123 A.R. Gutiérrez, J. Friedrich, D. Haarer, H. Wolfrum: IBM J. Res. Dev. **26**, 198 (1983)
5.124 T.P. Carter, B.L. Fearey, G.J. Small: unpublished results
5.125 J.M. Hayes, G.J. Small: J. Lumin. 18+19, 219 (1979)
5.126 H.W.H. Lee, C.A. Walsh, M.D. Fayer: J. Chem. Phys. **82**, 3948 (1985)
5.127 B.L. Fearey, T.P. Carter, G.J. Small: Chem. Phys. **101**, 279 (1986)
5.128 Specifically, when a sample which had been burned at 1.7 K, was raised rapidly to some higher temperature (T_h) and then promptly recooled to 1.7 K, significant broadening was observed when hole filling was comparable to LIHF. For example, for filling of $\simeq10\%$ ($T_h \simeq 10$ K), a broadening of $\simeq20\%$ occurred. This is in marked contrast to the results for LIHF.
5.129 R.M. Macfarlane: R.M. Shelby: In *NATO Workshop on Coherence and Energy Transfer in Glasses*, ed. by P.A. Fleury, B. Golding (Plenum, New York 1982)
5.130 J.M. Hayes, B.L. Fearey, T.P. Carter, G.J. Small: Int. Rev. Phys. Chem. **5**, 175 (1986)
5.131 T.P. Carter, G.J. Small: J. Phys. Chem. **90**, 1997 (1986)
5.132 B. Golding, M.V. Schickfus, S. Hunklinger, K. Dransfeld: Phys. Rev. Lett. **43**, 1817 (1979)

5.133 L. Léger, P.-G. de Gennes: Ann. Rev. Chem. 33, 49 (1982)
5.134 P.-G. de Gennes: *Entangled polymers*, Phys. Today 36, 33 (June 1983)
5.135 J. Friedrich, D. Haarer, R. Silbey: Chem. Phys. Lett. 95, 119 (1983)
5.136 R. Jankowiak, L. Shu, M.J. Kenney, G.J. Small: J. Lumin. 36, 293 (1987)
5.137 R.M. Facfarlane, R.M: Shelby: J. Lumin. 36, 179 (1987)
5.138 C.A. Walsh, M. Berg, L.R. Narasimhan, M.D. Fayer: Chem. Phys. Lett. 130, 6 (1986)
5.139 C.A. Walsh, M. Berg, L.R. Narasimhan, M.D. Fayer: J. Chem. Phys. 86, 77 (1987)
5.140 C.A. Walsh, M. Berg, L.R. Narasimhan, K.A. Littau, M.D. Fayer: Chem. Phys. Lett. 139, 66 (1987)

6. Persistent Infrared Spectral Hole-Burning for Impurity Vibrational Modes in Solids

A. J. Sievers and W. E. Moerner

With 27 Figures

One unusual group of excitations in which persistent spectral hole-burning has been observed consists of infrared vibrational transitions of impurity molecules in solids. Examples include 1,2-difluoroethane in rare gas matrices, perrhenate ions in alkali halide crystals, and most recently, cyanide and nitrite ions in KBr. The hole formation mechanisms involve molecular reorientation or conformational changes at low temperatures induced by excitation of an internal vibrational mode of the impurity molecule.

6.1 Introduction

6.1.1 Matrix-Isolated Molecules in Van der Waals and Ionic Solids

As was described in the last chapter, the formation of persistent spectral holes by nonphotochemical or photophysical mechanisms has been reported for several inhomogeneously broadened electronic transitions of molecular and ionic impurities in solids in the visible region [6.1 – 3]. The phrase "persistent infrared photophysical hole burning" completely specifies the theme of this chapter. By itself "photophysical" is too general to identify the area of interest here because this phrase is generally used to cover all electronic energy transfer processes in which a chemical bond is not created or destroyed [6.4] whereas, the topic to be considered in this chapter pertains to the small subset of persistent photophysical systems in which no electronic excitation is involved and only vibrational degrees of freedom are photoexcited with infrared radiation. In particular, cases where low temperature infrared (IR) photochemistry (bond-breaking) occurs and entire inhomogeneous lines disappear, are not reviewed here [6.5].

What are the most general requirements for the formation of persistent IR spectral holes (PIRSHs) that last much longer than excited vibrational state lifetimes at low temperatures? In a fundamental sense, there are three basic requirements for PIRSH formation. First, there must be several ground state configurations of the total system, and the IR absorption energies from these ground states must differ by more than the laser linewidth. Second, there must exist an IR pumping pathway that connects these ground state configurations. Finally, the relaxation among the ground states must

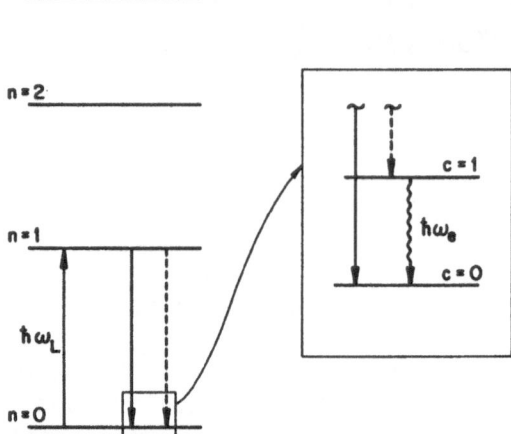

Fig. 6.1. Schematic diagram showing the generalized mechanism producing persistent spectral hole-burning in a vibrational mode

be slower than the excited state decay rate. If all of these conditions are met, persistent spectral holes may result. A schematic picture of such an IR-induced vibrational process for a molecule at low temperatures is shown in Fig. 6.1. The monochromatic laser source of frequency ω_L induces transitions between the $n=0$ and $n=1$ vibrational states as represented by the solid vertical lines. In addition to this allowed process there also exists a weakly allowed decay channel (the dotted line) to a different configuration of the molecule-lattice system which is nearly degenerate in energy with the configuration which is being pumped. These two ground states are labeled in the expanded view of the ground state region by the quantum numbers $c=0$ and $c=1$. After many laser-induced cycles some molecules end up in the metastable $c=1$ configuration, out of equilibrium with the lattice at low temperatures. The large energy barrier separating the two elastic configurations prevents the molecule-lattice system from simply emitting the elastic energy $\hbar\omega_e$ and returning to the ground state.

To date, only a few such persistent IR spectral holes have been reported for low temperature matrix-isolated molecules, but they have been found in solids with two different kinds of bonding: van der Waals [6.6 – 8] and ionic [6.9 – 13]. These new high resolution hole-burning measurements are attracting widespread interest not only because they complement earlier results obtained with other spectroscopic techniques on the vibrational spectra of trapped molecular species, but also because related persistent effects may occur in the vibrational spectra of solids with other types of bonding.

Spectroscopic research on molecules isolated in the two classes of hosts has been driven forward by different potential applications. Typically, van der Waals matrices of rare-gas atoms have been utilized as hosts when the interest is centered mainly on the properties of the dopant molecule itself [6.14,15], such as in the spectroscopic determination of the structure of free radicals [6.16]. This sample preparation technique, which is designed to make the internal vibrational modes of unstable molecules stand out, sometimes obscures the details of the molecule-lattice interaction. The material to be examined is formed by condensing the molecular and matrix gases on a cryogenically cooled substrate so that the host is necessarily disordered and some ambiguity in the possible arrangements of the nearest neighbors of the dopant molecule results. Although initial measurements did focus on the molecular vibrational modes, it was soon appreciated that a good understanding of the local environment and molecule-lattice interaction would be essential before the static and dynamic properties of the vibrating molecule could be characterized properly [6.14]. Each new advance in spectroscopic technique has continued to resolve more fine structure in the vibrational stretch transitions demonstrating that IR can probe precisely the normal modes of the coupled system [6.17].

The discovery [6.18] of vibrational fluorescence from matrix-isolated CO marked the beginning of a new era in which IR laser probes played a central role in the determination of vibrational energy lifetime and transfer processes for both single and multicomponent molecular species at low temperatures [6.19−27]. An important result of this work was that molecular rotation plays a key role in determining the excited state vibrational lifetime of diatomic molecules with small moments of inertia [6.19]. Although fluorescence can be measured for most simple molecules, it is much weaker or nonexistent for complex ones, no longer because of rotation but instead due to energy exchange between the anharmonic normal modes. Because complex molecules always contain low frequency modes, nonradiative processes from these low frequency internal modes to external lattice modes of similar frequencies dominate the relaxation, and fluorescence methods cannot readily be used to determine the vibrational lifetime; instead, the more demanding techniques of incoherent saturation and infrared hole-burning must be employed [6.28,29].

Alkali halide host crystals doped with molecules are prepared by diffusion near the crystal melting temperature, by pulling a crystal from the doped molten salt (Czochralski-growth) or by dropping the encapsulated doped melt through a steep temperature gradient (Bridgman-growth). All of these sample preparation techniques limit the possibility of isolating large complex molecules in these single crystal hosts; however, simple molecules can be readily embedded in nine simply related fcc lattices and three bcc ones. Because the electrostatic potential produced at the impurity site by an array of point charges does not contribute to the vibrational frequency of the molecule, one can compare the spectroscopic results obtained for molecules embedded in these matrices with those found for van der Waals solids.

The first infrared measurements on simple molecules in alkali halides were made in 1928 [6.30], but the field did not begin to develop until the late 1950's [6.31 – 33]. By the early 1970's the complex vibrational spectra obtained for these simple systems had been interpreted in terms of rotational, librational, and tunneling motion of the vibrating molecule as well as local, gap, resonant, and tunneling mode motion of the impurity center of mass in conjunction with the nearby lattice [6.34 – 37]. A comparison of the spectroscopic results for diatomic molecules embedded both in ionic and van der Waals hosts has demonstrated that the vibrational dynamics are similar even though the crystal bonding strength is very different [6.37].

The first attempt to determine the energy transfer properties of molecules in alkali halides developed from the desire to make a room temperature saturable absorber for a specific high power CO_2 laser system using perrhenate (ReO_4^-) ions in alkali halides [6.38]. Low temperature studies on the ReO_4^- molecule were initiated when it became evident that long-lived vibrational states in matrix-isolated molecules could occur in the CO_2 laser frequency region. *Gethins* [6.39] predicted that the local vibrational mode for the H^- ion in KBr would have an intrinsic linewidth of 3 MHz due to three- and four-phonon decay processes. At about the same time *Moos* [6.40] showed that the multiphonon decay rate between electronic states in rare earth ions was exponential in the energy gap to the next lowest level. By analogy with these optical results, the molecular multiphonon relaxation frequency should decrease exponentially with increasing ratio of vibrational frequency to maximum phonon frequency while the radiative relaxation rate should increase as the cube of the vibrational frequency. Because of this, a long relaxation time window could occur in the 10 μm wavelength region [6.41].

High-resolution diode-laser spectroscopy and the nonequilibrium techniques of laser saturation and transient hole-burning spectroscopy over a wide temperature range can be used to identify the microscopic mechanisms responsible for T_1 (the energy relaxation time) and T_2 (the dephasing time) for the ReO_4^- ion in various alkali halides [6.42 – 46]. The intrinsic symmetry and simplicity of this hole-burning system have allowed many features of the anharmonic decay process to be identified: the decay channel consists of multistep emission of lower-energy internal modes, localized modes, and band phonon modes [6.46]. It was in the course of these transient hole-burning measurements that the production of persistent spectral holes at extremely low laser powers was first observed in a vibrational mode in a crystal lattice [6.9].

6.1.2 Persistent IR Hole-Burning in Vibrational Modes

The three major sections in this chapter describe in turn three examples of persistent infrared hole-burning in vibrational modes. This division is natural because the spectra for the three molecule-host systems are strikingly different although the underlying process appears to involve molecule reorientation in each case. The description of the salient features of these systems should provide a framework for the interpretation of future studies of vibrational modes in solids showing PIRSH formation.

In the next section the persistent spectral properties of the molecule 1,2-difluorethane (CH_2F-CH_2F, or DFE) in van der Waals matrices are described. At low burning powers such as those available from Pb-salt semiconductor diode lasers, the v_{17} mode of the *trans* conformation of DFE matrix-isolated in Ar, Kr or N_2 shows PIRSHs due to molecular reorientation in the host. At the higher laser powers available from CO_2 lasers, however, a different hole-burning mechanism becomes observable: conformer interconversion.

Section 6.3 is devoted to a discussion of the PIRSH dynamics of the v_3 vibrational mode of tetrahedral ReO_4^- molecules substitutionally doped into alkali halide single crystals. The holes and antiholes are produced with $\simeq 10.8$ μm radiation from CO_2 and semiconductor diode lasers also at intensities far below saturation intensity levels. The PIRSH mechanism is photon-induced 90° reorientation of the molecule in the crystal cage. This defect-lattice system also has the distinction of proving that PIRSHs can occur in infrared-active impurity modes in *crystalline* hosts.

For the CN^- molecule in alkali halides, a molecular reorientation of 180° occurs during PIRSH formation but the low symmetry of the molecule and the proximity of a second dopant yield hole-burning properties quite different from the ReO_4^- – alkali halide system. These results are described in Sect. 6.4. Current research is uncovering even more vibrational mode-lattice systems that show PIRSH production without requiring transitions among TLSs or other degrees of freedom associated with the host lattice.

6.2 Molecules in Van der Waals Matrices

6.2.1 1,2-Difluorethane (DFE)

a) Diode Laser Measurements

IR hole-burning with a low intensity (10 mW/cm^2) tunable diode laser has been demonstrated for one of the matrix-split components of the v_{17} mode of the *trans* conformation of DFE matrix-isolated in Ar [6.6–8] or Kr [6.7,8] and for both of the components in N_2 [6.8]. These investigations have attracted widespread interest because this ultrahigh resolution probe, if applicable to matrix-isolated molecules in general, would provide an analytical tool of considerable power. The central issue which was addressed with these first sets of measurements is the identification of the mechanism behind the persistent spectral effects. The experimental results which are reviewed here indicate that molecular reorientation in the matrix cage during vibrational de-excitation is the most likely cause of the low intensity hole-burning effect.

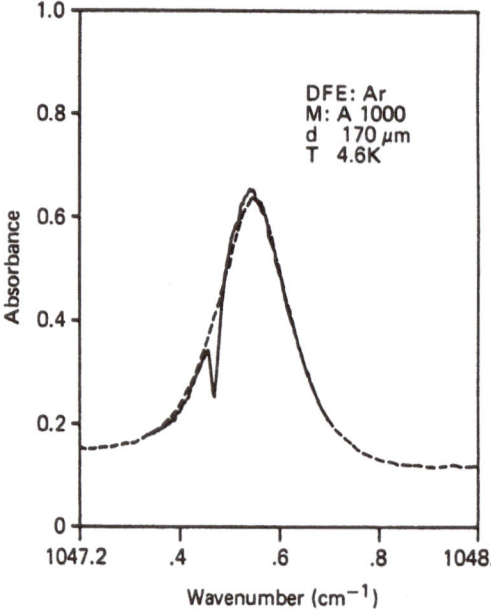

Fig. 6.2. Diode laser spectrum of the matrix-split v_{17} mode of *trans* DFE isolated in Ar. Dashed line: before irradiation. Solid line: after 60 sec of irradiation at 1047.46 cm^{-1} with an intensity of 10 mW/cm^2. A persistent spectral hole is produced by the laser light. (After [6.8])

i) DFE : Ar. Upon deposition of 1,2-difluorethane (DFE) in Ar (M/A = 1000) from room temperature onto a 17 K cold plate, approximately 20% of the molecules remain trapped in the less stable *trans* form [6.47]. The dashed trace in Fig. 6.2 shows one of the matrix-split components of the inhomogeneously broadened v_{17} absorption band of the *trans* conformation in Ar at 4.6 K which is centered at 1047.54 cm^{-1} with a linewidth of 0.17 cm^{-1}. (The other two components are centered at 1049.5 cm^{-1} and 1045.6 cm^{-1}, respectively.) The solid trace shows the same feature with a spectral hole burned at 1047.46 cm^{-1} [6.8] . By shifting the diode frequency a hole can be produced in a few minutes at any position in this band. The hallmark of this process is that for fixed laser frequency the transmittance of the sample is observed to increase as a function of time, rapidly at first and then finally approaching a limiting value at longer times. After the burn is completed, the laser intensity is attenuated and the frequency swept continuously to produce a spectrum such as that represented by the solid trace in Fig. 6.2. The difference between the two traces identifies both the hole and anti-hole (region of *increased* absorption) positions. Note that the anti-hole is broader than the hole and its frequency position is between that of the hole and band center. This result is obtained independent of where the hole is placed. An experimental reason accounts for the difference in area between the hole and antihole shown in Fig. 6.2: even with the laser attenuated some additional burning occurs during the frequency sweep. Thus the key experimental finding which will help in the identification of the underlying mechanism is that the integrated area of this band does not change after hole-burning.

An experimental problem has been that the hole strength at low temperature decays with time with a decay constant which depends on the radiation shielding in the cryostat. When the optical access ports are blocked by the 50 K shield which surrounds most of the sample, the decay constant decreases by an order of magnitude [6.8]. With increasing sample temperature above 10 K the decay constant increases rapidly due to a thermal deactivation process so that by 17 K the decay is so fast that hole-burning is no longer detectable. Keeping this difficulty in mind, an estimate of the low-temperature quantum efficiency of the hole-burning process, η, (defined as the number of molecules burned divided by the number of absorbed laser photons) can be obtained. For the Ar host the measured value is quite large: $\eta = 0.03$.

From the width of the low-temperature hole an estimate can be obtained for the relaxation time of the excited vibrational state [6.8]. For short irradiation times a Lorentzian line shape is observed. In this limit the measured width of the hole is twice the homogeneous linewidth. By extrapolation the measured width at 0 K is obtained and identified with the vibrational relaxation time, T_1, under the assumption that the dephasing is dominated by excited state decay. For the Ar matrix the estimate for $T_1 = 3.5$ ns.

Hole-burning has been attempted on the other strong matrix-split component (with no effect [6.48]) and also on several bands of the *gauche* and *trans* conformation which were brought into coincidence with the laser modes by using fully deuterated DFE. None of the four deuterated bands investigated (*trans* (980 cm^{-1}) nor *gauche* (974 cm^{-1}, 989 cm^{-1}, and 1239 cm^{-1})) showed any signs of hole-burning.

ii) DFE : Kr. The hole-burning results on one of the matrix-split v_{17} mode absorption lines (1050.4 cm^{-1}) for the Kr host are similar to those obtained with Ar except that the holes are more stable. By shielding the sample completely from room temperature background radiation the lifetime of the holes at liquid helium temperatures is measured to be about 24 hours [6.7]. With the optical

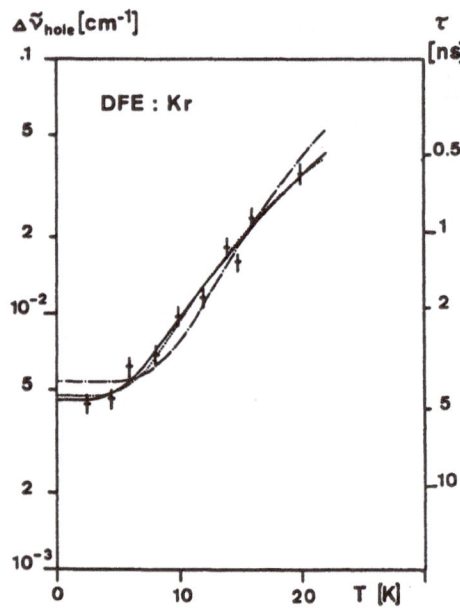

$\Delta \tilde{\nu}_{hole}$ [cm⁻¹]

DFE : Kr

τ [ns]

Fig. 6.3. Temperature dependence of the hole linewidth of the matrix split ν_{17} mode in *trans* DFE:Kr. The crosses identify the experimental points which have been taken at short irradiation time and have been corrected for the laser linewidth. The lines identify phonon-induced vibrational dephasing as given by the Raman mechanism: Solid curve, Debye model with $\theta_D = 38$ K; Dot-dash, Debye model with θ_D(Kr) = 72 K; Dotted, Einstein model with $\theta_E = 35$ K. (After [6.8])

access ports open the lifetime is reduced by an order of magnitude. For this system $\eta = 0.01$.

The thermal deactivation process becomes important at temperatures greater than 15 K and the holes are erased rapidly at 28 K. Nevertheless, the hole profile can be measured up to 22 K. The temperature dependence of the homogeneous linewidth for short irradiation times is shown in Fig. 6.3. The observed width is independent of the laser frequency in the band. Note that the laser linewidth is about an order of magnitude smaller than the smallest width shown in Fig. 6.3. To interpret these results the authors [6.8] assume that the temperature dependence of the linewidth comes from dephasing (T_2) processes which vanish as T→0. The three curves superimposed on top of the data in Fig. 6.3 are appropriate for the Raman dephasing process associated with phonon scattering. For the Debye model, the solid curve is for a best fit with $\theta_D = 38$ K and the dashed curve for θ_D (Kr) = 72 K. The dotted curve describes a similar process but for a local mode (Einstein oscillator) centered at $\theta_E = 38$ K. The homogeneous width as T → 0 gives $T_1 = 4.3$ ns.

Burning and analyzing the hole strength with polarized radiation demonstrates that the molecules are not freely rotating at low temperature. The experimental data supporting this conclusion are shown in Fig. 6.4 with the spectrum of the virgin sample displayed at the top. The notations V and H denote vertical and horizontal polarization directions for either the burn or probe beams. The hole intensity ratio for perpendicular and parallel polarization directions is found to be 1/3. If the molecules are rotating with a period small

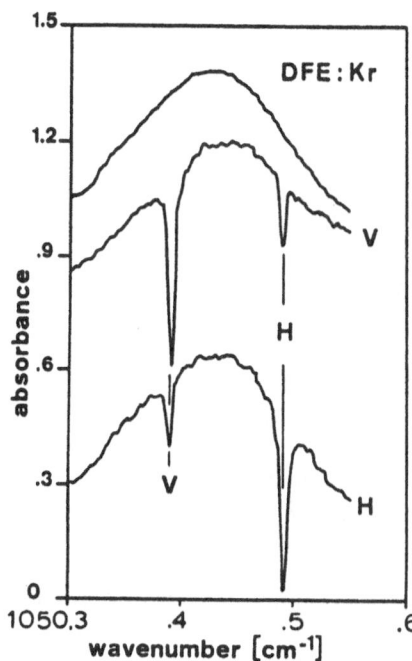

absorbance

DFE : Kr

V

H

V

H

1050.3   .4   .5   .6

wavenumber [cm⁻¹]

Fig. 6.4. Polarization properties of holes in the matrix-split v_{17} mode in *trans* DFE:Kr. The top trace shows the spectrum before irradiation. A hole is burned at one frequency with the polarization vertical (V) and at a second with the polarization horizontal (H). The spectra are then recorded with the two polarizations (V) and (H) as shown. (After [6.8])

compared to the measuring time (a few minutes) then no difference in hole strength between the two polarizations in a given trace should occur, in disagreement with the experiment.

iii) DFE : N₂. In the diatomic matrix N_2, the v_{17} mode of DFE is split by the matrix into two components, a strong band labeled A at 1045.39 cm⁻¹ and a weaker band, B, at 1046.17 cm⁻¹. Fortunately, both bands are covered by two adjacent diode laser modes [6.8]. Hole-burning can be seen in either of these bands and, in addition, the antihole or product state from each appears spread over a new inhomogeneously broadened band, C, centered at 1046.21 cm⁻¹ which is resolved from band B. This redistribution of oscillator strength into a frequency region separated from the original inhomogeneously broadened band is very different from that found for the other two matrices; furthermore, once band C is populated, then sharp holes can be burned in it as well. The T→0 vibrational relaxation time is comparable to that observed for the other two hosts, namely, $T_1 = 2.2$ nsec. The temperature dependence of the erasing of the holes is similar to that found in the Ar host.

b) CO₂ Laser Measurements

i) Molecular Monomers in Various Hosts. Site conversion for the *trans* v_{17} doublet for DFE : N₂ was first observed by *Frei* et al. [6.49] with a moderate intensity cw CO₂ laser. Because of the

211

displacement of band C from B or A, the pump could be a fixed frequency laser and the probe beam, a high-resolution spectrometer. This experimental technique with cw CO_2 laser intensities 10^4 times larger than that for the earlier diode laser studies was used to probe DFE isolated in Ar, Kr and Xe in the high-intensity regime [6.50].

Due to the accidental coincidence of the CO_2 laser lines 9R(6) and 9R(8) with the v_{16} doublet band of *gauche* DFE : Ar, earlier measurements had already produced a negative result [6.49] for PIRSH production. Unfortunately, the v_{17} triplet (1049.5, 1047.6, and 1045.6 cm^{-1}) associated with monomeric *trans* DFE in Ar is not coincident with any CO_2 laser line. In addition, irradiation of the sample for several hours with unfiltered radiation from a glowbar source produced no spectral changes in band strengths so this host-defect system will be set aside temporarily.

IR irradiation of DFE in Kr and Xe hosts gives positive results: radiation coincident with the v_{17} absorption of *trans* DFE induces a frequency selective *trans* → *gauche* conversion. Since the exper-imental results are similar for both defect-host systems only the DFE : Xe results will be outlined here [6.50]. Figure 6.5 shows the IR

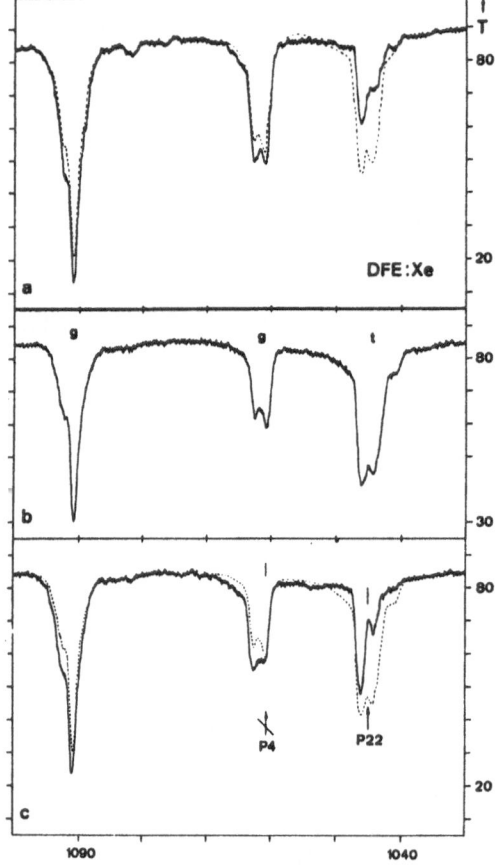

Fig. 6.5. Spectra demonstrating IR-induced interconversion of *gauche* and *trans* DFE:Xe. (a) Solid line: immediately after deposition, dashed line: after 90 min exposure to unfiltered globar source; (b) sample exposed during deposition and for 120 min to give photostationary trace, "g" and "t" stand for *gauche* and *trans*, respectively; (c) Solid line: after 100 min CO_2 laser irradiation (9P(22) at 130 mW/cm^2), dashed line: spectrum from (b) before irradiation. (After [6.51])

spectra of *trans* and *gauche* DFE : Xe in the region of the C-F stretch modes for different conditions of irradiation after the sample had been fabricated in the dark. Figure 6.5b displays the labeling of the different bands, two of which are coincident with CO_2 laser lines. The v_{16} mode of *gauche* DFE is matrix-split into two lines near 1062 cm^{-1} which overlap the 9P(4) laser line and the v_{17} mode of *trans* DFE is split into three components near 1045 cm^{-1} which overlap with the 9P(22) line. The laser frequencies are identified in Fig. 6.5c.

The solid line in Fig. 6.5a shows the line strengths as measured with a filtered glowbar source immediately after the matrix was deposited in the dark. The dashed trace in the same figure indicates how the strengths have changed after the sample had been exposed to unfiltered IR source radiation for 1.5 hrs. Clearly, the IR radiation facilitates the conversion of the more stable *gauche* to the less stable *trans* DFE. Figure 6.5b shows the spectrum for a sample in a photostationary state. It had been illuminated both during deposition and also for the next 2 hrs with unfiltered glowbar light. Changing to the filtered source induces a slow back-conversion from *trans* to *gauche* DFE. The effect of the 9P(22) laser light on this conversion is shown in Fig. 6.5c. The dashed trace identifies the initial state and the solid trace, the spectrum after 100 min. of laser illumination at an intensity of 130 mW/cm^2. A dramatic decrease in the strength of the lower frequency *trans* site is observed; in addition, the laser-induced production of *gauche* DFE appears in the higher frequency component of the v_{16} absorption doublet. A similar irradiation procedure with 9P(4) in the *gauche* band does not cause hole-burning, *gauche* → *trans* conversion or any other change whatsoever in the spectrum.

A close examination of the *trans* region with a spectrometer resolution of 0.4 cm^{-1}(FWHM) at intermediate laser illumination times shows that persistent hole-burning occurs at the laser frequency. The hole depth, as determined by this low-resolution probe beam, is 50% after irradiation for 30 min. This observation demonstrates that a frequency-selective *trans* → *gauche* conversion is taking place.

ii) Molecular Aggregates in Ar. When the molecular concentration in Ar is increased so that M/A = 850 then the three site components of the v_{17} mode of *trans* DFE are superimposed on top of a broad background feature [6.51]. Irradiation of this feature with 9P(18) light induces persistent hole-burning which can be detected with an IR spectrometer with a resolution of 0.25 cm^{-1} after laser illumination for 10 min with an intensity of 750 mW/cm^2. After exposure to unfiltered source light for 30 min the hole is erased and the band recovers its original shape.

New broad absorption features appear when the concentration is increased to M/A = 90. Hole-burning can now be identified in both the *gauche* and *trans* absorption bands as well as in a new aggregate band which appears at 1060 cm^{-1}, but no evidence of conformational interconversion between *gauche* and *trans* is observed. Again the presence of unfiltered source light erases the holes in about 30 min [6.51].

6.2.2 Interpretation of Persistence

The low-intensity diode and high-intensity CO_2 laser measurements probe two different regimes of hole-burning in matrix-isolated DFE. For low intensities the main findings are as follows: (1) spectral persistence is produced by a one-photon process, (2) the low concentrations ensure that the process involves a single molecule, (3) for the Ar and Kr matrices the matrix-split modes do not show the same hole-burning behavior, (4) the hole polarization property demonstrates that the molecule is not rotating freely nor tunneling rapidly around a fixed site, (5) for the Ar and Kr matrices the antihole (product state) remains within the inhomogeneous line while for the N_2 matrix the oscillator strength is transferred among three different inhomogeneous lines, and (6) the erasing of the hole by room temperature thermal radiation and by modest temperature cycling indicates that the process contains a thermal activation energy and is reversible.

The measurements on DFE:Kr and DFE:Xe with a CO_2 laser in conjunction with a thermal globar source provide the following additional results: (7) unfiltered source radiation induces a *gauche* → *trans* conversion, (8) laser radiation coincident with the v_{17} mode of *trans* DFE induces hole-burning and a *trans* → *gauche* conversion, (9) laser radiation coincident with the v_{16} mode of *gauche* DFE induces no change, (10) filtered source radiation (< 1450 cm^{-1} transmitted) induces a very slow *trans* → *gauche* conversion, and (11) the thermal source radiation induces no change in the spectrum of DFE:Ar (excluding for the moment the single aggregate study).

Three fundamentally different kinds of vibrationally induced structural rearrangements could all be sources of the observed persistent spectral changes for this matrix-isolated molecule. The first type (photochemical) involves the shape change of the molecule itself (the *trans* - *gauche* conversion), the second (photophysical) involving rearrangement of the neighboring matrix, and the third (photophysical) involving reorientation of the rigid molecule in the matrix.

Items (7) through (10) listed above definitely show that photochemical conversion is taking place, and the frequency selectivity mentioned in item (8) illustrates that the effect is not produced by local heating. Item (11), on the other hand, demonstrates that the photochemical rate is dependent on the host in a crucial way. There is additional evidence from the low intensity work that photochemical conversion is not the sole mechanism behind the observed hole-burning.

At low intensity the oscillator strength of the inhomogeneously broadened v_{17} mode in *trans* DFE:Kr is independent of the burning (item (5)) in marked contrast with the high-intensity results. This behavior is consistent with a photophysical process involving motion of the molecule and/or rearrangement of the local lattice. *Dubs* et al. [6.8] have completed model calculations with the DFE molecule in a substitutional site in an fcc lattice and conclude that the photophysical hole-burning is caused by a reorientation of the molecule across a large barrier to an energy configuration which would be equivalent except for the occurrence of nearby lattice defects. The nonspherical shape of the N_2 host molecule eliminates the orientational degeneracy of DFE so that the inequivalent directions which are still connected by a large energy barrier have vibrational transition energies separated by more than the inhomogeneous linewidths. Presumably the photophysical hole-burning for the DFE aggregates in Ar is produced by a similar process.

It is possible to distinguish both photophysical and photochemical hole-burning with excitation of the same vibrational mode for Kr isolated DFE because the efficiencies for the two processes are very different: for the photophysical process the efficiency $\eta \simeq 10^{-2}$ while for the photochemical hole-burning $\eta \simeq 10^{-6}$. The v_{17} mode of *trans* DFE in Kr is characterized by a very complex matrix-split spectrum with six absorption lines at 1051.4 cm^{-1}, 1050.4 cm^{-1}, 1048.5 cm^{-1}, 1046.9 cm^{-1}, 1045.0 cm^{-1}, and 1042.9 cm^{-1}, respectively but so far diode laser measurements have been reported only for the 1050.4 cm^{-1} transition. A continued exploration of DFE in this host is to be encouraged. Probing the sample simultaneously with the diode and one of the four coincident CO_2 laser lines is an approach which would enable one to examine with some precision the interplay between these different processes.

6.2.3 Molecular Aggregates of Methyl Nitrite or Methanol

Methyl nitrite (MEN) has been studied [6.50] at a relatively high concentration (M/A = 300) in a nitrogen matrix with a nominal layer thickness of 25 μm. The experimental procedure is first to record the spectral region (about 15 cm^{-1}) around the v_7 *trans* band (FWHM = 2.8 cm^{-1}), then to irradiate the sample with 170 mW/cm^2 of cw CO_2 laser radiation at 9P(28) for about 10 min and, finally, to block the laser beam and record the spectral region a second time. A spectral hole is observed in the second scan at the laser frequency with a depth that decays with time. This decay time increases when an IR band pass filter is placed in front of the sample indicating that the broad band thermal radiation from the source erases the hole. From the hole width, a lower limit of 100 ps is estimated for the vibrational relaxation time from the excited state. The line shapes of other vibrational absorption bands of *trans* MEN:N_2 at different frequencies are found to be independent of this hole-burning at 9P(28). The location of the displaced oscillator strength has not been identified.

Analogous measurements [6.50] on *trans* MEN:Ar show the same behavior at the laser frequencies 9P(28) and 9P(26) but not at 9P(24). Since the monomer component has a peak at this last frequency, the persistence does not appear to be associated with isolated molecules but instead with dimers, trimers, etc.

Methanol in an argon matrix (M/A = 200, layer thickness = 20 μm) also shows persistent hole-burning in the C - O stretch vibrational mode region [6.50]. Laser irradiation with 9P(38) or 9P(40) in the broad band associated with methanol dimers generates holes, but no holes are produced when the frequency is tuned to the strong band produced by isolated methanol or the transitions due to other aggregates. These results indicate that although some degree of molecular complexity is important, aggregation, in general, cannot be relied upon to guarantee persistent spectral effects.

Felder and *Günthard* [6.50] have speculated on the possible causes of these persistent effects. They eliminate the *trans* → *cis* isomerization of MEN:Ar,N_2 as a possibility because of the large barrier between the two conformers. In addition, one would expect that if a photon-assisted conformational change had occurred, then holes should have appeared in many of the normal mode bands of the complex, but the experiments show that a hole only appears at the burn frequency. They propose that the inhomogeneously broadened bands are produced by the many geometric arrangements available to the aggregates and that the persistence stems from the laser-induced rearrangement of those resonant complexes which happen to coincide with the driving frequency. Although the model is plausible, it is unlikely that only one normal mode would be changed after such a collective rearrangement. A precise measurement of the difference in band shapes before and after irradiation should be made to pinpoint how the oscillator band shapes and strengths are altered.

6.3 ReO$_4^-$ in Alkali Halide Crystals

6.3.1 Background and Spectroscopic Information

In this section we describe the persistent spectral hole dynamics for a high-symmetry photostable molecule in several crystalline alkali halide hosts. The ReO$_4^-$ molecule (perrhenate ion) was originally selected for study because one of the four vibrational modes of this tetrahedral molecule, the ν_3 mode, lies in the wavelength range spanned by the CO_2 laser (9 – 10 μm). The molecule is stable at the temperatures required for the growth of alkali halide crystals and enters the crystal substitutionally in appreciable concentrations for those crystals with lattice constants larger than that of NaCl. Because the molecule has a much larger ionic radius than the halogen ions for which it substitutes, rotational motion is suppressed and steric effects are important. There are only two distinct symmetry-preserving orientations of the tetrahedral molecule in the octahedral anion site. Thus, perrhenate ions in alkali halides are ideally suited for spectroscopic study of a zero-phonon vibrational transition at low temperatures.

Before tunable semiconductor-diode lasers in the infrared became available, a major experimental problem was to produce a coincidence between any of the fixed frequency CO_2 laser lines and the ν_3 mode absorption at liquid helium temperatures. The experimental difficulty is illustrated in Fig. 6.6 which shows a diode laser trace in the 10μm wavelength region of the transmissivity of both a sample of ReO_4^- molecules in KI at low temperatures and a hot CO_2 gas cell [6.43]. The vibrational absorption line can be as narrow as 0.016 cm^{-1} at 1.4K [6.43]. With this type of measurement the frequency interval between the nearest CO_2 laser line and the ν_3 mode can be determined. Inspection of this figure shows that the two frequencies do not coincide and that the chance of an accidental coincidence between the comb of CO_2 laser lines (spaced almost 2 cm^{-1} apart) and the vibrational mode absorption is very small.

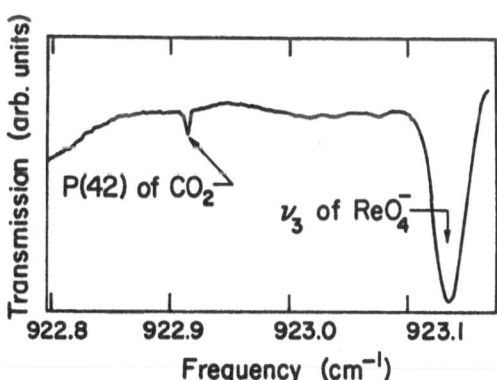

Fig. 6.6. Diode laser transmission spectrum of 0.2 cm thick KI + 0.001 mole % KReO$_4$ at 2K. The 10P(42) absorption of hot CO_2 gas serves as a frequency marker. (after [6.43])

One way to circumvent this problem is to use "poor-man tuning", or double-doping, to intentionally induce excess inhomogeneous broadening of the zero-phonon line. This broadening is accomplished by adding a second cation dopant to the melt [6.42,52] when the crystal is grown. The second impurity, depending on its position in the lattice, inhomogeneously strain broadens or splits the ν_3 absorption line producing an overall linewidth on the order of $1-2$ cm^{-1}. Line coincidences can also be produced simply by hot-forging the ReO$_4^-$–doped alkali halide sample before cooldown [6.11]. With either method of sample preparation it becomes relatively easy to find a CO$_2$ laser line coincidence with some part of the v_3 absorption spectrum and therefore perform saturation and hole-burning measurements.

6.3.2 Measurements of Relaxation Times T$_1$ and T$_2$

The relaxation dynamics of the v_3 mode provide an important framework for the interpretation of persistent spectral hole-burning results. The excited state lifetime T$_1$ and the dephasing time T$_2$ have been measured at low temperatures for perrhenate ions in alkali halides by a combination of spectroscopic techniques [6.44–46], including incoherent laser saturation and transient hole-burning spectroscopy.

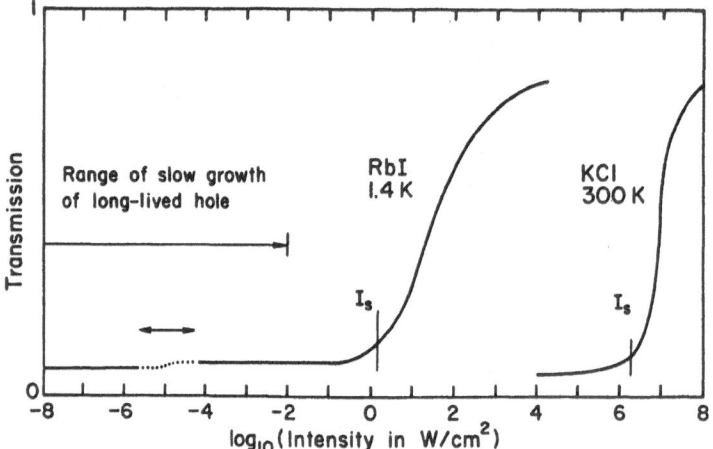

Fig. 6.7. Illustration of incoherent laser saturation measurements for matrix-isolated ReO$_4^-$ at two different temperatures. Persistent (long-lived) spectral features are observed with very small laser intensities. (after [6.52])

Incoherent laser saturation studies of the v_3 mode absorption at two different temperatures are illustrated in Fig. 6.7. The room temperature data follow the shape expected for a homogeneously broadened system while the 1.4 K data agree with the shape expected for an inhomogeneously broadened system. In the first case the saturation intensity, I$_s$, is directly related to the energy relaxation time T$_1$ through the usual relation

$$I_s = \frac{h\nu}{2\sigma T_1} \qquad (6.1)$$

where v is the vibrational mode frequency, h is Planck's constant, and σ is the absorption cross section. In the low-temperature case, I_s is directly related to the product of T_1 T_2 through

$$I_s = \frac{nc\hbar^2}{8\pi T_1 T_2 L(|\mu|^2/3)},$$

(6.2)

where n is the index of refraction, c is the velocity of light, L is the local field factor, and μ is the dipole moment of the transition. In this second case T_2 can be obtained independently from transient (two-level) hole-burning spectroscopy since the hole width is directly connected to the homogeneous linewidth through the relation $\Delta v_{hole} = (2/\pi T_2) = 2\Delta v_{HOM}$ (in the absence of spectral diffusion effects).

Using (6.1) and (6.2), values of T_1 and T_2 have been determined for the ReO_4^- v_3 mode in a variety of alkali halide hosts [6.45,46]. Table 6.1 shows the results of these measurements. Within experimental error, T_2 is approximately equal to $2T_1$ indicating that the v_3 homogeneous line width is lifetime-limited at liquid helium temperatures. Thus, in this case I_s is proportional to $(T_1)^{-2}$. Since $(1/T_1)$ is a measure of the excited state decay rate, then according to the simple energy-gap law for multiphonon decay [6.40], a plot of log I_s versus $(v_3/v_{lattice})$ where $v_{lattice}$ is a dominant lattice phonon frequency should produce points that lie on various straight lines with each line corresponding to a different alkali or halide series.

Figure 6.8 shows one plot of such data. The ratio of the v_3 frequency to the maximum frequency in the longitudinal acoustic

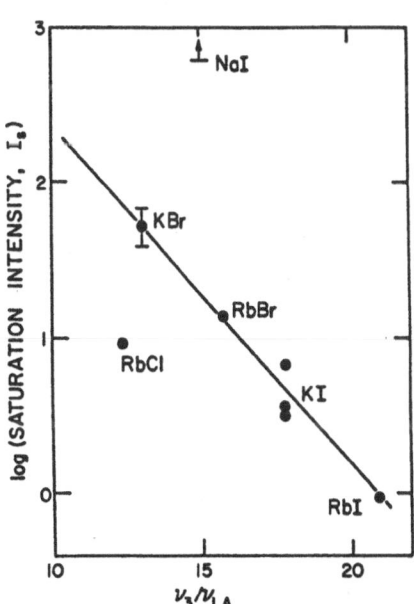

Fig. 6.8. Saturation intensity I_s versus v_3 mode frequency divided by the LA-phonon frequency at the zone boundary. (after [6.52])

Table 6.1. Measured vibrational relaxation times for the ν_3 mode for ReO_4^- in alkali halide lattices (T < 10 K)

Host	RbI	KI	KI	KI	RbBr	KBr	RbCl	NaI	KCl
ReO_4^- [mole %]	0.8	0.005	0.05	0.05	0.05	0.3	0.1	0.2	0.3
Other [mole %]	$0.8\ K^+$	$2.5\ Rb^+$	$0.2\ Cs^+$	$2\ Na^+$	$0.05\ K^+$	$2.5\ Rb^+$	$0.1\ K^+$	$0.2\ K^+$	$2.2\ Rb^+$
$\bar{\nu}_3$ [cm^{-1}]	922.9 P(42)	922.9 P(42)			927.0 P(38)	931.0 P(34)	933.0 P(32)	938.7 P(26)	939.0 P(26)
Lattice constant a [Å]	7.34	7.07			6.85	6.60	6.58	6.47	6.29
I_s [W/cm^2]	0.9 ± 0.2	3.6 ± 0.8	6.8 ± 1.4	3.2 ± 0.6	14 ± 3	53 ± 10	9.3 ± 2	> 700	22 ± 5
α_n [cm^{-1}] (± 25%)	0.80	0.13	0.55	0.12	1.1	0.41	1.4		0.1
$T_1 T_2$ [10^{-16} s^2]	24 ± 5	6.7 ± 1.5	3.6 ± 0.8	7.6 ± 1.5	1.8 ± 0.4	0.49 ± 0.09	2.9 ± 0.6	< 0.05	1.2 ± 0.3
$\Delta\nu_{HOM}$ [MHz, FWHM]	5.0 ± 1.5	7.8 ± 0.5	9.0 ± 1.3	10 ± 1.5	18 ± 1.5	30 ± 1.5	17 ± 1.3		
T_2 [10^{-8} s]	6.4 ± 2.0	4.1 ± 0.3	3.5 ± 0.5	3.2 ± 0.5	1.8 ± 0.2	1.1 ± 0.1	1.9 ± 0.2		
T_1 [10^{-8} s]	3.8 ± 1.4	1.6 ± 0.5	1.0 ± 0.3	2.4 ± 0.5	1.0 ± 0.3	0.45 ± 0.15	1.5 ± 0.3		

spectrum (the number of LA phonons required to relax the v_3 mode) is plotted along the abscissa. Since for the assumed multiphonon decay mechanism, the data for NaI, KI, and RbI should lie on one straight line while the data for RbCl, RbBr, and RbI should line on another, one can conclude that the actual decay scheme must be more intricate than simple multiphonon decay. A systematic study of these data suggests that the low temperature decay channel actually consists of multistep emission of other internal molecular modes, localized phonon modes, and band phonons [6.46]. Although this decay scheme is more complex, it is also more probable because it is of lower order than simple multiphonon decay. Further, this lower-order decay scheme also accounts for the relatively small value of T_1 ($\simeq 10^{-8}$ s) and the correspondingly large value of I_s $(1-10$ W/cm$^2)$ at low temperatures.

6.3.3 Persistent Spectral Holes for ReO_4^- in Alkali Halides

a) Summary of Characteristics

By far the most interesting part of Fig. 6.7 is the persistent change in sample transmission which appears at very small laser intensities. This is due to the formation of persistent infrared spectral holes under the action of the incident laser light. Such holes can be produced with $\simeq 10.8$ μm radiation from CO_2 and semiconductor diode lasers at intensities far below the two-level saturation intensity, I_s. The hole formation in this system appears to involve reorientation of the excited impurity molecule during nonradiative vibrational de-excitation. These holes have a variety of novel properties, and a brief description of the hole dynamics forms the remainder of this section. (For more detail, the reader is referred to [6.10].) As a result of these measurements, PIRSH formation at low laser intensity is shown to be a general solid-state phenomenon which may be reasonably expected to occur whenever the complete ground state of the system has configurational degeneracy.

Figure 6.9 shows the growth and detection of a persistent spectral hole in RbI: ReO_4^- using a single CO_2 laser (called the probe). The fixed frequency probe laser burns a hole, and time-varying probe laser transmission serves as a measure of hole depth. After cooling the sample to 1.4K in the dark, the laser is unblocked at time t_1 (Fig. 6.9a). The vertical axis in Fig. 6.9 shows the infrared power transmitted through the sample as a function of time. The initial sample transmission is T_i; it then slowly grows over many seconds to a steady-state value T_{ss}. If the laser is now blocked for a period of, say, 10 minutes and then unblocked the sample transmission is observed to remain at the larger T_{ss} value indicating that the spectral hole has not decayed away during the dark period.

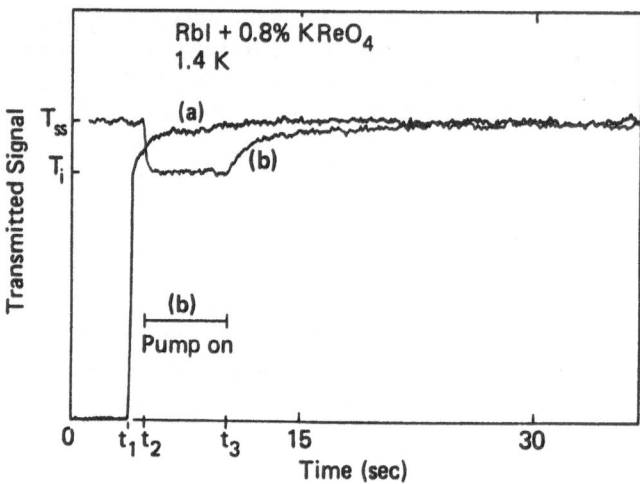

Fig. 6.9. Examples of PIRSH growth and erasing. (a) After cooling a sample of RbI + 0.8 mole % KReO$_4$ from 77 K to 1.4 K in the dark, the CO$_2$ laser probe beam was unblocked at t_1. (b) After long exposure to the probe beam, a second CO$_2$ laser displaced by 10 MHz is unblocked at t_2, erasing the probe hole, and blocked again at time t_3. (after [6.10])

One way to erase the hole is to use a second CO$_2$ laser which is tuned roughly 10 MHz away from the first laser. When this second laser irradiates the sample from t_2 to t_3, as shown in Fig. 6.5b, then the hole burned by the first laser is erased. When the second laser is turned off the transmission again rises to T_{ss} signifying reburning of the first hole by the probe laser. This is a general phenomenon: laser light in the wings of a previously burned hole causes hole erasure, signifying that the product absorption for the initial hole is only slightly displaced from the initial state absorption. Another way to erase the hole is to simply cycle the sample temperature up to 20 K and back down. Thus the spectral holes in this system only have long lifetimes at temperatures less than 10 K.

To measure the hole lineshape, either a tunable diode laser or a second cavity-length tuned CO$_2$ laser can be utilized to scan the hole directly. Figure 6.10 shows the hole lineshape as determined by the latter technique. The hole spectrum consists of a single, central hole of width 10±2 MHz in RbI and 20±4 MHz in KI, which is essentially identical to the hole lineshape for transient saturation holes [6.44]. The hole width is independent of temperature below 10 K, and grows very quickly above 10 K. In addition, the hole depth $\Delta\alpha/\alpha_i \equiv (\alpha_i - \alpha_f)/\alpha_i$ appears to saturate at a shallow value in the range from 0.05 to 0.3 independent of the burning laser power.

Curves like those presented in Fig. 6.9 also illustrate the kinetics of hole growth for the ReO$_4^-$ – alkali halide systems. The hole growth is nonexponential, showing a fast rate at small burning time and slower and slower rate as hole-burning continues. Nevertheless,

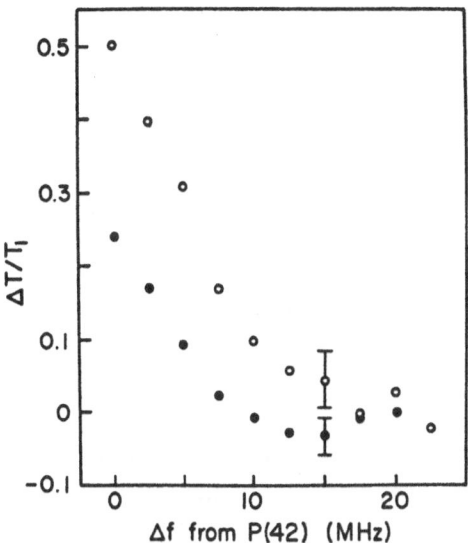

Fig. 6.10. Persistent-hole lineshape measurements with two CO_2 lasers for RbI: ReO_4^-. The initial probe transmissions at $\Delta f = 20$ and 22.5 MHz were averaged to define T_i. The open and closed circles correspond to measurements at different sample locations. (after [6.10])

an effective quantum efficiency at t = 0, η_e, can be determined from the data to be roughly equal to 10^{-3}. Values in this range are typical of the ReO_4^- system, and are to be contrasted with the 10^{-5} to 10^{-7} efficiencies that have been reported for NPHB in electronic transitions (see Chap. 5).

Since the inhomogeneously broadened v_3 mode absorption is $1-2$ cm^{-1} in width in the RbI and KI hosts, attempts were made with a tunable diode laser to burn holes at frequencies other than that for the 10P(42) CO_2 laser line at 10.8 μm. The 10P(42) laser line is $\simeq 2.2$ cm^{-1} from the $^{185}ReO_4^-$ line center in RbI, and 0.22 cm^{-1} from the $^{187}ReO_4^-$ line center in KI. Detectable persistent holes could not be burned with the diode laser in the immediate region of the v_3 mode line center in either KI + 0.001 mole % $KReO_4$ or RbI + 0.001 mole % $RbReO_4$. These results suggest that the long-lived hole-burning phenomenon is somehow associated with the larger values of strain found in the wings of the inhomogeneously broadened line.

b) Model for the PIRSH Process

A natural configuration for the ReO_4^- molecule substituted for the halide ion in an alkali halide crystal is with the four tetrahedral Re-O bonds directed along four of the eight available <111> directions. Indeed, careful studies of the infrared and Raman spectra of ReO_4^- molecules in alkali halides [6.53] indicate that the symmetry of the ReO_4^- molecule in the crystal is still tetrahedral to high accuracy. Examination of an ionic model for the lattice with the ReO_4^- at the halide ion site shows that the steric pressure from nearby ions is somewhat relieved if the Re-O bonds lie along <111> body diagonals.

The molecular ion may thus be expected to have two configurations in the lattice related by a $\pi/2$ rotation about a <100> axis with a high barrier to rotation between the two configurations. Infrared spectroscopy of the v_3 mode absorption (over a range of several cm^{-1}) [6.10] suggests that there are no strong librational absorptions at low temperatures, in contrast to the situation for CN$^-$ molecules in alkali halides [6.58], for instance.

The most reasonable first model for the persistent hole formation mechanism involves a photostimulated reorientation of the molecule between these two normally equivalent orientations. The model that originally evolved in the early work was very similar to that used by *Hayes* et al. [6.1] to explain NPHB in glasses in the visible. The principal difference was that in the infrared case, the configurational degeneracy is not provided by the presence of the nearby tunneling systems in the host matrix, but rather by the two possible orientations of the molecule in the crystalline host.

Two possible vibrational deexcitation pathways of the v_3 mode which can give rise to optically induced reorientation are indicated in Fig. 6.11. Pathway P1 is that proposed previously for NPHB in glasses; photoexcitation in well A and configurational tunneling to the new configuration B while in the optically excited state, with subsequent relaxation to the ground state of this second configuration. The alternative pathway is P2, photoexcitation in configuration

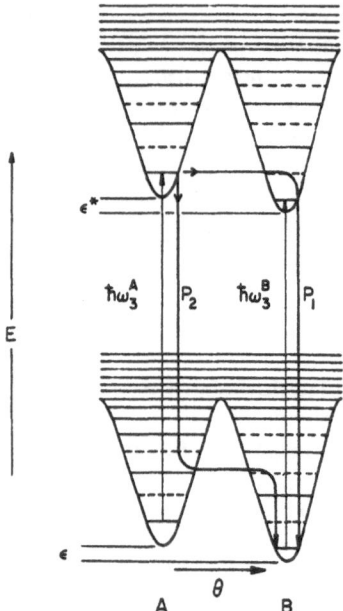

Fig. 6.11. Potential energy diagrams for the ReO$_4^-$ spherical-top molecule in the ground and excited v_3 mode vibrational states. Two possible configurations A and B occur for each state. (after [6.10])

A, return to an excited librational state within well A, subsequent tunneling in this librational state to well B and finally relaxation to the new ground state. The consequence of the resultant molecular reorientation is to shift the absorption at any particular active site from the frequency corresponding to a vertical transition between wells in configuration A to a different frequency corresponding to a vertical transition between wells in configuration B. Because there is a distribution of both A and B sites, molecules which initially have vibrational frequencies coincident with the laser frequency end up being reoriented to the alternate configuration which absorbs at a slightly different energy, producing a spectral hole.

Using the level structure in Fig. 6.11, a phenomenological rate equation model can be developed that can be used to compute the hole growth characteristic and the hole lineshape [6.10]. Although the model predicts the correct experimental behavior in most re-gimes, a fundamental problem has been that it also predicts libra-tional states in the far infrared absorption spectrum of the material, yet none have been observed. This mystery has been resolved with the discovery of the antihole, or "peg".

6.3.4 Persistent Spectral Pegs

Figure 6.12 shows a transmission spectrum of a thick alloy sample (KI + 0.05 mole % $KReO_4$ + 2 mole % NaI) at 6 K as measured with a high resolution Fourier transform spectrometer. The doublet structure of the v_3 mode comes from the two Re isotopes. The two doublets farthest removed from the unperturbed mode at 923.5

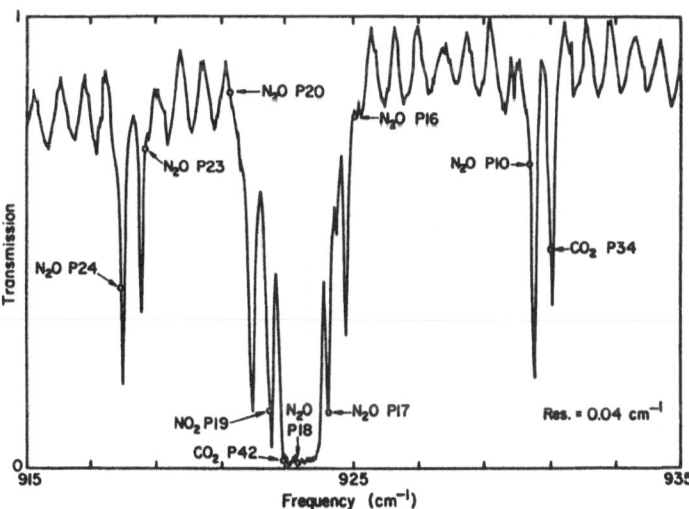

Fig. 6.12. High resolution spectrum of KI + 0.05 mole % $KReO_4$ + 2 mole % NaI showing laser coincidences at T = 6 K

cm^{-1} appear because of a nearby Na$^+$ impurity. The close-in doublets come from Na$^+$ impurities further away from the ReO$_4^-$ ion. The laser coincidences for both N$_2$O and CO$_2$ lasers for all perturbed v_3 frequencies are also indicated in Fig. 6.12. The important observation is that now a number of laser transitions are coincident with vibrational modes of the perturbed and unperturbed ionic absorptions.

The persistent holes which have been burnt using N$_2$O, P(19) are very similar in behavior to those produced by CO$_2$, 10P(42) in the main band. The surprising result is that at the P(17) line of N$_2$O, a persistent antihole (peg) is produced [6.11].

An enlargement of the important spectral region is shown in Fig. 6.13a. A first derivative spectroscopic measurement of the peg has been made with a tunable low-power diode laser in Fig. 6.13b. The line shape corresponds to a central line of *enhanced* absorption with a full width at half maximum of 28 MHz, but surrounded by broad wings of decreased absorption. The overall shape is a negative rep-

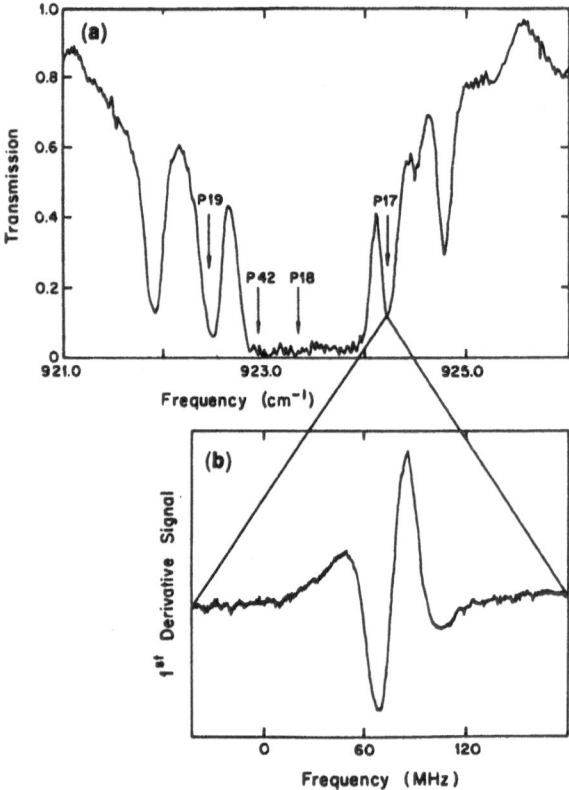

Fig. 6.13. (a) Infrared transmission spectrum of KI + 0.05 mole % KReO$_4$ + 2 mole % NaI in the region of the v_3 mode. (b) Diode laser first derivative spectrum of the peg at N$_2$O P17. The central feature corresponds to enhanced absorption. (after [6.11])

lica of the persistent hole which had been observed at the 10.8 μm P(42) CO_2 line. A reexamination of other ReO_4^- –host lattice combinations at a number of N_2O frequencies has shown that this peg formation phenomenon is quite general. Persistent holes or persistent pegs occur depending upon the particular spectral position and the particular sideband in which irradiation occurs.

By now one should be intuitively comfortable with the idea that narrow-band laser radiation within an inhomogeneously broadened absorption line might lead to a burned spectral hole through saturation, photodecomposition, photoinduced reorientation or perhaps other processes which remove absorbers from the spectral region under irradiation. An *increased* absorption at the frequency of irradiation is not so easy to explain. A simple but subtle variation in the model described earlier (Fig. 6.11) appears to account qualitatively for the persistent hole and peg phenomena [6.11]. Once again we imagine that the local lattice strain gradients make the two ground state orientations, A and B, of the tetrahedral molecule in the cubic site inequivalent (see Fig. 6.14). In addition it is assumed that the energies of the two orientations in the excited vibrational states A' and B' are different from the ground state. Although the random nature of the strain field produces distributions of all of these energy levels, the important features with respect to hole and peg production can be demonstrated with the four levels shown in Fig. 6.14. In arrangement (1) the potential-well asymmetry in the ground state is larger than the asymmetry in the excited state while the converse is true for arrangement (2). For both cases, well asymmetry is assumed to be large compared to the tunnel splitting

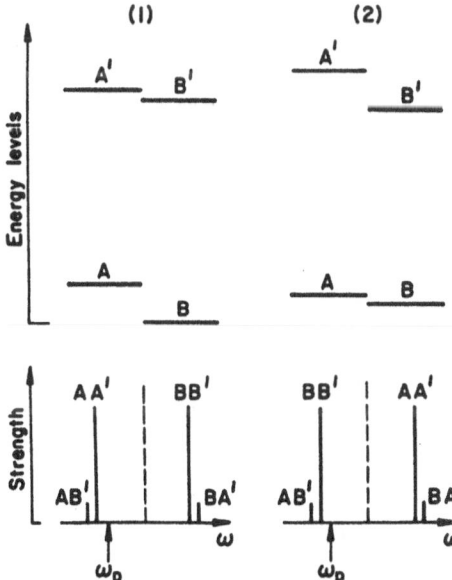

Fig. 6.14. Photoinduced reorientation model for ReO_4^- in alkali halides. Two possible configurations denoted A(A') and B(B') occur for each zero libration state. Arrangement (1) has the double-well asymmetry of the two ground-state configurations larger than for the two excited states and arrangement (2) has the opposite. The bottom figure shows the line strengths vs frequency for each of the transitions in the two arrangements. (after [6.11])

227

so the infrared selection rule strongly favors left well→left well or right well→right well transitions. These transitions are labeled AA' and BB' in the lower part of the figure. The nearly forbidden transitions between configurations, labeled AB' and BA', are the pathways that lead to persistent elastic polarization of the resonant molecules.

When the laser frequency ω_p is not symmetrically located with respect the AA' and BB' transition frequencies in Fig. 6.14, persistent spectral effects are produced. For the energy level arrangement (1) the laser pump drives centers from A to B via the overlap of the laser with the homogeneous width of the AB' transition. Since the number in A decreases, the strength of AA' decreases and a hole appears at the pump frequency. For the energy level arrangement (2) the laser pump again drives centers from A to B via the AB' transition but increased absorption now occurs near ω_p since the dominant absorption results from the overlap of the homogeneous width of the BB' transition with the laser line instead of the AA' transition.

6.3.5 Ultrasonic Studies of Multiple Ground State Configurations

The models represented by Fig. 6.11 and by Fig. 6.14 have one common feature, namely, that the tetrahedral molecular ion has two rotationally equivalent configurations at the substitutional lattice site which are separated by a large energy barrier due to steric effects and which, in fact, have slightly different energies due to local strain gradients in the crystal. For the first model, IR-active librational sidebands should occur at frequencies near the vibrational mode value while for the second model low-lying librational states are not required.

A measurement of the temperature-dependent absorption of 10 MHz ultrasound in RbBr, KI and RbI crystals doped with ReO_4^- has produced the first direct experimental identification of a two-configuration energy barrier associated with this defect-lattice system[6.54]. The loss peak which is shown in Fig. 6.15 stems from an activation process with a barrier height of 60 meV for RbBr, 50 meV for KI and 30 meV for RbI. The decreasing barrier height with increasing lattice constant may account for the observation that spectral holes are burned and erased most easily in the RbI host.

The combination of ultrasound with the persistent spectroscopic effects has led to the development of a new phase-insensitive optical detector for ultrasonic waves in solids [6.55]. When ultrasonic waves are applied to an alkali halide which contains persistent holes or pegs the hole depth or peg height is modulated. This effect occurs because the local ultrasonic strain introduces a variety of frequency shifts which add up to a broadening of the persistent feature. Since, at the hole's (peg's) center frequency, the absorption can only in-

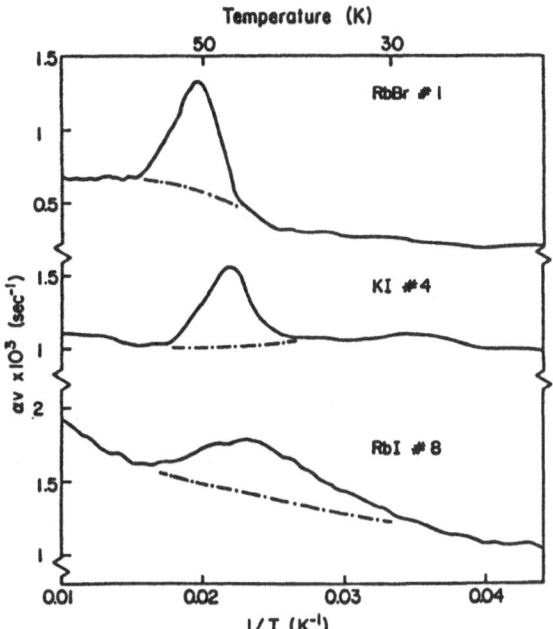

Fig. 6.15. Ultrasonic absorption at 10 MHz versus temperature for three alkali
halides doped with $KReO_4$.
(after [6.54])

crease (decrease) due to these frequency shifts; the transmitted signal
reflects the rectified displacement. Such a process is completely independent of the phase of the ultrasound, provided that the acoustic
wavelength is smaller than the laser beam diameter.

An estimate of the sensitivity of this new optical method of ultrasonic detection has shown that for the ReO_4^- molecule an ultrasonic wave amplitude of 1 Å should be easily detectable. Even
more sensitivity can be obtained by using the persistent photochemical hole of the F_3^+ center in NaF where it was predicted that
for 10 MHz sound an amplitude of 10^{-2} Å could be detected
[6.55]. A recent study of this defect-lattice system [6.56] has confirmed and extended these ideas to the problem of hole detection.
Phase-sensitive modulation of spectral holes at MHz frequencies has
also been observed [6.57].

6.3.6 Conclusions on the ReO_4^- System

The view that the vibrational modes of matrix-isolated molecules
should have long relaxation times in the 10 μm wavelength region
is not generally valid because other relaxation processes dominate
multiphonon decay. The intrinsic symmetry and simplicity of the
matrix-isolated ReO_4^- molecule in alkali halide crystals has enabled
the determination of many aspects of the anharmonic decay process.
At low temperatures the decay channel consists of multistep emission
of other internal modes, localized modes and band phonons.

The burning, at extremely low laser powers, of persistent spectral holes and antiholes in inhomogeneously broadened vibrational mode absorptions in crystals indicates that tunable diode lasers can be used to measure the homogeneous width of vibrational modes of matrix-isolated molecules throughout the IR. From this work it appears that the burning of such holes and antiholes at low laser intensity is a general solid state phenomenon which can occur whenever the complete ground state of the system to be excited has configurational degeneracy.

Early models for persistent nonphotochemical hole-burning have relied on configurational change during deexcitation to move centers from one configuration to another. The dynamics of persistent nonphotochemical spectral hole and peg formation in the ν_3 vibrational mode absorption of ReO_4^- molecules in alkali halide crystals can be understood qualitatively with a simple energy level picture in which nearly forbidden interconfiguration transitions are important. Whether or not such forbidden transitions also play a central role in explaining the earlier results on NPHB for electronic transitions in organic glasses remains to be seen.

6.4 Persistent Spectral Hole–Burning for CN⁻ Molecules in Alkali Halide Crystals

6.4.1 Background Information on Matrix-Isolated CN⁻

The IR spectrum of CN^- molecules matrix-isolated in alkali halide crystals has attracted interest for many years because of its intrinsic simplicity. Depending on the particular matrix, the room-temperature vibrational spectrum takes on one of two characteristic shapes: for KCl, KBr, KI and RbCl the band shows a predominant double hump structure with a total width of approximately 50 cm⁻¹ (FWHM) while for NaCl and NaBr a single band is observed with a width of $\simeq$ 20 cm⁻¹ (FWHM) [6.58 – 61]. The temperature-dependent properties of the double hump are identical to those observed at low resolution for the stretch region of a polar diatomic in the gas phase; thus the spectra have been identified with the P and R branches of freely rotating CN^- in the solid. The absence of any room temperature P and R structure at the stretch frequency for the Na salt hosts indicates that the molecule is not free to rotate in these lattices.

A temperature-dependent absorption spectrum [6.58] of the first overtone band of KBr:CN⁻ which illustrates this high temperature free-rotor-like behavior is presented in Fig. 6.16. The separation between the two maxima varies as $T^{1/2}$ between 60 K and 300 K giving a rotational constant B $\simeq$ 1.2 cm⁻¹ which is quite a bit smaller than the undistorted CN^- free rotor value, B = 1.94 cm⁻¹ [6.36] . At temperatures below 60 K the spatial anisotropy produced by the

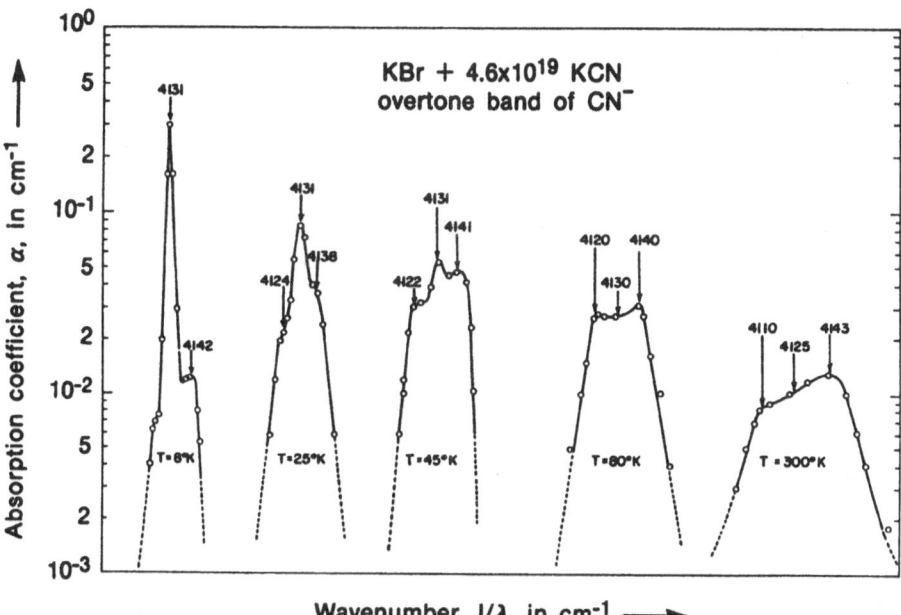

Fig. 6.16. Temperature dependence of the first overtone vibration of CN⁻ in KBr. (after [6.58])

local crystalline electric field becomes observable. The interpretation of the lowest temperature spectrum (ignoring tunneling for the moment) is as follows: the strong transition is a zero-libron transition, the weak high-frequency satellite, a one-libron sum band and the faint low-frequency satellite, a one-libron difference band [6.35]. Within the zero-libron band of width $\simeq$ 2.5 cm⁻¹ a tunnel splitting of $\simeq$ 1 cm⁻¹ was predicted but not observed [6.58] when measurements were attempted with a spectral resolution of 0.6 cm⁻¹. Later stress dependent studies [6.62,63] demonstrated that the equilibrium orientations of the molecule in the potassium halide salts are directed along the equivalent {111} axes of the face centered cubic lattice.

In the most recent spectroscopic investigation with a grating instrument which was carried out at a spectral resolution of 0.1 cm⁻¹, the tunneling structure for CN⁻ in KCl, KBr and KI hosts has been resolved [6.63]. A doublet split by 2.5 cm⁻¹ in KCl is interpreted as the Stokes and anti-Stokes components of combined vibration-tunneling transitions [6.63]. With increasing lattice constant of the host (KCl → KI) the band shape shifts from two satellite peaks to a central line. The relative strengths of the central line compared to the two satellites vary from sample to sample for a given host depending on growth procedures; hence, it is concluded that the satellites identify those molecules which can tunnel and the central line, those which cannot. Apparently, some CN⁻ molecules are always pinned in the lattice because of interactions with nearby impurities, internal strains, or other crystal defects [6.63].

6.4.2 High-Resolution FTIR Spectroscopy in the CN⁻ Stretch Region

Figure 6.17a shows the low temperature absorption coefficient in the stretch mode region at a resolution of 0.04 cm⁻¹ for a KBr crystal doped with 0.05 mole % KCN [6.12,13]. The vertical dashed line in the figure divides the spectrum into two frequency intervals. The absorption lines which occur at frequencies larger than 2100 cm⁻¹ are associated with different isotopes of the linear molecule NCO⁻, a common impurity in the KCN dopant. The linewidths of these transitions are all about 0.05 cm⁻¹ (FWHM) and the frequencies are given in Table 6.2. None of these features show spectral hole-burning.

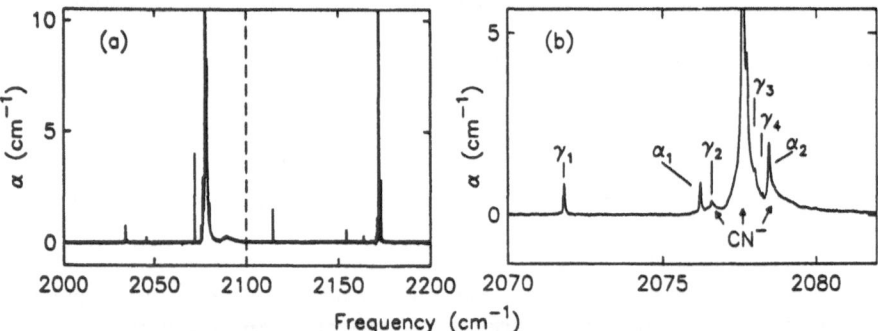

Fig. 6.17. High-resolution absorption spectra of CN⁻ in KBr. (a) Impurity-induced absorption coefficient in the CN⁻ stretch region for KBr + 0.05 mole % KCN. The vertical dashed line divides the spectrum into two parts: right, isotope lines of NCO⁻; left, lines of CN⁻. The center frequencies are recorded in Table 6.2. (b) Expanded view of the CN⁻ region for a 0.01 mole % dopant concentration. Hole burning has been observed on the lines α_1, α_2, γ_1, γ_2, γ_3, and γ_4. The sample temperature is 1.7 K and the FTIR resolution is 0.04 cm⁻¹. (after [6.13])

The absorption lines at frequencies below 2100 cm⁻¹ in Fig. 6.17a are all associated with the CN⁻ stretch vibration in some way or other and the complexity of this spectrum is quite surprising [6.13]. Again the center frequencies, line identifications and hole-burning results are summarized in Table 6.2. The fine structure near the main ($^{12}C^{14}N^-$) absorption line can be seen more clearly in Fig. 6.17b where the spectrum of a lower concentration sample (0.01 mole % KCN) is displayed. This spectrum is characterized by one strong central line with a large linewidth ($\simeq 0.2$ cm⁻¹ FWHM) surrounded by lines which have much narrower linewidths ($\simeq 0.04$ cm⁻¹ FWHM). In the figure these satellite lines are labeled α_1, α_2, γ_1, γ_2, γ_3 and γ_4.

The spectral shapes in Fig. 6.17b are somewhat different from those reported by *Beyeler* [6.63] for the same impurity concentration. His three components of equal strength (frequency positions labeled

Table 6.2. Center frequencies and hole-burning properties of the different vibrational modes in the KBr:CN⁻ spectrum at 1.7K

Vibrational center	Frequency [cm⁻¹]	Hole burns?
$^{14}N^{12}C^{16}O^{-}$	2171.57	no
$^{14}N^{12}C^{18}O^{-}$	2163.45	no
$^{15}N^{12}C^{16}O^{-}$	2154.23	no
$^{14}N^{13}C^{16}O^{-}$	2114.43	no
$^{10}B^{16}O_2^{-}$	2030.96	–
$^{11}B^{16}O_2^{-}$	1960.71	–
$^{12}C^{14}N^{-}$	2077.60	no
$^{12}C^{15}N^{-}$	2045.80	no
$^{13}C^{14}N^{-}$	2034.46	no
$^{12}C^{14}N^{-}:Na^{+}$ (γ_1)	2071.83	yes
(γ_1')	2077.75	yes
(γ_2)	2076.59	yes
(γ_3)	2077.99	yes
(γ_4)	2078.22	yes
$^{12}C^{15}N^{-}:Na^{+}$ (γ_1)	2040.22	–
$^{13}C^{14}N^{-}:Na^{+}$ (γ_1)	2028.79	–
$^{12}C^{14}N^{-}:Cl^{-}$ (α_1)	1076.21	yes
(α_2)	2078.45	yes
$^{12}C^{14}N^{-}:Li^{+}$ (β_1)	2066.81	no
(γ_1)	2074.07	no
(β_2)	2078.14	no
(β_3)	2078.39	–
(γ_1')	2083.57	no
(β_1')	2084.43	–
$^{12}C^{14}N^{-}:Rb^{+}$ (γ_2)	2076.84	no
$^{12}C^{14}N^{-}:^{12}C^{14}N^{-}$	2079.98	no
$^{12}C^{14}N^{-}:^{12}C^{14}N^{-}$	2075.22	no

by the arrows in the figure) are to be compared with the single strong and weak satellite transitions shown here. The hole-burning measurements reported by *Spitzer* et al. [6.12,13] indicate that both the central line and related broad structure should be attributed to tunneling CN⁻ molecules and the ultrasharp lines, to CN⁻ molecules frozen in orientation by other nearby impurities in the crystal. With a broad band source an FTIR resolution of $\simeq$ 0.04 cm⁻¹ is required to spectrally resolve these two types of centers in the region of overlap.

6.4.3 Hole-Burning in the CN⁻ Stretch Mode Region

No hole-burning has been detected in the main line shown in Fig. 6.17b but the sharp lines labeled α_1, α_2, γ_1, γ_2, γ_3 and γ_4 all show PIRSHs when irradiated with a diode laser [6.12,13]. The signatures are very different from that found for the spherical top molecule ReO₄⁻ (see Sect. 6.3). Here, for each absorption line, the oscillator

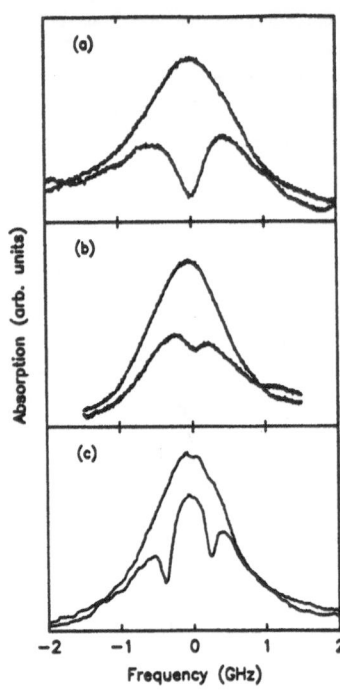

Absorption (arb. units)

Frequency (GHz)

Fig. 6.18. Three examples of persistent hole-burning spectra. Each case shows both before and after burning: (a) α_1 band, (b) α_2 band, and (c) γ_1 band. All holes are power broadened. Sample temperature is 1.7 K. (after [6.13])

strength removed from the laser frequency does not reappear at another frequency within the inhomogeneously broadened line. Examples of hole-burning in the three absorption lines α_1, α_2, and γ_1 are shown in Fig. 6.18. Figure 6.18a shows a hole burned in the α_1 line which can made 100% deep but which washes out after about three minutes. The lines γ_2, γ_3 and γ_4 (not shown) display a similar erasing behavior. Only a shallow hole can be burned in the α_2 line (shown in Fig. 6.18b) which erases in about 30 sec. Figure 6.18c shows the γ_1 mode before and after two holes have been burned in it. These holes which can be made 100% deep have the interesting property that they are permanent as long as the sample is maintained at liquid helium temperatures.

The temperature stability of the holes burned in the γ_1 band has been determined by sweeping the spectrum with a low intensity probe beam while the sample temperature is varied [6.13]. Scans at four different temperatures are shown in Fig. 6.19: trace (a) of a sample at 1.7 K displays a spectrum which contains two holes, trace (b), the same sample at 15 K, trace (c), at 25 K and trace (d), at 35 K. At this highest temperature the spectrum appears nearly featureless yet upon recooling to 1.7 K the original spectral shape is recovered. Erasing does not become noticeable until the sample temperature is cycled up to about 48 K.

A systematic study of crystals double-doped with CN$^-$ and with different cation and anion dopants demonstrates that the vibrational

234

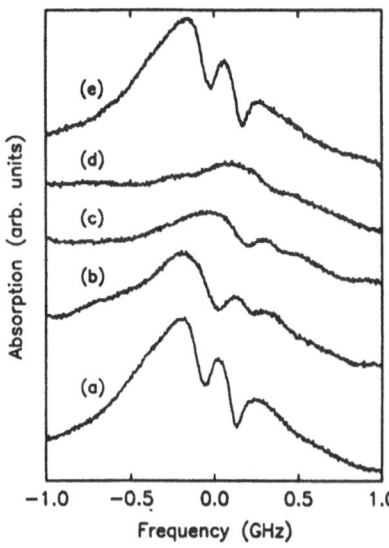

Absorption (arb. units)

(e)

(d)

(c)

(b)

(a)

-1.0   -0.5   0.0   0.5   1.0

Frequency (GHz)

Fig. 6.19. Stability of the persistent holes in the γ_1 band. The temperatures for the four cases shown are: (a) 1.7 K, (b) 15 K, (c) 25 K, (d) 35 K and (e) back to 12.7 K again. (after [6.13])

center which produces the γ_1 band is a CN⁻ molecule with a near neighbor Na⁺ impurity replacing one of the K⁺ host ions [6.63].

6.4.4 A Study of the CN⁻: Na⁺ Center Dynamics

a) Fluorescence

Solid-state vibrational fluorescence was first observed in 1972 for CO matrix-isolated in van der Waals solids [6.18] and somewhat later for isomorphic CN⁻ in ionic crystals [6.64–67]. Subsequent direct vibrational excitation studies of the fluorescence from matrix-isolated CN⁻ at low temperatures have shown the importance of an efficient vibrational energy exchange process when the concentration exceeds $\simeq 0.05$ mole % which ultimately led to the first solid-state vibrational laser [6.67–70]. This same vibrational energy exchange process has been used to determine the excited-state vibrational lifetime of the CN⁻: Na⁺ center [6.13].

For low CN⁻ concentration at 1.7 K the time-resolved fluorescence signal from the $1 \to 0$ transition gives an excited state decay time of 27 ms; however, when the fundamental vibration of a high-concentration sample is pumped, the fluorescence signature is completely different. Figure 6.20 shows the emission spectrum for KBr + 0.12 mole % KCN + 0.2 mole % NaBr at 1.7 K. At this high CN⁻ concentration a series of sharp emission lines appear at frequencies below the pump frequency. These emission features result from the electric-dipole mediated V-V transfer process between CN⁻ molecules. This V-V transfer not only leads to population of highly excited vibrational states and isotope-shifted states of the

235

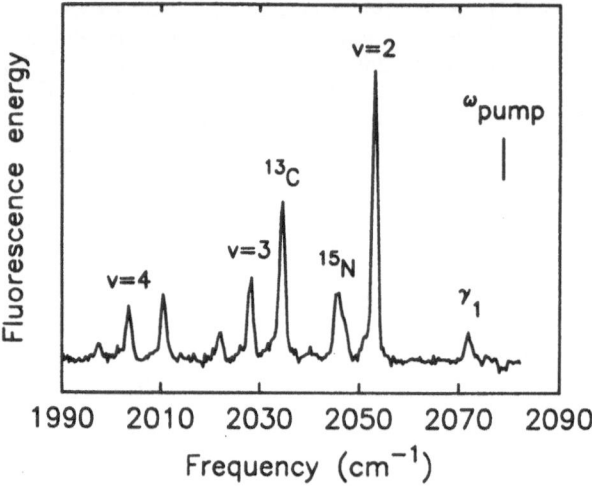

Fig. 6.20. Vibrational emission spectrum of CN$^-$ in KBr. The concentration is 0.12 mole % KCN + 0.2 mole % NaBr. (after [6.13])

CN$^-$ molecule but also to a large population in the first excited state of the CN$^-$: Na$^+$ center. To conserve energy in the transfer from a singly excited CN$^-$ donor to the CN$^-$: Na$^+$ acceptor a lattice phonon is emitted at a frequency equal to the energy difference. Since this energy deficit is much larger than the thermal energy kT, the back reaction is not allowed and the vibrational energy is funneled from the laser-pumped CN$^-$ molecules into the γ_1 centers.

Measurement of the time-resolved fluorescence signal from the γ_1 center and from the isolated ^{13}C^{14}N$^-$ center give lifetimes of 10 ms and 25 ms, respectively. The shorter lifetime for the γ_1 center could stem from its higher concentration so the best estimate of its excited state lifetime, T$_1$, is between 10 ms and 25 ms; hence, the vibrational mode lifetime of the CN$^-$ molecule is not radically changed by the presence of an impurity ion [6.13].

b) Hole-Burning and γ_1 Center Geometry

i) Polarization Studies. Polarized spectral hole-burning spectroscopy has established the crystallographic orientation of the γ_1 center in the lattice. The main results are summarized in Fig. 6.21. The top two traces for the probe beam show the γ_1 line before and after hole-burning with the laser beam propagating along a [001] axis and the probe and burn polarizations parallel to the [100] crystal axis. When the probe polarization is rotated by 90° to be perpendicular to the burn polarization as shown in the bottom two traces, no absorption change is observed. From these experimental results it is clear that the γ_1 center cannot be oriented along the <111> crystal axes which are the equilibrium directions for the molecule in the

236

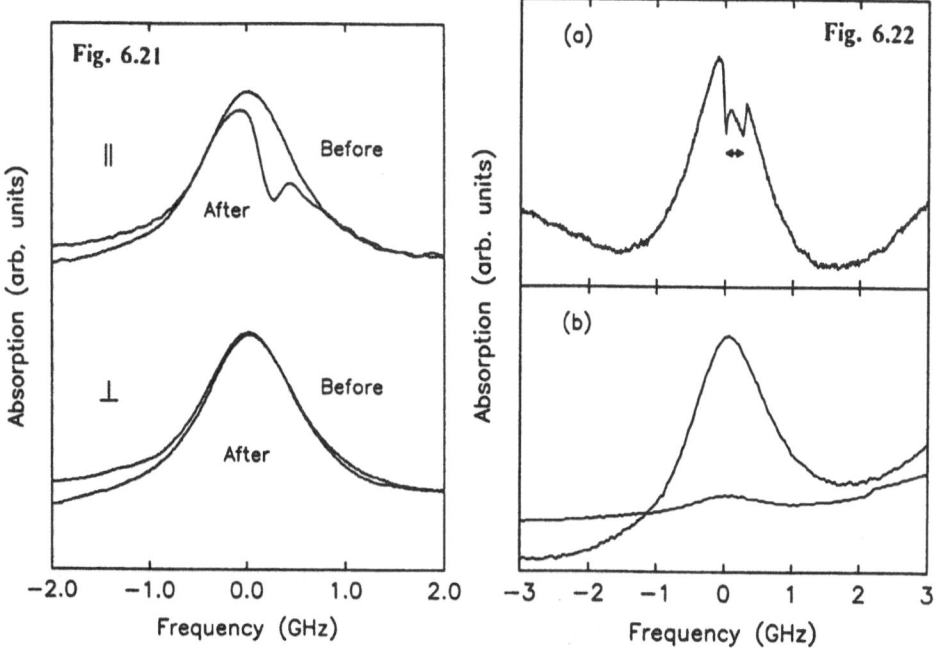

Fig. 6.21. Polarized hole-burning study of the γ_1 band. Top: Before and after a hole is burned with probe laser polarization parallel to burn polarization. Bottom: Probe perpendicular to burn polarization. The burn polarization is along the [100] axis. Sample temperature is 1.7 K. (after [6.12])

Fig. 6.22. Milling and erasing the γ_1 band. **(a)** Sweeping the diode frequency over a fixed frequency interval (arrow width) mills a slot. **(b)** Sweeping the diode frequency over 4 GHz for 80 min erases the entire γ_1 band. (after [6.13])

unperturbed lattice. The polarization results are consistent with the molecule being directed along one of the six $<100>$ axes with the particular axis determined by the location of the near-neighbor Na^+ impurity [6.13].

ii) Union of Broad-Band and Hole-Burning Techniques. The frequency shift of the hole-burned product state to a region outside of the γ_1 inhomogeneous line makes the product absorption difficult to locate directly because of the small continuous tuning range of a diode laser. The solution of this problem has been to use another unusual property of the γ_1 mode hole burning which is shown in Fig. 6.22. Figure 6.22a demonstrates that by sweeping the diode laser frequency over a fixed interval for about 2 min at an intensity of 500 mW/cm^2 a slot can be milled in the absorption line. If the sweeping interval is increased to include the entire absorption line then the inhomogeneous line is erased (Fig. 6.22b) after 80 min.

To locate the vibrational absorption of the product state the diode laser is used as the pump in conjunction with a FTIR interfer-

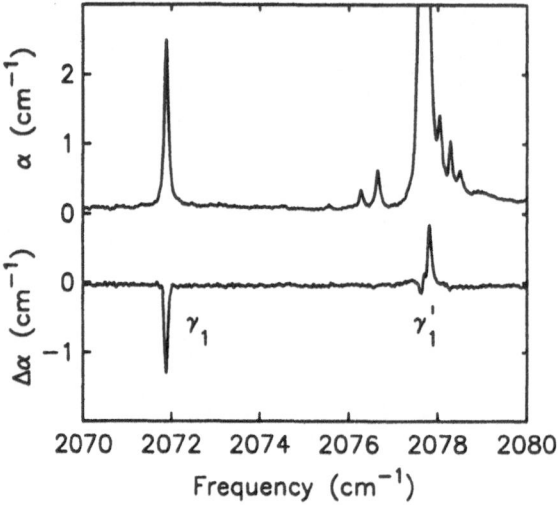

Fig. 6.23. Identification of the persistent hole product state, γ_1'. Top: The FTIR spectrum before hole-burning on the γ_1 band. Bottom: FTIR difference spectrum after most of the γ_1 band is erased by the diode. The product state, γ_1', has the same polarization as the hole burned in the γ_1 band. FTIR resolution is 0.04 cm^{-1} and the sample temperature is 1.7 K. (after [6.13])

ometer which provides a broad-band probe beam [6.13]. Since the FTIR resolution is only 0.04 cm^{-1} the diode is swept over the 0.04 cm^{-1} wide γ_1 line to burn away most of the absorption for one polarization. After erasing the γ_1 mode an FTIR spectrum over a 10 cm^{-1} range is obtained, then the sample is warmed to 50 K to recover the γ_1 mode, cooled back down to 1.7 K, and another spectrum obtained. The spectra are presented in Fig. 6.23: the top trace displays the FTIR absorption spectrum before hole-burning and the bottom trace shows the difference in absorption coefficients [γ_1(after) $-$ γ_1(before)]. Erasing the γ_1 mode produces the γ_1' mode at a higher frequency which is nearly coincident with that of the unperturbed CN$^-$ molecule; in addition, the polarization of the γ_1 and γ_1' modes is the same as are the integrated absorption strengths. Note that none of the other sharp CN$^-$ complex lines has changed so the particular geometrical arrangement for CN$^-$:Na$^+$ which produces the γ_1 mode is not the same as those which produce γ_2, γ_3 and γ_4 absorption lines. These latter spectral lines must be caused by centers with distinctly different locations of the Na$^+$ relative to the CN$^-$.

With the frequency position of the γ_1' mode identified it is now possible once again to use high-resolution diode-laser hole-burning to explore its properties [6.13]. After the γ_1' mode has been populated a persistent spectral hole (FWHM $\simeq$ 100 MHz) can be burned in it. It is also possible to sweep the diode frequency and burn away the entire absorption line with the oscillator strength returning to the

γ_1 mode. Another interesting property is that the entire γ_1' mode can be erased by tuning the diode anywhere in the main CN^- stretch region. This nonresonant erasing is a consequence of the efficient V-V transfer process which can not only transfer an excitation in the crystal but also at the appropriate place transfer from an isolated CN^- to the higher frequency γ_1' center because the energy deficit is less than kT.

iii) Two-Configuration Model for Hole Production. The simplest hole-burning model consistent with the polarization and photoproduct data is that the CN^- molecule rotates by 180° between two configurations [6.13]. Since the CN^- has a large elastic dipole moment it will preferentially align in the lattice to minimize the elastic energy. The CN^- molecular axis points at the nearest neighbor Na^+, and the four nearest neighbor K^+ ions in the plane perpendicular to the molecular axis relax inward to produce the large barrier to rotation. The two states correspond to the two equivalent directions for the elastic dipole axis with the inequivalence produced by the interaction of the CN^- electric dipole and the Na^+ impurity.

In the two-configuration model, which is drawn to scale in Fig. 6.24, only one state is populated at low temperature because the energy separation between the two configurations, $\varepsilon = 150$ K. If, at low temperature, some fraction of centers are put in the high energy state they remain there because of the large energy barrier, $V = 1500$ K, which prohibits classical reorientation and makes re-

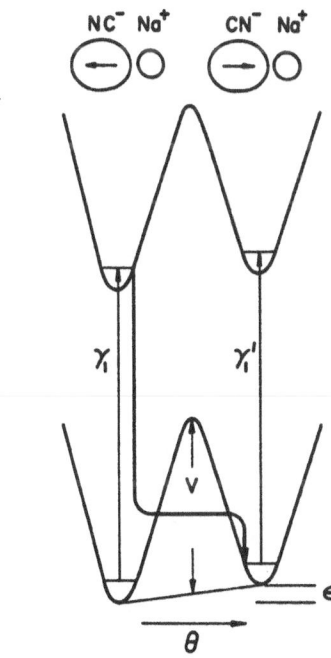

Fig. 6.24. Two-configuration diagram for the γ_1 and γ_1' centers. The measured energies are $\gamma_1 = 2071.83$ cm^{-1}, $\gamma_1' = 2077.75$ cm^{-1}, $V = 1040$ cm^{-1} and $\varepsilon = 104$ cm^{-1}. (after [6.13])

orientation by tunneling extremely unlikely. At high temperatures thermal activation over the barrier becomes possible, thermal equilibrium is restored and the relative population in the two configurations is governed by the Boltzmann factor.

The energy mismatch between the ground state energies in the two configurations has been obtained from high resolution broadband measurements of the absorption profile at elevated temperatures [6.13]. By 50 K all of the sharp lines near the stretch mode have melted into the characteristic P and R bands associated with free rotor behavior of the isolated CN^-; however, the γ_1 mode and the NCO^- modes remain distinct and narrow in width: the former because it is pinned by the Na^+ impurity and the latter because the long linear molecule is pinned by the lattice. The fact that NCO^- does not fluoresce while both the broad and sharp modes of CN^- do, clearly demonstrates that the vibrational mode lifetime cannot readily be estimated from the measured widths of spectral lines.

The γ_1' mode becomes observable in normal FTIR spectroscopy at about 50 K which is the same temperature at which holes in the γ_1 mode erase rapidly. Hole refilling data at elevated temperatures together with the classical Arrhenius relaxation approximation provide a convenient way for determining the barrier height shown in Fig. 6.24.

6.4.5 Other CN^- Complexes

The interesting behavior obtained for the CN^-: Na^+ center has not been duplicated when other cations are added to the host [6.13].

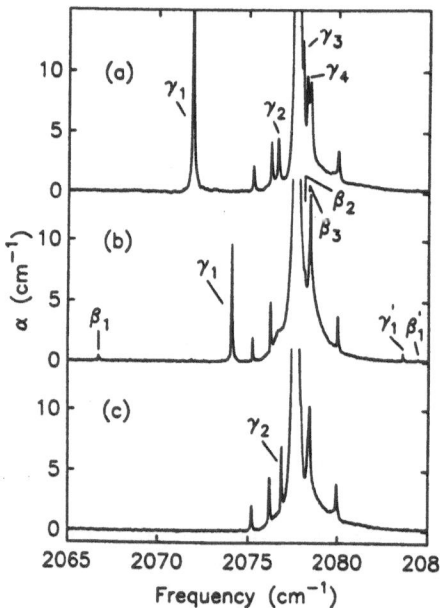

Fig. 6.25. Influence of alkali ion impurities on CN^- vibrational spectrum in KBr. (a) 0.12 mole % KCN + 0.2 mole % NaBr. The labeled lines involve the Na^+ ion and the unmarked ones are either pair modes or Cl^- perturbed modes (see Table 6.2 for frequencies). (b) 0.12 mole % KCN + 0.2 mole % LiBr. The labeled lines involve the Li^+ ion. (c) 0.12 mole % KCN + 0.2 mole % RbBr. The labeled line involves the Rb^+ ion. Sample temperature is 1.7 K and the resolution is 0.04 cm^{-1}. (after [6.13])

240

Figure 6.25 shows the influence of different alkali impurity ions on the CN^- stretch mode region. Since the same concentration of alkali dopant has been used in each case, the spectral features can be compared directly with the Na^+ case which is shown in Fig. 6.25a. Figure 6.25b indicates that the perturbations produced by double-doping with LiBr cause six new lines to appear. Sharp lines are observed at all temperatures and the line positions for 1.7 K are tabulated in Table 6.2. The γ lines involve CN^-: Li^+ and the β lines, CN^-: Li^+ (pairs). Double-doping with RbBr produces a single new line close to the main CN^- stretch mode frequency (Fig. 6.25c) which disappears by a temperature of 25 K.

None of the sharp spectral lines induced by Li^+ or Rb^+ show persistent spectral effects. This null result is surprising because both an alkali and a halide (Cl^-, see [6.13]) impurity ion have already been used successfully to obtain persistence and the two-configuration model illustrated in Fig. 6.24 should apply to all of these isoelectronic dopant cases. The fact that the CN^- ion in conjunction with monovalent ions in KBr may or may not give rise to persistent effects in the vibrational spectrum suggests that the physical process behind the hole-burning is not yet clearly identified.

6.5 Conclusion

6.5.1 Comparison of the Three Types of Vibrational Hole-Burning Systems

The low-power diode laser source, which in principle can be fabricated and tuned to cover any region of the vibrational infrared, is an effective hole-burning probe for the v_{17} mode of the *trans* conformation of DFE matrix-isolated in Ar, Kr or N_2. From the temperature dependence of the spectral holes together with some assumptions about the dephasing mechanism, an estimate of the vibrational excited state lifetime has been made while from polarization studies molecular tunneling or free rotation of the molecule in the host can be ruled out. Fixed-frequency CO_2 laser measurements have provided valuable additional information on the hole-burning properties of this molecule in Xe as well as in the hosts listed above. At a laser intensity which is about 10^4 times larger than obtained from the diode laser but still small compared to the values required to produce nonlinear effects, vibrational photochemical hole-burning in the form of *trans - gauche* conversion is observed for the Kr and Xe hosts but not for Ar. Finally, IR photophysical hole-burning has been found for three different molecular aggregates in Ar and N_2 matrices.

The holes and antiholes burned in the ReO_4^- v_3 vibrational mode have a variety of novel properties, some similar to the PIRSHs in the DFE system detailed in Sect. 6.2. Because both T_1 and T_2 have been determined for this vibrational mode from transient techniques, it has been possible to demonstrate, at least in this defect-lattice system, that the transient and persistent hole-burning spectral widths are the same. The persistent properties of this combination appear to involve reorientation of the impurity during IR excitation to the excited vibrational state. A photon-induced reorientation model accounts for most of the observed properties. This study demonstrates that persistence is possible even for a spherical-top molecule located in a cubic site. Ultrasonic measurements have shown that because of steric effects the two rotationally equivalent configurations for this tetrahederal molecule in a substitutional lattice site are separated by a large energy barrier and are made inequivalent by the local strain gradient at the molecular site.

A different kind of PIRSH production has been observed for CN^- complexes in KBr crystals. These complexes consist of the CN^- molecule together with an alkali or halogen ion impurity at a nearby lattice site. Of the fourteen new centers identified, six of them show persistent hole-burning when the perturbed CN^- stretch mode is probed with a low-power diode laser. One extremely stable defect complex: $CN^-:Na^+$, which both fluoresces and hole burns, has been investigated in some detail. By combining broad-band Fourier transform interferometry with narrow-band diode-laser hole-burning it has been demonstrated that this particular persistent effect occurs when, during vibrational de-excitation, the CN^- molecule flips by 180° between inequivalent energy configurations generated by the presence of the nearby Na^+ ion. For this system the measured hole width has no connection to the excited state lifetime, T_1, which is determined from a fluorescence measurement. This result is completely different from that found for the more complex ReO_4^- and DFE molecules and demonstrates that, in general, the hole width cannot be used to find the excited vibrational state lifetime. Although the persistent effect can no longer by itself be used to extract information about the decay process, it still provides a powerful high-resolution probe of the center and its environment.

6.5.2 Systems Which do not Exhibit PIRSH Formation

a) Derivatives of the CN^- Molecule

The identification of the essential components for persistent IR spectral hole-burning in solids requires cataloging those systems which do not produce PIRSH is well as those which do. Inspection of Table 6.2 shows that, for 21 molecular centers in alkali halide crystals in which diode-laser hole-burning has been attempted, PIRSHs have not been found in two-thirds of these cases. The absence of hole-burning for NCO⁻ can be rationalized on the grounds that the linear triatomic molecule is too long to fit easily into a substitutional lattice site and these large steric effects from the lattice inhibit the reorientation of the molecule between equivalent sites when it is excited by a vibrational quantum.

The isolated CN⁻ molecule represents the opposite limit. The molecule is small enough to tunnel rapidly between the equivalent {111} orientations so that the thermal ground state is composed from a symmetric combination of this configuration set. PIRSHs would not be expected to form unless another configuration set, say {100} orientations, was separated by a large energy barrier but nearby in ground state energy. The experimental results show that this is not the case for the KBr host.

Both the Na⁺ and Cl⁻ ions are effective in lifting this tunneling degeneracy for the CN⁻ molecule so that PIRSHs can occur while, on the other hand, PIRSHs are not found for crystals similarly doped with Li⁺ or Rb⁺. Given the simple geometric model described in Sect. 6.4 in which the CN⁻ molecule during vibrational de-excitation simply rotates by 180° with respect to the monatomic impurity site, this null result is quite surprising. Presumably, a transition from the excited vibrational and ground rotational state to a ground vibrational but excited rotational state in the first configuration puts the system in a region of low barrier height so that it can reach the second configuration. The picture should be the same whatever the monatomic impurity. The fact that CN⁻ ions in conjunction with monovalent impurity ions in KBr may or may not give rise to PIRSHs indicates that the physical process behind the cross-relaxation effect is not yet clearly defined.

Null results also have been obtained for another group of CN⁻ complexes in KCl [6.71]. When O_2^- is diffused into a previously CN⁻ doped KCl or KBr crystal the absorption lines above 2100 cm⁻¹ increase in strength indicating that some CN⁻ is converted to NCO⁻; in addition, a large number of new lines appear below 2100 cm⁻¹ in the CN⁻ stretch region indicating that CN⁻:O_2^- complexes have formed. The KCl:CN⁻ low temperature spectrum (before heating the sample in O_2 gas) is shown in Fig. 6.26a. The same crystal after heat treatment is shown in Fig. 6.26b. All of the new absorption lines which must stem from a variety of defect geometries have spectral widths (FWHM) of about 0.05 cm⁻¹. Although these absorption lines appear in the same frequency region as the γ absorption lines described in Sec. 6.4, PIRSHs have not been found for any of these new centers. Since the statics and dynamics of these centers are unknown the only conclusion which can be drawn from these measurements together with Table 6.2 is that most CN⁻ complexes do not produce persistent spectral holes.

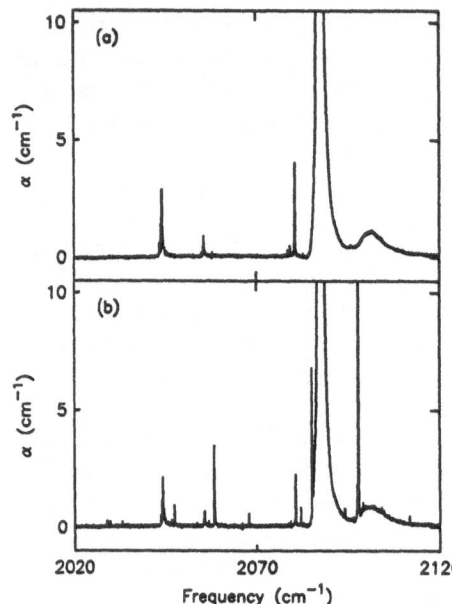

Fig. 6.26. Spectrum of oxygenated CN^- complexes in KCl (a) before diffusion of oxygen (b) after diffusion. (after [6.71])

b) Spherical-Top Molecules Which Contain Hydrogen

The systematic investigation of the transient and persistent hole-burning properties of the v_3 mode for ReO_4^- molecules in alkali halide crystals has produced a detailed understanding of the microscopic mechanisms responsible for the decay of the excited vibrational state and for the appearance of PIRSHs. The key component for the relaxation process is that the v_3 mode can decay into the low lying v_4 mode of this spherical-top molecule which is near in energy to the phonon spectrum so that a low-order multi-phonon process is probable.

A recent study [6.72,73] has focused on the vibrational relaxation of a different spherical top molecule, BH_4^-, so that the lowest frequency v_4 mode occurs in the CO_2 laser wavelength region. Since, at liquid helium temperature, the v_4 mode frequency is slightly above the 9R branch of the CO_2 laser and the BD_4^- mode is slightly below the 10P branch, both BH_4^- and BD_4^- molecules are diffused into the alkali halide at the same time. Because of isotope exchange, molecules of BH_3D^-, $BH_2D_2^-$, and BHD_3^- also appear in the crystal and the lower frequency modes of these molecules are in the CO_2 laser tuning range.

Of interest here is the v_{4b} vibrational mode of $BH_2D_2^-$ in KBr at low temperatures which is coincident with the strong CO_2 laser line, 10P22 (942 cm^{-1}). A comparison of the vibrational energy-level diagrams for ReO_4^- and $BH_2D_2^-$ is shown in Fig. 6.27. The lower-case letters in the figure identify the three non-degenerate v_4

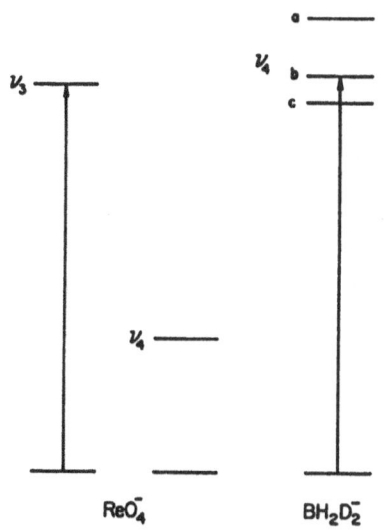

ν_3

ν_4 a ——

b —

c —

ν_4 ——

ReO$_4^-$ BH$_2$D$_2^-$

Fig. 6.27. Vibrational energy-level diagram of ReO$_4^-$ and BH$_2$D$_2^-$ at low frequencies. The three-fold degenerate ν_4 mode of ReO$_4^-$ which is important in the anharmonic decay of the three-fold degenerate ν_3 mode is indicated. The BH$_2$D$_2^-$ modes are nondegenerate. The arrows identify the CO$_2$ laser frequency used in the saturation experiment

modes of BH$_2$D$_2^-$, while both the ν_3 and ν_4 modes of ReO$_4^-$ are threefold degenerate. The arrow indicates the transition pumped in each system by the CO$_2$ laser. At low laser intensities no PIRSH formation is observed for the BH$_2$D$_2^-$ ν_{4b} mode; furthermore, at high laser intensities of 1 MW/cm^2 no bleaching of the absorption band is observed. These null results are to be contrasted with the positive results obtained for the ReO$_4^-$ ν_3 mode which have already been described in Sect. 6.3.

We now consider some of the possible reasons for this optically inert behavior. Because of the small moment of inertia for BH$_2$D$_2^-$, free rotation or rapid tunneling among equivalent molecular orientations could eliminate the PIRSH possibility; however, the absence of transient bleaching is not so easy to explain. The nonradiative vibrational relaxation time of both HCl and DCl matrix-isolated in Ar [6.21], which is consistent with a relaxation process in which the rate determining step is vibration-rotation transfer, is relatively long ($\simeq$ 100 μs) when compared to the relaxation time for BH$_2$D$_2^-$ estimated from the high intensity CO$_2$ laser measurement ($<$ 20 ps), yet in both cases the ratio of the mode frequency to the rotational constant is about the same.

Another difference between ReO$_4^-$ and BH$_2$D$_2^-$ is that the former has a very large mass (250 amu) and the latter a very small mass (15 amu) when compared to the lattice ions. Although the ν_{4b} mode is far removed from the normal phonon energies the small mass defect would be expected to produce a local mode with frequency above the normal phonon band mode region. If the ν_{4b} decay channel made use of this high-frequency external mode then the decay would occur in lower order, and hence be faster. Such a local mode has been observed for BH$_4^-$ in RbI [6.74]. If this feature represents the important decay channel then four local modes plus one band mode are required to relax the excited vibrational state. This is the same order of multi-phonon decay that is required to explain the ReO_4^- data in Sect. 6.3 so the two relaxation times should be comparable. The experimental results show that the v_{4b} mode must relax at least three orders of magnitude faster than does the ReO_4^- vibration.

An important difference between the two systems is that a Fermi resonance exists between $2v_4$ and v_3 for the $BH_2D_2^-$ molecule [6.74] but not for the ReO_4^- system. Because of the strong coupling between these two vibrational ladders the high laser intensity may simply populate both ladders so that saturation would be much harder to attain; however, a search in the infrared and visible has shown that no vibrational fluorescence occurs for this defect. In order to account for these data both a Fermi resonance and either a free-rotor-like degree of freedom or a local lattice mode are required to account for the rapid decay. The local mode or rotor states insure that the decay process is as fast as ReO_4^- ($\simeq 5$ nsec) and the Fermi resonance insures that a large reservoir of levels is available to suppress bleaching. Since the vibrational radiative lifetime is $\simeq 1$ ms, fluorescence is completely quenched by the much faster nonradiative decay channels.

6.5.3 Future Prospects

a) NO_2^- in Alkali Halides

The nitrite molecule has a bent O-N-O structure with C_{2v} symmetry, and optical and infrared measurements [6.75 – 80] have determined that when it is doped into the alkali halide crystal the symmetry axis of the NO_2^- molecule lies in a <110> direction with the molecular plane in a (100) plane. The vibrational modes of this asymmetric top are: v_1 (the high-frequency symmetric stretch), v_2 (the low-frequency bending mode), and v_3 (the intermediate-frequency asymmetric stretch). For KCl, KBr and RbCl hosts molecular energy levels which correspond to free rotation, libration, and tunneling have been identified [6.75 – 79]. The KI results are very different probably because the molecule occupies an off-center position in the lattice so that translation and rotation are strongly coupled together. It has been proposed that KI:NO_2^- may represent the most general case for motion of molecules in solid solution [6.77].

A preliminary report has appeared which describes PIRSH production in KBr:NO_2^- [6.81]. Probing the vibrational modes with tunable diode lasers reveals that PIRSH's occur in two of the three modes. The results are that: a) the v_1 mode does not burn, b) the v_3 mode burns with a hole width $\simeq 100$ MHz, and c) both the zero-libron line and the sidebands of v_2 burn with hole widths of

350 MHz and 1 GHz, respectively. Additional experiments have shown that PIRSHs appear in the same modes for KCl and KI hosts as well. In all cases the missing oscillator strength relocates to a different polarization inside the inhomogeneous linewidth, unlike either the spherical-top, ReO_4^-, or the CN^-:Na^+ center. Similar studies may be possible on the isomorphic defect, PO_2^-.

The observation of PIRSHs in the internal modes of KI:NO_2^- opens up the possibility of generating holes in the external vibrational mode spectrum produced by the defect. Impurity-induced spectral features, which occur in the gap of the lattice vibration frequencies for KI, have been attributed to translational modes of the NO_2^- ion in its cavity, high-frequency librational modes, or perturbed lattice modes of the host crystal [6.37]. By pumping one of the internal modes of the molecule with a diode laser the molecule can be transferred from one elastically polarized configuration to another. Since this new configuration will have its own polarized external mode spectrum it may be possible to produce persistent holes in the far infrared by burning in the IR. Generation of such far IR spectra would provide the information required to properly interpret the present gap mode results and also to determine if external modes are inhomogeneously broadened.

b) Disordered Solids

The burning of PIRSHs in the modes of molecular impurities in alkali halides indicates that similar effects should be observable with some molecules embedded in any crystal lattice. In a more general sense one need not rely only on the molecule to produce more than one ground state configuration to support PIRSH dynamics since it is the total system – defect plus lattice – which is important. For example, in glasses it is known that at low energies the dynamics are dominated by the two-level tunneling states [6.82]. If the vibration of an impurity molecule in a glass is excited with a diode laser and the two-level states (TLS) of the host change during the IR transition, once again a persistent hole could appear. Measurements on impurity electronic transitions in organic glasses have already demonstrated that visible energy photons cause such a process (See Chap. 5 and [6.3]).

Once it is realized that either reorientation of the impurity molecule and/or rearrangement of the host can give rise to persistent hole-burning features in the vibrational spectrum then complex molecular solids which have a large number of low-lying conformational states should be found particularly susceptible to this high-resolution technique. Organic polymers appear to satisfy all of the necessary conditions. If holes could be produced in these materials then one would have high-resolution information about the vibrational dynamics at specific polymer sites, information which is currently buried in the inhomogeneously broadened spectrum.

Acknowledgements

The authors acknowledge the fruitful collaboration over the years with W. P. Ambrose, A. R. Chraplyvy, T. R. Gosnell, H. Lengfellner, and R. Spitzer on some of the results reported in this chapter. The authors also thank H. H. Günthard for permission to reprint Figs. 6.2 – 6.5 and V. Narayanamurti for permission to reprint Fig. 6.16. This work has been supported in part by U.S. Army Research Office Grants DAAG-29-79-C-0170, DAAG-29-83-004, and DAAG-29-84-0034; by National Science Foundation Grants 79-24008 and DMR-80-08546; by National Science Foundation Grant No. DMR-76-81083 through the Cornell Materials Science Center; and by the U. S. Office of Naval Research.

References

6.1   J. M. Hayes, R. P. Stout, G. J. Small: J. Chem. Phys. **74**, 4266 (1981)
6.2   F. G. Patterson, H. W. H. Lee, R. W. Olson, M. D. Fayer: Chem. Phys. Lett. **84**, 59 (1981)
6.3   G. J. Small: "Persistent Nonphotochemical Hole Burning and the Dephasing of Impurity Electronic Transitions in Organic Glasses", in *Spectroscopy and Excitation Dynamics of Condensed Molecular Systems*, ed. by V. M. Agranovitch and R. M. Hochstrasser, (North-Holland, Amsterdam 1983), pp. 515-554
6.4   J. T. Yardley: "Dynamic Properties of Electronically Excited Molecules" in *Chemical and Biochemical Applications of Lasers*, Vol. I, ed. by C. B. Moore (Academic Press, New York 1974), pp. 231-279
6.5   See for example, M. Poliakoff, N. Breedon, B. Davies, A. McNeish, J. J. Turner: Chem. Phys. Lett. **56**, 474 (1978)
6.6   M. Dubs, H. H. Günthard: Chem. Phys. Lett. **64**, 105-107 (1979)
6.7   M. Dubs, H. H. Günthard: J. Mol. Struct. **60**, 311-316 (1980)
6.8   M. Dubs, L. Ermanni, H. H. Günthard: J. Mol. Spectr. **91**, 458-491 (1982)
6.9   W. E. Moerner, A. J. Sievers, R. H. Silsbee, A. R. Chraplyvy, D. K. Lambert: Phys. Rev. Lett. **49**, 398 (1982)
6.10 W. E. Moerner, A. R. Chraplyvy, A. J. Sievers, R. H. Silsbee: Phys. Rev. B **28**, 7244 (1983); Phys. Rev. B **29**, 4791 (1984)
6.11 T. R. Gosnell, A. J. Sievers, R. H. Silsbee: Phys. Rev. Lett. **52**, 303 (1984)
6.12 R. C. Spitzer, W. P. Ambrose, A. J. Sievers: Opt. Lett. **11**, 428-430 (1986)
6.13 R. C. Spitzer, W. P. Ambrose, A. J. Sievers: Phys. Rev. B **34**, 7307 (1986)
6.14 See for example, *Vibrational Spectroscopy of Trapped Species*, ed. by H. E. Hallam (Wiley, New York 1973)
6.15 B. M. Chadwick: "Matrix Isolation", in *Molecular Spectroscopy*, ed. by R. F. Barrow, D. A. Long, D. J. Millen (The Chemical Society, London 1975), p. 281ff
6.16 See for example, *Cryochemistry*, ed. by M. Moskovits and G. A. Ozin (Wiley, New York 1976)
6.17 B. I. Swanson, L. H. Jones: "High Resolution Infrared Studies of Site Structure and Dynamics for Matrix-Isolated Molecules" in *Vibrational Spectra and Structure, A Series of Advances*, Vol. 12, ed. by J. R. Durig (Elsevier, Amsterdam 1983) pp. 1-67
6.18 H. Dubost. L. Abouaf-Marguin, F. Legay: Phys. Rev. Lett. **29**, 145 (1972)
6.19 F. Legay: "Vibrational Relaxation in Matrices", in *Chemical and Biochemical Applications of Lasers*, Vol. II, ed. by C. B. Moore (Academic, New York 1977) pp. 43-86

6.20 H. Dubost: Ber. Bunsenges. Phys. Chem. **82**, 112 (1978)
6.21 J. M. Wiesenfeld, C. B. Moore: J. Chem. Phys. **70**, 930-946 (1979)
6.22 H. Dubost, R. Charneau: Chem. Phys. **41**, 329 (1979)
6.23 H. Dubost, A. Lecuyer, R. Charneau: Chem. Phys. Lett. **66**, 191 (1979)
6.24 L. Abouaf-Marguin, B. Gauthier-Roy: Chem. Phys. **51**, 213 (1980)
6.25 B. Gauthier-Roy, L. Abouaf-Marguin, F. Legay: Chem. Phys. **46**, 31 (1980)
6.26 L. Abouaf-Marguin, P. Boissel, B. Gauthier-Roy: J. Chem. Phys. **75**, 495 (1981)
6.27 H. Dubost: "Spectroscopy of Vibrational and Rotational Levels of Diatomic Molecules in Rare Gas Crystals" in *Inert Gases: Potentials, Dynamics, Energy Transfer in Doped Crystals*, ed. by M. L. Klein (Springer Verlag, New York 1984), pp. 145-257
6.28 M. Hartig, H. Dubost: J. Lumin. **24/25**, 643 (1981)
6.29 M. Hartig, H. Dubost: Phys. Rev. Lett. **49**, 715 (1982)
6.30 I. Maslakowez: Z. Physik **51**, 696 (1928)
6.31 J. C. Decius, A. Maki: J. Chem. Phys. **28**, 1003 (1958)
6.32 W. C. Price, W. F. Sherman, G. R. Wilkinson: Proc. Roy. Soc. **A255**, 5 (1960)
6.33 W. C. Price, W. F. Sherman, G. R. Wilkinson: Spectrochim. Acta **16**, 663 (1960)
6.34 F. Lüty: J. de Physique, Coll. **C4**, 120 (1967)
6.35 V. Narayanamurti, R. O. Pohl: Rev. Mod. Phys. **42**, 201 (1970)
6.36 W. F. Sherman, G. R. Wilkinson: 'Infrared and Raman Studies on the Vibrational Spectra of Impurities in Ionic and Covalent Crystals', in *Vibrational Spectroscopy of Trapped Species*, ed. by H. E. Hallam (Wiley, New York 1973), pp. 246-318
6.37 A. S. Barker, A. J. Sievers: Rev. Mod. Phys. **47**, Supp. 2, 1-179 (1975)
6.38 R. K. Ahrenkiel, J. F. Figueira, C. R. Phipps, D. J. Dunlavy, S. J. Thomas, A. J. Sievers: Appl. Phys. Lett. **33**, 705-707 (1978)
6.39 T. Gethins: Can. J. Phys. **48**, 580 (1970)
6.40 H. W. Moos: J. Lumin. **1-2**, 106 (1970)
6.41 A. J. Sievers: Cryst. Latt. Def. and Amorph. Mat. **12**, 441 (1985)
6.42 A. R. Chraplyvy, A. J. Sievers: Opt. Lett. **3**, 112 (1978)
6.43 A. R. Chraplyvy, W. E. Moerner, A. J. Sievers: Opt. Lett. **6**, 254 (1981)
6.44 A. R. Chraplyvy, W. E. Moerner, A. J. Sievers: Opt. Lett. **6**, 431 (1981)
6.45 W. E. Moerner, A. J. Sievers, A. R. Chraplyvy: Phys. Rev. Lett. **47**, 1082 (1981)
6.46 W. E. Moerner, A. R. Chraplyvy, A. J. Sievers: Phys. Rev. B **29**, 6694 (1984)
6.47 P. Huber-Walchli, H. H. Günthard: Chem. Phys. Lett. **30**, 347-351 (1975)
6.48 M. Poliakoff, J. J. Turner: 'Infrared Laser Photochemistry in Matrices', in *Chemical and Biochemical Applications of Lasers*, Vol. 5, ed. by C. Bradley Moore (Academic, New York 1980), pp. 175-216
6.49 H. Frei, L. Fredin, G. C. Pimentel: J. Chem. Phys. **74**, 397-411 (1981); H. Frei, G. C. Pimentel: J. Chem. Phys. **78**, 3698 (1983)
6.50 P. Felder, H. H. Günthard: Chem. Phys. Lett. **88**, 473-476 (1982)
6.51 P. Felder, H. H. Günthard: Chem. Phys. **85**, 1 (1984)
6.52 W. E. Moerner: Ph.D. Thesis, Cornell University, Ithaca, New York 1982 (unpublished)
6.53 M. R. Mohammad, W. F. Sherman: J. Phys. C **14**, 283 (1981)
6.54 H. Lengfellner, A. J. Sievers: Phys. Rev. B **31**, 2591 (1985)
6.55 H. Lengfellner, T. R. Gosnell, R. Tkach, A. J. Sievers: Appl. Phys. Lett. **43**, 437 (1983)
6.56 A. L. Huston, W. E. Moerner: J. Opt. Soc. Amer. **B1**, 349 (1984)
6.57 W. E. Moerner, A. L. Huston: Appl. Phys. Lett. **48**, 1181 (1986)
6.58 W. D. Seward, V. Narayanamurti: Phys. Rev. **148**, 463 (1966)
6.59 G. R. Field, W. F. Sherman: J. Chem. Phys. **47**, 2378 (1967)
6.60 M. A. Cundill, W. F. Sherman: Phys. Rev. **168**, 1007 (1968)

6.61 R. Callender, P. Pershan: Phys. Rev. A **2**, 672 (1970)
6.62 F. Lüty: Phys. Rev. B **10**, 3677 (1974)
6.63 H. U. Beyeler: Phys. Rev. B **11**, 3078 (1975)
6.64 Y. Yang, F. Lüty: Phys. Rev. Lett. **51**, 419 (1983)
6.65 K. P. Koch, Y. Yang, F. Lüty: Phys. Rev. B **29**, 5840 (1984)
6.66 T. R. Gosnell, R. W. Tkach, A. J. Sievers: Solid State Commun. **53**, 419-421 (1985)
6.67 R. W. Tkach, T. R. Gosnell, A. J. Sievers: Opt. Lett. **9**, 122 (1984)
6.68 T. R. Gosnell, R. W. Tkach, A. J. Sievers: J. Lumin. **31-32**, 166 (1984)
6.69 T. R. Gosnell, R. W. Tkach, A. J. Sievers: Infrared Phys. **25**, 35 (1985)
6.70 T. R. Gosnell, A. J. Sievers, C. R. Pollock: Opt. Lett. **10**, 125 (1985)
6.71 R. C. Spitzer, Ph.D. Thesis, Cornell University, 1987 (unpublished)
6.72 D. M. Kammen, T. R. Gosnell, R. W. Tkach, A. J. Sievers: Bull. Amer. Phys. Soc. **29**, 502 (1984)
6.73 D. M. Kammen, T. R. Gosnell, R. W. Tkach, A. J. Sievers: J. Chem. Phys. **87**, 4371 (1987)
6.74 M. I. Memon, W. F. Sherman, G. R. Wilkinson: Spectrochim. Acta, **A37**, 461 (1981)
6.75 T. Timusk, W. Staude: Phys. Rev. Lett. **13**, 373 (1964)
6.76 C. D. Lytle, A. J. Sievers: Phys. Rev. Lett. **14**, 271-273 (1965)
6.77 V. Narayanamurti, W. D. Seward, R. O. Pohl: Phys. Rev. **148**, 481-494 (1966)
6.78 R. Avarmaa, L. Rebane: Phys. Status Solidi **35**, 107 (1969)
6.79 A. R. Evans, D. B. Fitchen: Phys. Rev. B **2**, 1074 (1970)
6.80 K. K. Rebane, L. A. Rebane: Pure and Appl. Chem. **37**, 161-181 (1974)
6.81 W. P. Ambrose, A. J. Sievers: Bull. Amer. Phys. Soc. **31**, 278 (1986)
6.82 See for example, *Amorphous Solids*, ed. by W. A. Phillips, Topics Curr. Phys., Vol. 24 (Springer, Berlin, Heidelberg 1981)

7. Frequency Domain Optical Storage and Other Applications of Persistent Spectral Hole-Burning

W. E. Moerner, W. Lenth, and G. C. Bjorklund

With 32 Figures

This chapter describes several possible applications of persistent spectral hole-burning (PSHB) with particular emphasis on the application that has received the most attention to date, frequency domain optical storage (FDOS). Engineering and systems issues influencing the form a FDOS system might take are summarized, and optimal properties of single-photon materials are derived. The limitations of monophotonic mechanisms underscore the need for photon-gated processes, and the examples of photon gating known to date are summarized. Other methods of organizing an optical storage system based on PSHB are also described, including holographic, time domain, and electric field schemes. Applications of PSHB to spectral filtering and optical waveform processing are briefly mentioned.

7.1 Introduction

The previous chapters have described the fundamental physical and chemical changes that can lead to the formation of persistent spectral holes in inhomogeneously broadened optical or infrared transitions in solids at low temperatures. Persistent spectral hole-burning (PSHB) has proven to be a powerful tool for high-resolution spectroscopy of impurities in solids and can be used to study dephasing mechanisms, microenvironments in crystals, amorphous host dynamics, low-temperature solid-state photochemistry, and microscopic perturbations due to external fields. At the same time, PSHB has the potential for technological applications to optical data storage, pulse shaping, and optical signal processing. This has resulted in engineering and materials research on PSHB at a variety of laboratories around the world.

In this chapter, several potential technological applications of persistent spectral hole-burning will be reviewed. The application that has received the most attention so far is based on the concept of using persistent spectral hole-burning to form a frequency domain optical storage (FDOS) system. In FDOS, the optical frequency or wavelength at which holes are burned is used to encode digital information where, for instance, the presence of a hole at a particular optical frequency may be used to encode a digital "1" and the absence of a hole a digital "0". Figure 7.1 shows an example of this

251

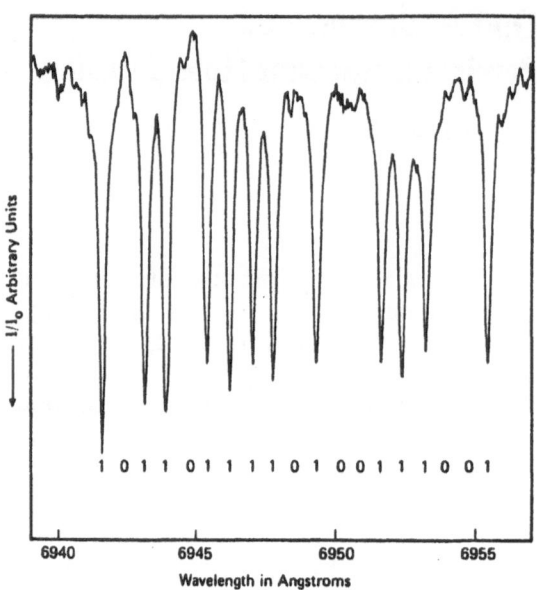

Fig. 7.1. A sequence of spectral holes burned in the inhomogeneously broadened 0-0 absorption band of free-base phthalocyanine in PMMA at 4.2 K. A possible decoding of the hole pattern into binary data is shown. (after [7.6])

type of digital encoding in the frequency domain for the case of free-base phthalocyanine in poly(methyl methacrylate) (PMMA). Due to the additional use of the frequency dimension, multiple bits can be stored in one laser focal volume, yielding areal densities several orders of magnitude higher than those for conventional optical or magnetic storage. Moreover, such a storage system does not require rotation of the storage medium and the use of complex voice-coil operated actuators. Thus, FDOS can offer considerable improvement in data random access time compared to either magnetic or optical rotating-disk technologies. Section 7.2 will consider configurations that can be used for data storage if frequency domain readout of the spectral holes is employed. Section 7.3 will review materials research efforts for FDOS, with particular emphasis on the mechanisms and microscopic dynamics required for a practical storage system with frequency-domain readout. Section 7.4 will briefly review configurations for FDOS that utilize alternative schemes for data readout, such as time-domain, electric-field, or holographic techniques. Section 7.5 will summarize other proposed applications of PSHB.

252

7.2 Systems Issues for Frequency Domain Optical Storage

7.2.1 General Remarks

Conventional laser-disk optical recording consists of focusing a laser beam to a diffraction-limited spot $\simeq 1~\mu$m in diameter on a recording material and writing a single bit of information in the volume irradiated by the laser by ablative, magneto-optic, or other means [7.1]. Areal storage densities can approach 10^8 bits/cm^2, fundamentally limited by optical diffraction. The recording material is rotated under an actuator, and a voice-coil moves a recording head in and out along a radius of the rotating disk. Such a rotating storage technology can suffer from two performance limitations on data access: latency and seek time. Latency refers to the time required on average to reach data stored on the opposite side of the rotating disk, and this number is in the $10-30$ ms range, depending predominantly upon the rotation rate and disk diameter. Seek time refers to the delay incurred when the voice-coil actuator must move in or out to find the actual track on which the desired data is stored. Both latency and seek time determine the random access performance of this type of data storage.

For future data storage technologies it is desirable to improve both storage density and data access time. FDOS seeks to use the phenomenon of persistent spectral hole-burning to achieve improvements in both areal density and random access performance. As described in Chap. 1 of this book, PSHB consists of particular photoinduced changes that can occur within inhomogeneously broadened lines of absorbing centers in solids at low temperatures. In one laser focal volume, many groups of absorbing centers are available, and different groups of centers can be selectively excited by tuning the laser frequency. Under these conditions, 1000 or 10,000 bits can, in principle, be stored in the frequency domain in the volume illuminated by a focused laser beam, offering the potential for storage densities as high as $10^{11}-10^{12}$ bits/cm^2. However, as will be explained in Sects. 7.2.1 and 7.3.2, materials and focusing optics considerations limit practical spot diameters to $\geq 10~\mu$m, resulting in areal densities in the 10^9-10^{10} bits/cm^2 range. Since the frequencies at which holes are burned add a new dimension for addressing data, this type of storage scheme is called "frequency domain optical storage". The emission wavelength of a semiconductor diode laser can be tuned rapidly, so that very high data rates at a given spatial location can be achieved. In addition, very fast random spatial access can be accomplished without rotation of the storage medium by using angular beam deflection techniques. The realization of a practical FDOS system represents an exciting application of laser spectroscopy of the solid state that would utilize many of the unique properties of laser radiation: tunability, spatial coherence, and high spectral resolution.

The idea of using transient saturation spectral holes for data storage was patented by *Szabo* in 1975 [7.2]. Since the holes and thus the data disappear within the excited state lifetime, such a storage system would involve excessive refresh. In order to configure a practical direct-access storage device, it is essential to consider materials in which the photoinduced change persists on time scales of months or years, as is often the case with PSHB. Usually this means that the mechanism for PSHB is photochemical, with high barriers to the reverse reaction at low temperatures, although non-photochemical mechanisms can also have long lifetimes (See Chaps. 3 and 5). The concept of using persistent spectral holes with lifetimes much longer than any excited state lifetime for FDOS appeared in a patent by *Castro* et al. [7.3] in 1978. Over the past few years, a number of review articles have appeared that relate to materials and configurations for FDOS [7.4–9].

The properties of the recording materials naturally place certain constraints on the system configuration for FDOS. At the same time, the system configuration places certain requirements on the recording material, and it must be capable of delivering the required signal-to-noise in an appropriate bandwidth for practical data storage and retrieval. At the present time, in spite of considerable progress no material exists that displays all the characteristics required for reliable frequency domain data storage, and materials research is actively being pursued in this area. In the rest of this section, several critical engineering considerations and system configurations will be discussed, without describing the details of the recording material. Then in Sect. 7.3 the central materials issues and their partial solutions will be described. It is a stimulating and interdisciplinary challenge for the solid-state spectroscopist, photochemist, recording engineer, and laser physicist to find suitable materials and devise recording configurations that make this application of high-resolution laser spectroscopy of solids possible.

Figure 7.2 illustrates one way to organize a frequency domain optical store. The goal from the outset is to configure a storage system that does not require rotation of the recording material. Starting at the top of the figure, a sequence of "1's" and "0's" has been burned into an inhomogeneously broadened line within one laser focal volume. Here it is assumed for definiteness that 1000 bits can be written in the frequency domain, although this number may vary depending upon the material. Furthermore, 10 μm diameter laser spots are assumed, because maintaining 1 μm diameter focused spots over a large field of view would require expensive and impractical optics. In addition, 1 μm diameter spots would also place unnecessarily severe constraints on the recording material (see Sect. 7.3.2). The real challenge as far as large areal densities are concerned is to access the various bits at the highest possible rate in a practical manner. PSHB addresses this challenge by utilizing fast spatial ac-

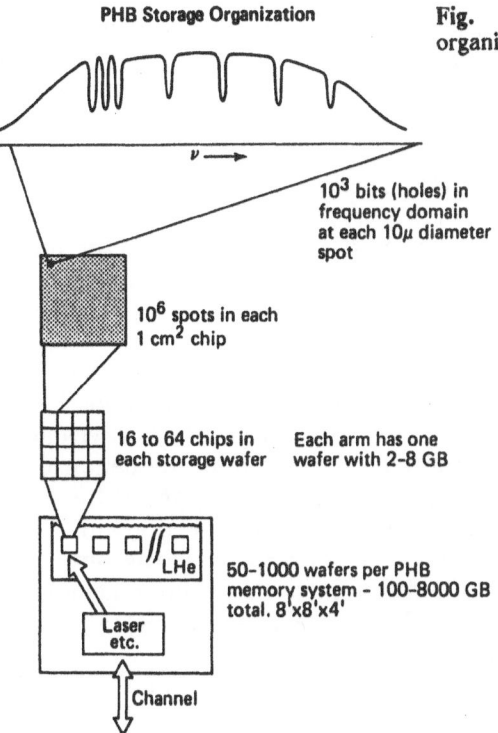

Fig. 7.2. Sketch of the overall organization of a PSHB storage system

10^3 bits (holes) in frequency domain at each 10μ diameter spot

10^6 spots in each 1 cm^2 chip

16 to 64 chips in each storage wafer

Each arm has one wafer with 2–8 GB

LHe

50–1000 wafers per PHB memory system – 100–8000 GB total. 8'x8'x4'

Laser etc.

Channel

cess based on angular beam deflection and fast spectral data access using laser frequency tuning. This high-speed access scheme, which does not involve rotation of the recording medium, reduces seek time and latency.

On one "chip" in the storage system, any one of a large number of laser spots (or focal volumes) could be accessed in the x-y plane using simple beam deflection. Sixteen to 64 chips could be arranged in a rectangular pattern to form one arm of the FDOS system, which would contain from $2-8$ gigabytes of storage. All of the chips in the arm might be accessed in parallel, using a bundle of laser beams. Of particular importance is the fact that moving the bundle of beams to a particular spot on each of the chips in the arm can be done using a pair of mirrors mounted on galvanometers. Galvanometer access can easily deliver fast (2 ms) seek times, which is $10-20$ times faster than access via voice coils. One limitation of angular beam deflection in general is that only a limited angular range (roughly 1000x1000 spots) can be accessed with a single pair of x-y actuators. This is where use of the additional frequency dimension produces a substantial advantage: with PSHB, $1000\text{x}1000\text{x}1000 = 10^9$ bits can be accessed with one x-y galvanometer pair and tuning of the laser frequency. Without the use of the frequency dimension, only 10^6 bits can be accessed by each x-y galvanometer pair and conse-

255

quently, a galvanometer-accessed, 1-bit-per-spot storage system such as magneto-optic recording would require too many actuators and would therefore be excessively costly. Similar considerations apply if an x-y acousto-optic deflector is used in a FDOS system: 1000x1000 angular positions can be accessed with 25 μs random access time, so that by using PSHB, 1 Gbit of data can be reached with one actuator having very small seek time.

The reading and writing laser should be a semiconductor diode laser because these devices are small, reliable, and inexpensive, and have the performance characteristics needed, i.e., single-mode operation, narrow linewidth, and rapid current-tunability. At present, single-mode GaAlAs diode lasers operating in the range $780-850$ nm have received the most development, but single-mode diode lasers are rapidly becoming commercially available in nearby wavelength ranges. The laser would be repetitively scanned in frequency over the inhomogeneously broadened line in a sawtooth fashion for writing and reading of the information in the frequency domain. The readout process then becomes a problem of laser spectroscopy: detection of the spectral features written into the inhomogeneously broadened line with sufficient signal-to-noise ratio (SNR) with low laser power in a short time period.

The recording material has to be kept in a liquid-helium cryostat at temperatures ≤ 10 K. This is essential in order to keep the absorption line inhomogeneously broadened. Zero-phonon lines broaden with temperature due to phonon scattering (Raman) processes and other phonon-assisted dephasing mechanisms, and at elevated temperatures the homogeneous width can become larger than the inhomogeneous width. (See Chaps. 3 and 5 for a detailed discussion of mechanisms determining the temperature dependence of the hole width.) Furthermore, for most materials the pattern of spectral holes is erased when the material is warmed up above a critical temperature. This can occur due to two effects: reversal of the photoinduced change that led to hole formation, or irreversible relaxation of the strain distribution that defines the inhomogeneous line. For this reason, the cryostat design for FDOS should be fairly conservative. Large-volume cryostats are available today that have boil-off times of a week or even a month in the absence of all electrical power [7.10]. On the other hand, some newly discovered materials have the property of retaining the pattern of spectral holes after cycling to room temperature (See Sect. 4.4.2 and [7.11]). However, low temperatures are still necessary for reading and writing of the spectral holes in all cases.

The schematic FDOS organization of Fig. 7.2 could be implemented as shown in Fig. 7.3. Only the wafers of the hole-burning material are kept at low temperature in a liquid-helium cryostat. The reading and writing laser beams could be brought into and out of the liquid helium bath through cryostat windows or fiber optics.

256

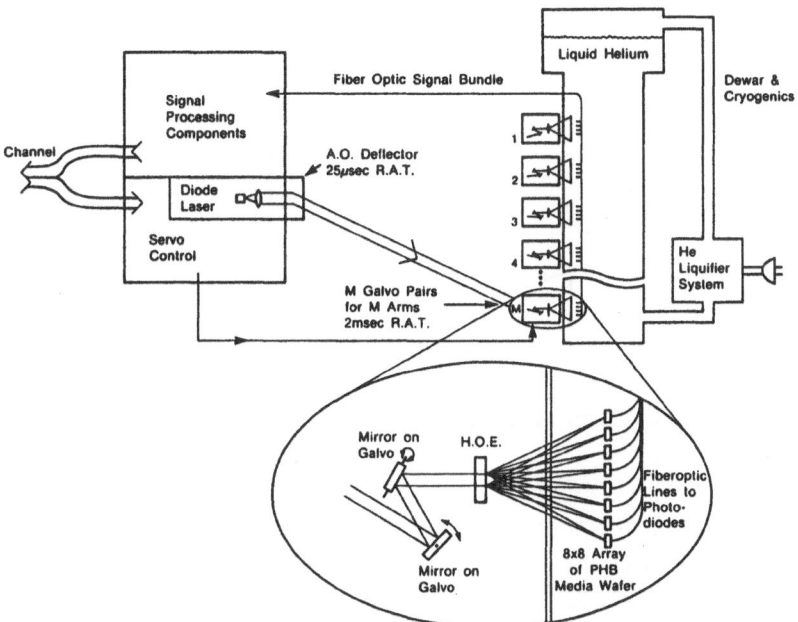

Fig. 7.3. Schematic of a PSHB storage system. Frequency controlled light from an array of diode lasers, each propagating at slightly different angles, is deflected acousto-optically to one of M arms. Each arm contains a pair of galvanometers for X-Y deflection, and a holographic optical element which divides the light from the individual diodes and focuses each beam onto PSHB media wafers. Fiber-optic lines collect the transmitted light from each wafer and carry the signals out of the cryostat for processing

Spatial data access is provided by an x-y deflection system consisting of a combination of acousto-optic deflectors and galvanometers, which permits fast random access to a large data volume. The system consists of several arms, which could be addressed in parallel by using several laser devices or optical beam-splitting to produce several laser beams. Such channel multiplexing can enable fast data access, and requires independent modulation capability for each data channel.

Figure 7.4 illustrates the basic timing of writing and reading. The laser frequency is tuned rapidly over the inhomogeneously broadened transition. The time allotted for reading and writing each bit is an important parameter determining the data rate of the system; 30 ns/bit is chosen to match the data rate of conventional magnetic and optical recording technologies. For 1000 bits in the frequency domain, the complete laser scan over the entire line requires 30 μs. This ramping of the laser frequency forms a clock. For writing, the laser power might be increased by an external modulator at those instants of time (i.e., frequencies) at which holes are to be burned. If a photon-gated material (see Sect. 7.3.3) is utilized, the gating light source would also have to be present to

257

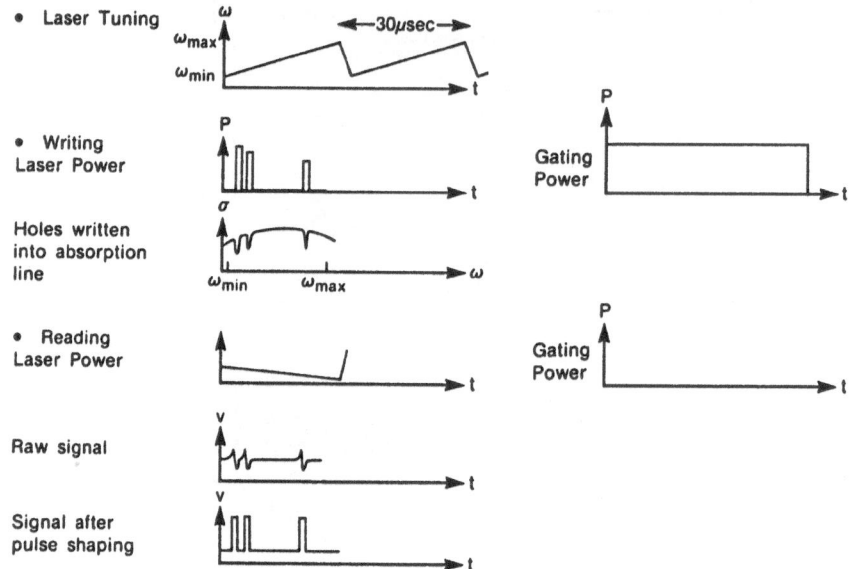

Fig. 7.4. Timing diagram for a hole-burning memory. The laser would continuously tune 100 GHz in frequency every 30 μsec. With the gating light flooding the PSHB material, the laser is modulated to write holes. The laser frequency is then scanned with the gating light off, and the heterodyne transmission signals from the holes are shaped into recognizable bits

burn the spectral holes. Writing could also be accomplished by modulating the gating power. Reading is performed with the gating field switched off, or in the case of single-photon materials, with the laser power dramatically reduced. This frequency domain optical storage system features fast spatial random access (due to use of beam deflection) and high data rates (due to the possibility for simultaneous readout by several beams in parallel with high-speed tuning of the laser frequency) as well as high areal density (due to the use of the frequency domain at each spot).

7.2.2 Engineering Studies

A FDOS system places a variety of demands on the engineering of the system and on the recording material in a coupled fashion. Table 7.1 lists some of the engineering requirements for the case of frequency-domain readout along with the references in which these properties have been demonstrated in recent research. The goal of most of what has been done has been to establish proofs of principle of the main system components, i. e., to observe the required characteristic in any convenient material without requiring that the complete system satisfy all the required properties.

258

Table 7.1: Basic engineering issues for frequency domain optical storage

Requirement	Technique	References
Tunable laser with high slew rate, wide tuning range	Diode laser: current and temperature tuning	[7.12-15]
Hole detection (30 ns/bit, high SNR)	FM, HUMPH, FREMPOLSPECT, λ modulation, ...	[7.12,16-32]
Focus/servo in liquid He (x-y access)	Ronchi grating, dithered tracer beam	[7.12,33]
Simultaneous spectral and spatial recording, low crosstalk	Measured: R' in LiF	[7.12,34,35]

The demands on laser performance are much more stringent in frequency domain optical storage than in laser optical-disk recording. A suitable diode laser should operate in a single transverse and longitudinal mode and exhibit narrow linewidth. Single-mode GaAlAs diode lasers are commercially available with a linewidth of typically 5-50 MHz. The laser also has to operate at the specific wavelength of the PSHB transition, and temperature stabilization of the device is needed. A very special requirement which is usually not found in other diode-laser applications is the capability of continuously tuning the laser frequency very rapidly over the entire inhomogeneously broadened transition of the recording medium to provide fast data access in the frequency dimension.

In principle, ramping of the injection current can be used for fast changes of the lasing frequency. The measured current-tuning response of two commercial GaAlAs diode lasers is shown in Fig. 7.5 as a function of current ramping rate [7.12]. The actual variable

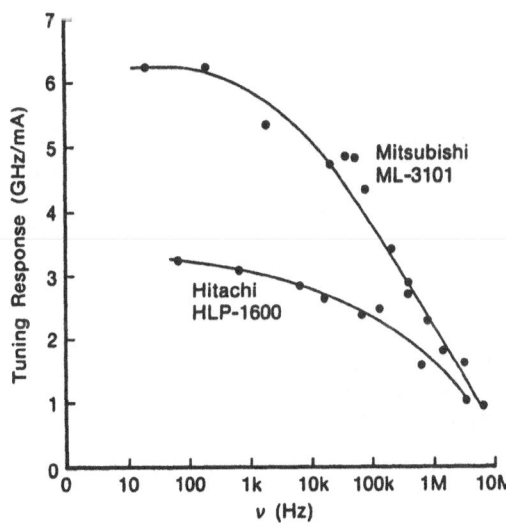

Fig. 7.5. Current tuning response of two single mode laser diodes; ν is the frequency of a 1 mA peak-to-peak triangle wave added to the DC bias current

governing the tuning response is the change of current with time, $\Delta I / \Delta t$. The laser frequency change induced by a triangle wave with an amplitude of 1 mA and different slopes $\Delta I / \Delta t$ was measured with a Fabry-Perot interferometer. The frequency change induced by current variations is actually caused by the associated change of the junction temperature, which, through refractive index changes, causes the laser mode frequency to shift. However, at very fast tuning rates the tuning response decreases due to the thermal inertia of the device (see Fig. 7.5). Small-signal frequency modulation can be induced by small carrier-density variations at rates up to several gigahertz [7.13]. However, this effect is small and does not lead to tuning over a wide frequency range. Nevertheless, both investigated diode lasers have a tuning response of 1 GHz/mA at the required scan rate of 3 GHz/μs (Fig. 7.5). For materials with hole widths on the order of 100 MHz, at least several hundred bits can be addressed by tuning the current and thus the frequency over a practical range. In the case of materials with very wide inhomogeneously broadened absorption lines and/or broader holes, several lasers could be used to lower the frequency span required from each laser. Advances in the development of coupled two-segment lasers (e.g., of the cleaved-coupled-cavity design) may result in devices with improved high-speed and wide-range tuning [7.14,15].

Hole detection is also a crucial requirement, and holes must be detected at a rate on the order of 30 ns/bit with a signal-to-noise ratio (SNR) adequate for digital data retrieval. A value of SNR proven to be adequate for conventional data storage technologies is $50-60$ dB in a 30 kHz bandwidth [7.1]. With white Gaussian noise, this translates approximately into 26 dB SNR in the 16 MHz detection bandwidth corresponding to a data rate of 30 ns/bit. In this section a variety of techniques that have been proposed for detecting spectral holes will only be mentioned briefly, and the reader is urged to consult the references for more detail. Modulation techniques at frequencies of tens to hundreds of megahertz are typically used, which provides the potential for fast temporal response and permits coherent detection in a frequency range where diode lasers have very little excess noise above the quantum limit. These techniques also offer zero-background detection, which helps in the reading of shallow absorption features.

Laser FM spectroscopy [7.16,17] is a zero-background method that involves phase modulating the incident laser beam at a fixed rf frequency and using the spectral holes to convert the phase modulation into amplitude modulation at the rf frequency. The technique can detect a hole in a time on the order of several rf periods, which would be near 10 ns for a 200 MHz rf frequency. Although the method has been demonstrated to be shot-noise-limited under double modulation conditions [7.18,19], only recently has shot-noise-limited performance and removal of residual amplitude modulation been realized without secondary modulation of the spectral features [7.20,21]. HUMPH (for High Resolution Ultrasonic Modulation of Persistent Holes) is an internal modulation technique that also shows promise [7.22-24]. In this technique, an ultrasonic field generates a time-varying strain field that directly modulates the shape of the spectral hole. This can be done at rf frequencies (tens of MHz) compatible with fast data rates.

FREMPOLSPECT (for Frequency Modulated Polarization Spectroscopy) [7.25,26] utilizes the dichroism and birefringence of spectral holes for sensitive, zero-background detection of narrow, polarization-anisotropic spectral features. This scheme may be regarded as a combination of both FM and polarization spectroscopy. One particularly simple technique, direct wavelength modulation of the diode laser, has also been utilized for zero-background sensitive detection [7.27,28]. Time-domain, electric-field, and holographic methods have also been proposed and demonstrated, but since these detection schemes require a different overall configuration of the storage system, their discussion will be postponed until Sect. 7.4.

One frequency domain measurement technique that has been studied in somewhat more detail is high-frequency heterodyne spectroscopy with current-modulated diode lasers [7.29 – 31]. This method is based on the coherent detection of the heterodyne beat signal that arises when the frequency-modulated light spectrum induced by current modulation probes a narrow absorption line. It is an important property of this measurement concept that the additional amplitude modulation associated with the current modulation cancels in one measurement quadrature, which is readily selected with phase-sensitive detection electronics [7.30]. Thus the desired measurement signal can be detected against zero background. The sensitivity limits of this spectroscopic method were recently investigated by measuring weak NO_2 absorption lines, and nearly shot-noise-limited detection was achieved using fast reading with 20 MHz detection bandwidth [7.32].

The suitability of this type of heterodyne spectroscopy for fast data readout in FDOS was studied in a simulation experiment [7.12]. A specially modified Fabry-Perot etalon in reflection was used to simulate sharp absorption features periodically separated by the free-spectral-range of 2.14 GHz. These Fabry-Perot resonances, which can be thought of as representing recorded frequency bits, had an absorption of 2.6% and a spectral width of 240 MHz. Figure 7.6 shows detection of these resonances using heterodyne spectroscopy with frequency modulation at 250 MHz using only 100 μW of laser power. The laser frequency was scanned at ~3 GHz/μs. Note that the laser power that can be used for detecting spectral holes written in small spots in PSHB recording materials can often be limited by a low saturation intensity of the optical transition. The (voltage) signal-to-noise ratio in Fig. 7.6 is SNR≈8 in a 20

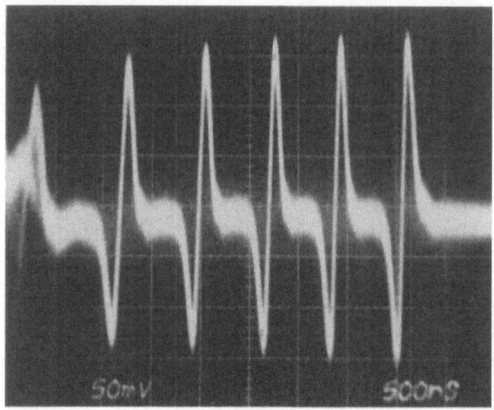

Fig. 7.6. Measurement of 240 MHz wide Fabry-Perot resonances representing an optical absorption of 2.6%. Coherent heterodyne detection was achieved by modulating the laser frequency at 250 MHz

MHz detection bandwidth, somewhat lower than what would probably be required in a practical frequency domain optical storage system. Nevertheless, this result demonstrates that heterodyne spectroscopy with current-modulated diode lasers is a promising reading technique for optical storage applications. Further improvements of the signal-to-noise ratio are possible by optimization of the photodetector, the wide-bandwidth amplifier, and the impedance matching.

A photolithographic pattern written on the medium could be used for servo-controlled spatial tracking and precise positioning of the focused laser beam. To keep the focusing optics relatively inexpensive and outside of the liquid helium environment, f/10 focusing to 10 μm spots at a working distance of 10 cm is desirable. Thus, to access different spots, the laser beam must be deflected to different off-axis positions at the focusing lens.

The operation of such a beam-positioning system has been demonstrated using 50 μm wide chrome-on-glass Ronchi fringes as tracking lines and a He-Ne tracer beam coincident with a GaAlAs diode laser beam [7.33]. A servo-control signal was derived from the transmission of the He-Ne beam by sinusoidally varying the position of the beam at a high frequency. The tracking feedback signal was supplied to a one-dimensional galvanometer that was used for directing the laser beam onto the material. It was easily possible to remain locked on any fringe indefinitely, as well as to count fringes during motion and settle to a final position in a time scale on the order of 5 − 10 ms, which indicated the possibility of fast random spatial access by this technique.

A further requirement, simultaneous high-density spatial and spectral recording with low crosstalk was demonstrated using a specially designed liquid-helium cryostat [7.12]. Large, optically flat windows with antireflection coatings permitted focusing to very small spots over 1 cm² of storage material. As shown in Fig. 7.7, a mirror

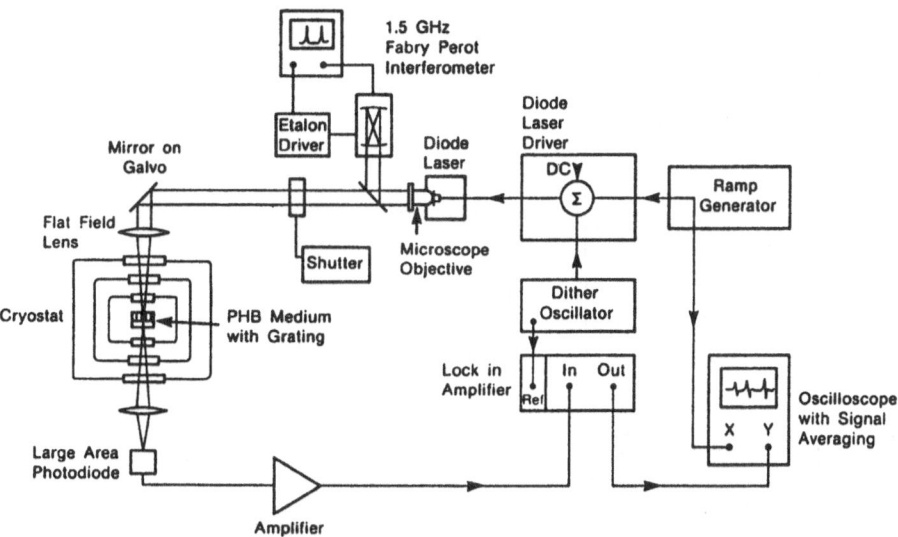

Fig. 7.7. Experimental setup for data storage experiments with diode lasers

mounted on a galvanometer was used to direct the laser beam to well-defined spatial locations on the storage medium. The 833 nm zero-phonon transition of the R′ color center in LiF exhibits single-photon hole-burning [7.34,35] and was chosen as the PSHB system, because photon-gated materials (see Sect. 7.3.3) with non-destructive reading were not then available at GaAlAs diode laser wavelengths. A $10 \times 10 \times 0.6$ mm³ LiF crystal was mounted against a chrome-on-glass Ronchi ruling, whose 50 μm wide fringes served as tracking lines. A 5 mW GaAlAs diode laser with a linewidth of ∼50 MHz was used for writing and reading of the spectral holes.

The hole-burning efficiency of the R′ color center material is rather low [7.35]. Due to the high cross section of the strong electric-dipole transition at 833 nm the saturation intensity is also low (∼9 W/cm²). In order to avoid saturation effects, the power of the tightly focused laser beam was kept below a few microwatts during writing and reading. Because of these limitations of the recording material, long exposure times of 60 s were needed for hole-burning and the need for low probing power required slow reading with signal averaging for signal-to-noise improvement. Figure 7.8 shows strings of frequency bits written at 19 different spatial locations. The spectral holes that represent the frequency bits are 600 MHz wide and separated by 900 MHz. The first three bits in each sequence are $1-1-1$, indicating the beginning of a string. Presence or absence of holes at 900 MHz intervals from these reference marks correspond to 1 and 0, respectively, to form 3 words of 8 bits stored in each spatial location. The uncertain bits indicated by question marks are due to irregular variations in bit spacing, which are caused

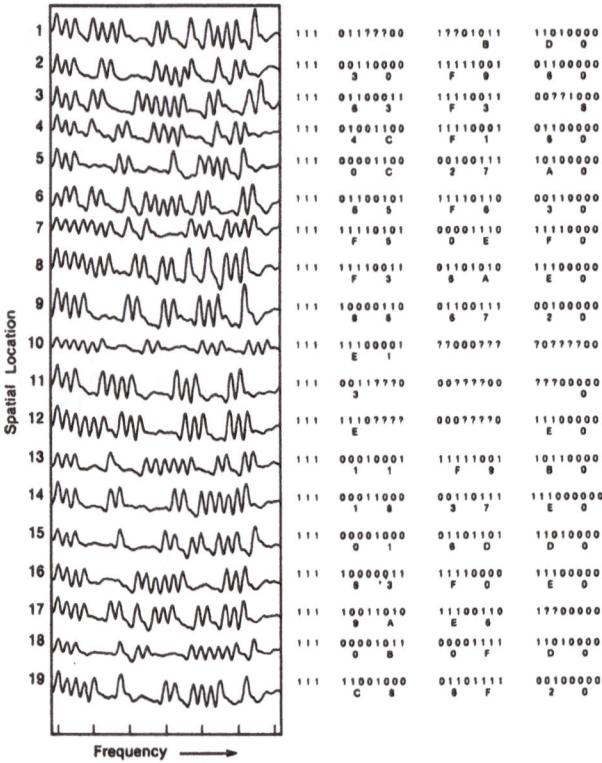

Fig. 7.8. Data written in both position and frequency domains. Each trace shows data from a different spatial location. Bits are read at 900 MHz intervals from the $1-1-1$ sequence, which begins each string. Three words of two hexadecimal letters each follow. Question marks indicate where a bit reading is uncertain due to irregular spacing of the holes; further detail is given in the text

by slow temperature drifts of the laser frequency over the relatively long time needed to write each string with this low-efficiency material.

In an actual storage system, data access will be very fast. Frequency drifts can be compensated by locking the laser frequency between certain time intervals to a frequency reference as has been successfully done with semiconductor diode lasers [7.36]. It is essential for simultaneous spatial and spectral recording that spatial crosstalk between adjacent tracks be avoided. In another experiment with improved spatial resolution of 25 μm, more than 30 dB of spatial isolation was observed [7.12].

One may conclude that investigations into the primary engineering issues for frequency domain optical storage have not uncovered to date any fundamental obstacles, although single-mode diode lasers with wider tuning ranges and higher tuning rates would be desirable. Despite considerable progress, the final problem of burning a hole in 30 ns and then detecting it in 30 ns with high signal-to-noise using a 10 μm laser spot still remains, and this problem depends heavily on the nature of the photoinduced process occurring in the recording material, as will be described presently.

7.3 Materials Research for Frequency Domain Optical Storage

7.3.1 General Materials Requirements

Any reasonable system design also places demands on the properties of the photoactive hole-burning material. It is assumed here that the reader is familiar with the basic process of PSHB as outlined in Chaps. 1 and 2 of this book. Table 7.2 lists some of the most basic materials requirements for practical frequency domain optical storage. Along with each requirement, examples of materials in which the property has been observed are listed.

Table 7.2: Basic materials requirements for frequency domain optical storage

Requirement	Material	Reference
$\Delta\nu_{hole} = 100-500$MHz	color centers, H_2Pc-PE,...	[7.37-40]
Operation at diode laser wavelengths	R' in LiF(8330 Å), H_2Pc-H_2SO_4, ...	[7.34,35,41, 42]
Reversible burning	H_2Pc, H_2P, color centers, ...	[7.6,46-49]
Long hole lifetime at low temperatures	Quinizarin in alcohol glasses	[7.50,51]
Thin-film compatibility	Aggregate color centers formed by electron irradiation, organic thin films, ...	[7.46,52,53]
Fast burning (30 ns/bit)	$(tBu)_4$-H_2Pc-PE, TZT-$CHCl_3$-PMMA	[7.56], Sect. 7.3.3
Fast burning, high SNR, fast reading, focused spot	Difficult for single-photon materials	[7.58]
Gated hole-burning mechanism	Sm^{2+} in BaClF, carbazole in boric acid glass, ...	[7.49,62]

Starting at the top of Table 7.2, the hole widths at liquid helium temperatures should be fairly small, ideally in the range $100-500$ MHz. This means that excited state lifetimes T_1 and dephasing times T_2 should not be much shorter than $\simeq 10$ ns. Holes much narrower than 100 MHz would be broadened by Fourier transform limitations during the required reading time of 30 ns/bit. If the holes are very broad, excessive demands will be placed on the tuning range of the diode laser. As shown in Table 7.2, a variety of systems have yielded hole widths in this range [7.37,38]. In general, for centers in crystalline hosts with small electron-phonon coupling and excited state lifetimes $\geq$ 10 ns, hole widths near 100 MHz occur fairly often at 1.5 K. Of particular interest several years ago was whether or not narrow holes could be observed for organic molecules in polymer hosts, where the inhomogeneous width is extremely large (typically 100's of cm^{-1}). Using careful low-power reading and shallow holes, hole widths of 120 MHz FWHM were observed for H_2-phthalocyanine (H_2Pc) in poly(ethylene) (PE) matrices at 1.4 K [7.37], and even narrower holes have been observed for free-base porphine (H_2P) in PE at 0.3 K [7.39]. This polymer, however, is partially crystalline, which may account for the relative ease of production of narrow holes at these temperatures. In H_2Pc-PE, the inhomogeneous line is $\simeq$ 100 cm^{-1} wide, and the ratio of inhomogeneous to homogeneous widths is on the order of 10^4. On the other hand, for other polymer hosts with high amorphicity such as poly(styrene) and poly(methyl methacrylate), the hole widths at superfluid helium temperatures can often approach 1 GHz with inhomogeneous widths of 300 cm^{-1} [7.37,40]. This larger hole width can be utilized in a practical FDOS system only if tunable lasers with wider tuning ranges can be found.

Since fast reading and writing must be achieved with a compact, low-cost tunable laser, the ultimate PSHB mechanism must be active at tunable single-mode diode laser wavelengths. PSHB in the GaAlAs diode laser wavelength range was first observed in 1982 in the R' aggregate color center in LiF crystals [7.34,35] using an infrared dye laser at 833 nm, and hole writing and reading were quickly demonstrated with a GaAlAs diode laser [7.27] as well. Hole-burning at GaAlAs laser wavelengths has also been demonstrated for an organic material composed of protonated H_2Pc in a sulfuric acid glass matrix [7.41] and for a high efficiency color center material, the F_4^- center in NaF:OH$^-$ crystals [7.42]. Continuing progress in diode laser technology [7.43] and in frequency doubling of diode lasers [7.44] will expand the wavelength range available for FDOS.

The problem of reversal of the hole formation process is very important because an archival, write-once low-temperature storage system would not be practical. Reversal occurs to a greater or lesser degree in almost all materials, with the obvious exception of the highly irreversible decomposition reactions that occur for s-tetrazine and dimethyl-s-tetrazine [7.45]. The proton tautomerization systems (H_2P and H_2Pc) offer a good example of a highly reversible PSHB mechanism: the educt and product are simply related and have localized excitations so that one may be transformed into the other and vice versa [7.46 – 48]. Materials that photoionize also show reversibility; for example, PSHB in color centers can be reversed by irradiation at high energies to release electrons from traps [7.38], and reverse electron transfer occurs for some samples of Sm^{2+}:BaClF material (see Chap. 4) as the electrons are transferred back and forth between Sm^{2+} and Sm^{3+} ions [7.49]. Rather than trapping the electrons at randomly distributed trapping sites somewhere in the host, a well-defined donor-acceptor pair with fixed spacing could permit reversible photoionization so that the electron could be transferred back and forth indefinitely without fatigue effects. Studies to determine the number of erase cycles and to optimize materials for reversibility are areas deserving further attention in future research.

Long hole lifetime at low temperatures is an important requirement for data storage applications. Most of the materials mentioned in this article have hole lifetimes at least of the order of days; hole lifetime measurements are usually limited only by the persistence of the experimenter in maintaining a cryostat at liquid helium temperatures for extended periods. One careful study of hole lifetimes has been performed for the case of quinizarin in alcohol glasses [7.50], and hole lifetimes of the order of 20,000 years were estimated for the deuterated system. Other effects such as time-dependent spectral diffusion can lead to a slow broadening of spectral holes and thus a degradation of the written information [7.51](See Chap. 3).

Since the laser spot for reading and writing must be on the order of 10 μm in diameter, the recording material must be fabricated in a thin film of thickness of the order of the Rayleigh range of the beam waist. This is to prevent overlap of the written information in adjacent spatial regions. The first PSHB experiments on inorganic thin film samples were performed with aggregate color centers in alkali halides formed by electron irradiation [7.52]. Later experiments utilized polycrystalline evaporated films and x-irradiation to produce color centers [7.53]. Many organic PSHB materials with allowed singlet-singlet transitions can be fabricated in thin film form fairly easily. The issue of which actual film thickness and optical density are required is intimately connected with other properties of the recording material, and a coupled analysis of the optimum parameters for FDOS applications will be presented in the next section.

One critical requirement, the burning of detectable holes in times of the order of 30 ns, proved fairly difficult to achieve. Prior to late 1983, no detectable holes had been burned in times less than 1 ms, and nanosecond burning times are obviously required for a data storage system with gigabyte capacities. Detectable fast burning was first achieved by using a high efficiency material with narrow holes, tetra-t-butyl-H_2Pc in poly(ethylene). The quantum efficiency, η, for this system is $\simeq 10^{-3}$ [7.37,54,55]. Hole-burning at constant energy in this material yielded smaller and smaller holes as the burning time was decreased, partially due to a bottleneck in the hole-burning process [7.37] resulting from the triplet decay time to the singlet ground state. Using a variation of FM spectroscopy with large

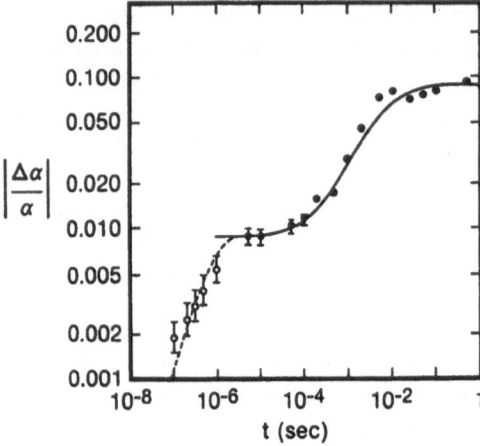

Fig. 7.9. Normalized change in absorption constant for holes written with different exposure times. The solid circles represent data taken with constant exposure energy; the open circles, with constant intensity. The point at 10^{-5} s corresponds to both the constant energy and constant intensity conditions. The solid and dashed lines are theoretical fits described in [7.56]. Both the vertical and horizontal scales are logarithmic

modulation index and detection with avalanche photodiodes to probe the sample with extremely low reading intensities, barely detectable holes were burned in less than 100 ns for tetra-t-butyl-H_2Pc in PE [7.56].

Figure 7.9 shows the observed hole depths for constant burning energy (closed circles) and for constant burning power (open circles). The falloff at 1 ms burn time is due to the triplet lifetime bottleneck; the plateau in the constant energy data just below 0.1 ms corresponds to unit probability of exciting each molecule into the triplet state during the burn time. The falloff at 10^{-6} s is due to the limited laser power available, but even if constant energy burning could have been maintained, the relative hole depth would fall to the 0.001 level near 10^{-8} s. These experiments showed that for any system, a fundamental bottleneck exists in the burning process defined by the excited-state lifetime, which should ideally be on the order of 10 ns [7.57] (see below). Since holes must be burned in 30 ns or so, a given center can at most absorb only a few photons during the typical burning time. Moreover, quantum efficiencies for PSHB are usually less than 1% to as small as 10^{-7}; thus the holes that are burned in short times can be very shallow. To detect shallow holes, high reading powers are required to increase the signal-to-noise, but high reading powers in small spots produce high intensities, destroying the written information by causing photochemistry over the entire laser scan range. This is why an extremely sensitive detection method with very low reading power and large 7 mm diameter laser spot sizes was required to detect the holes burned in 100 ns [7.56].

In fact, at the conclusion of the experiments leading to Fig. 7.9, it seemed fairly impossible that fast burning and detection could ever be accomplished in small focused laser spots unless some fundamental change were made in the hole-burning mechanism. The crucial problem with materials like H_2Pc-PE is the single-photon nature of the hole-burning process, which is the subject of the next section.

7.3.2 Limitations of Single-Photon Recording Mechanisms

This section addresses the entry in Table 7.2 that is second from the bottom: a material, in order to be useful in a practical storage system, must *simultaneously* show several required properties: the ability to form deep holes in short burning times (30 ns/bit) and yet allow fast reading (30 ns/bit) at high signal-to-noise ratios with focused (10 μm diameter) reading beams. To illustrate the central issue, a material with low hole-burning efficiency would be quite easy to read without serious destruction of the written holes, but such a system would be difficult to write with high contrast in short burning times. Conversely, a system that shows fast burning due to a high quantum efficiency for hole production would be difficult to read without burning of the unburned centers by the tightly focused reading beam.

To understand this problem more fully, a thorough analysis of the coupled fast reading and fast writing problem in small spots for materials with single-photon (or monophotonic) hole-burning mechanisms was performed [7.58]. The results of such a study are broadly applicable, because until very recently, essentially all materials undergoing PSHB did so via a single-photon process. The essential problem with such photoreactions is that there is no threshold in the hole formation mechanism. The process of hole detection requires the absorption of photons by the remaining unburned centers, and if high powers are necessary to detect the dip in the absorption line with adequate SNR, the hole pattern will be destroyed, since reading (and writing) must be done with a small diameter laser spot. This destructive reading problem was analyzed in detail [7.58] to determine whether any combination of single-photon materials parameters would yield acceptable reading and writing performance.

Any SNR analysis for single-photon materials must specify the detection technique used. Possible detection methods include any of the transmission methods summarized in Sect. 7.2.2, fluorescence detection, holographic methods, electric-field readout, and time-domain readout. As was remarked earlier, the last three methods will be discussed in Sect. 7.4. Fluorescence detection can be useful with materials with low oscillator strengths, but the detection of even a hole as large as 100% deep with an SNR of 26 dB in 30 ns requires

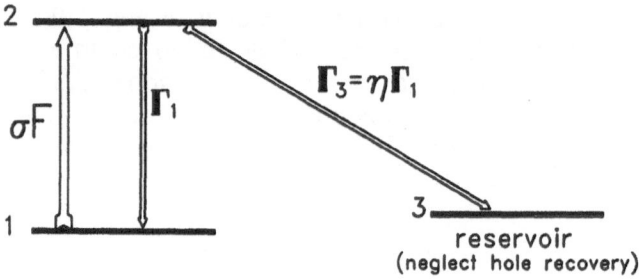

Fig. 7.10. Schematic energy level diagram for absorbing centers with a single-photon hole-burning mechanism. Level 1 is the ground state, level 2 is the excited state, and level 3 is a permanent reservoir ground state that schematically depicts the hole formation process. The various pumping and decay rates are defined in the text

count rates at the detector greater than 10^{10} per second [7.58], a value that can hardly be tolerated by low-noise photomultipliers. Shallower holes necessitate even higher count rates; and considering that optical losses are large in fluorescence detection, only quantum-limited transmission detection will be considered here.

Figure 7.10 shows the general level scheme for single-photon mechanisms. A photon flux F is incident on a general center with a peak absorption cross section σ at low temperatures. From the excited state, the center may then either decay back to the original ground state with rate Γ_1, or it may convert with probability (or quantum efficiency [7.54,55]) η to a new ground state, forming a spectral hole. The plan of the analysis involves using rate equations for the level diagram of Fig. 7.10 to compute the largest hole depth that can be produced in 30 ns, read the resulting hole in 30 ns and require that the SNR be large enough for practical digital data retrieval. The deepest hole that can be burned in 30 ns has relative depth near η, because with $T_1 \gtrsim 10$ ns, each center can absorb at most only a few photons. During reading, the detection process in assumed to be shot-noise limited, and the reading laser power is increased until the required SNR of 26 dB in 16 MHz bandwidth is reached, or until power broadening occurs due to saturation. The SNR can be expressed as

$$\text{SNR} = (\eta_Q F_R A \tau_R/2)^{1/2} \eta \alpha_o L \exp(-\alpha_o L/2), \qquad (7.1)$$

where η_Q is the detection quantum efficiency, F_R is the reading photon flux, A is the laser spot area, τ_R is the reading time, L is the sample length, and α_o is the initial absorption coefficient before burning. This is the SNR for the first read expressed in terms of a current ratio (or equivalently, voltage ratio across a fixed resistance).

Equation (7.1) can be used to identify combinations of material parameters that permit data readout with sufficient SNR. Figure 7.11 shows the results of the single-photon optimization problem for

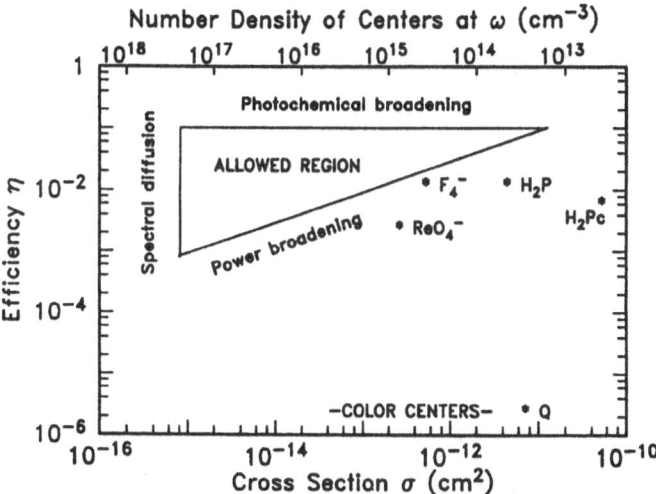

Fig. 7.11. Allowed region of efficiency η, number density of centers at ω, and cross section σ in order for the first read to yield acceptable SNR. The various abbreviations signify: F_4^- — 8892 Å color center in NaF:OH$^-$ ([7.42]); ReO$_4^-$ — perrhenate ion in alkali halides ([7.59]); H$_2$P — free-base porphine in n-alkanes ([7.47]); H$_2$Pc — free-base phthalocyanine in polymers ([7.6]); Q — quinizarin in boric acid glass ([7.54]); and color centers in general ([7.38]), with the exception of the F_4^- center

the standard conditions of 30 ns burn and read per bit, 10 μm diameter laser spot, 100 μm thick sample, and 10 mW maximum reading power (the reader is referred to [7.58] for details). A material with specific values of η and σ is represented by a point on the $\eta - \sigma$ plane shown in the figure. The analysis assumes that the sample absorption always satisfies $\alpha_0 L = \sigma N_\omega L = 2$ which can be shown to yield the optimum SNR [7.58]; hence the concentration of centers N_ω necessary to keep α_0 fixed (top axis) must decrease as the cross section increases. The concentration N_ω refers to the number density of centers within one homogeneous linewidth of the burning laser frequency ω; the total concentration of centers would be larger approximately by the ratio of the inhomogeneous to homogeneous widths. The triangular region in the figure represents the class of materials that would yield acceptable signal-to-noise ratios for the first read after burning a single hole. This means that a useful single-photon material must have *low* absorption cross section and *high* quantum efficiency, with sufficient solubility in the host to permit the concentrations shown on the upper axis. The boundaries of the allowed region are defined by an upper limit on the practical value of η that can be reached without hole broadening due to excessive photochemistry, an upper limit on the tolerable density of centers at ω without undesirable intercenter interaction, and the requirement that during readout the reading intensity be kept below the saturation intensity for the transition. The various symbols lo-

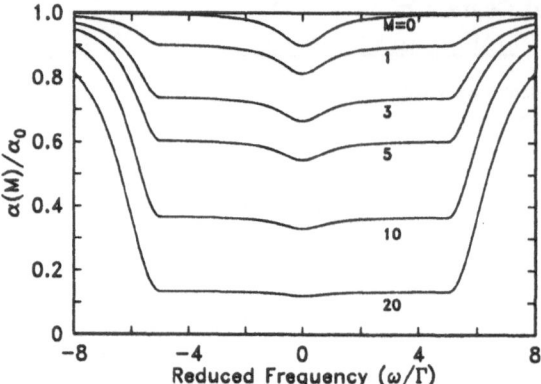

Fig. 7.12. Normalized absorption coefficient versus reduced frequency for a 10% deep hole. The hole lineshape is shown for several values of the number of reads, M. For this figure, $\eta = 0.1$ and the reading intensity is 1/3 the saturation intensity

cate a representative class of previously studied single-photon materials, none of which fall within the allowed region for the first read. This is mostly due to coincidence − since early investigators were unaware of the need to consider low-cross-section materials, only easily studied high-cross-section materials were considered.

So far, only the SNR requirements for the first read directly after writing have been considered. However, one realizes very quickly that at the power levels required to achieve usable SNR, the entire absorption line scanned by the reading laser is gradually burned away during each successive read. Figure 7.12 illustrates this with

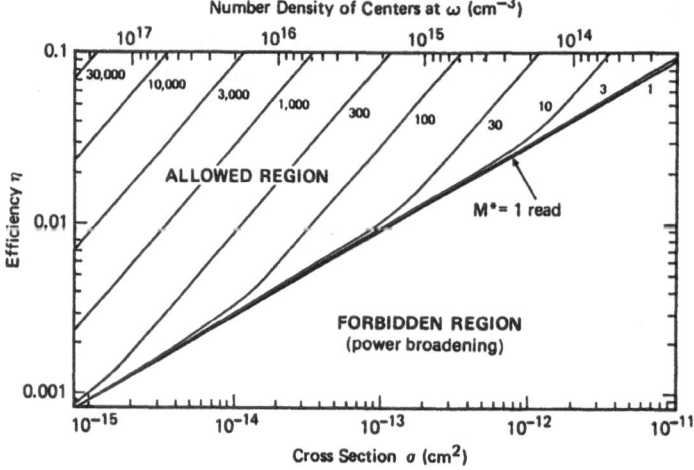

Fig. 7.13. Contour plot in the allowed region of the η-σ plane, showing contours of constant number of reads, M^*. The $M^* = 1$ line is the boundary of the allowed region defined by the constraint of no power broadening during reading. The upper density axis applies only to the case $\alpha_0 L = 2$; i.e., near the $M^* = 1$ line (see [7.58])

a computer simulation for a material with $\eta = 0.1$. This "trenching" of the line during reading is a general feature of single-photon materials when small laser spots are used as is required for data storage applications. Due to the accumulation effects of single-photon PSHB during successive reads, the SNR degrades as reading continues.

It is possible to determine M^*, the number of reads that can be achieved for each $\eta - \sigma$ pair before the SNR drops to unacceptable levels. Figure 7.13 contains an enlargement of the allowed region from Fig. 7.11 where contours of constant M^* have been added. Near the power broadening boundary line and in the upper right portion of the allowed triangle, the number of reads is fairly small and a practical storage system based on such materials would require excessive refreshing capability. Much better performance results in the upper left corner, i. e., at high quantum efficiency and low absorption cross section. From the physical point of view, high η is needed so that the hole produced in the 30 ns burning time is as deep as possible, thus less reading laser power is required. Low cross section means that more photons pass by a given center without absorption, and these photons can then improve the SNR at the detector.

For materials in the upper left region with η near 0.1 and σ near 10^{-15} cm^2, thousands to tens of thousands of reads can be achieved before refresh is required. Tens of thousands of reads would probably require refreshing of the written data. It is a matter of debate and engineering trade-offs at which values of M^* the refresh requirement would become impractical, and we assume for simplicity that with M^* greater than 1000, the overhead from the refresh capability would not be excessive.

One consideration not mentioned so far is the actual power required to detect the spectral holes. The analysis in [7.58] assumes that a maximum of 10 mW is available for reading, which is easily produced with currently available semiconductor diode lasers. The related question of sample heating is complex, and the amount by which a sample warms in superfluid helium depends upon the low-temperature thermal conductivity, heat capacity and Kapitza resistance of the sample in question. Excessive heating can erase the written data or broaden the spectral holes due to effects described in Chaps. 3 and 5. Preliminary modeling studies [7.60] indicate that sample heating effects are tolerable for crystalline hosts; however, in the case of polymeric hosts with potentially low thermal conductivity, data degradation can occur and measurements of sample heating must be performed for each specific material.

It may be concluded that single-photon materials with η-σ values in the upper left-hand corner of Fig. 7.13 might be practical materials that would allow fast reading and writing in small focused spots using thin films. Such materials must have high efficiency (near 0.1), low oscillator strength in the range 10^{-4} to 10^{-5}, and high solubility in the host material. (For 100 MHz homogeneous widths, the oscillator strength is equal to 1 for $\sigma = 1.7 \times 10^{-10}$ cm^2.) It is within the realm of possibility that single-photon materials exist with values in these ranges. One might look for hole-burning in partially forbidden transitions, such as $n - \pi^*$ transitions of organic molecules and d-f transitions of divalent rare earth systems, but no single-photon materials have been found to date that satisfy these requirements. This analysis provides future research with a clear direction as to which single-photon PSHB materials should be investigated for frequency domain optical storage applications.

7.3.3 Photon-Gated Mechanisms

The parameter space of useful $\eta - \sigma$ values for single-photon mechanisms yielding large numbers of reads presented in Figs. 7.11, 13 is very small. Consequently, it is important in future studies to consider materials that do not suffer from the destructive reading problems of single-photon processes with no threshold. One such class of materials are those with two-step PSHB mechanisms, called gated mechanisms in the last entry in Table 7.2. Figure 7.14 shows the general concept of gating. The wavelength excites a homogeneous packet within an inhomogeneously broadened line. If no external field (in the general sense of field) is present, the center returns to the original ground state without forming a spectral hole. However, in the presence of λ_1 plus some external field, the center undergoes the transformation leading to hole-burning and enters a new ground state or permanent reservoir. This is the origin of the term "gating": the external field acts as a gate on the hole-burning process. The hole may then be detected using λ_1 alone, since hole detection is merely measuring the ground state distribution of those centers

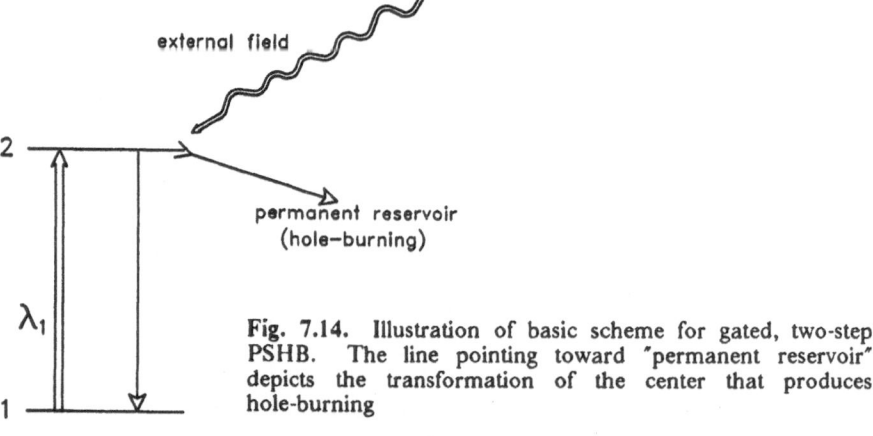

Fig. 7.14. Illustration of basic scheme for gated, two-step PSHB. The line pointing toward "permanent reservoir" depicts the transformation of the center that produces hole-burning

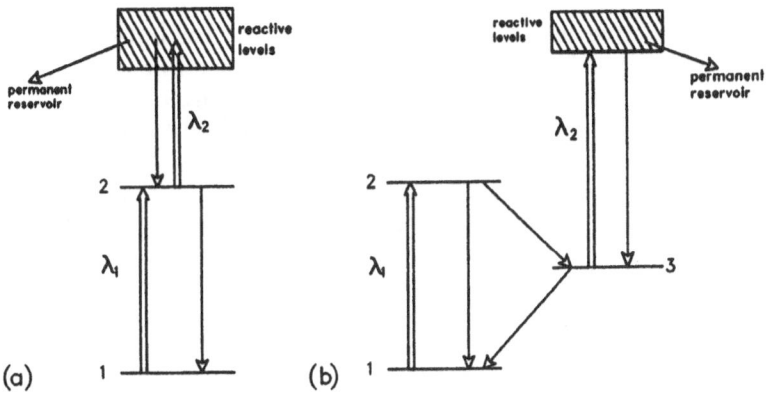

Fig. 7.15.   Two schemes for photon-gated PSHB.   (a) Three-level system.   (b) Four-level system

that did not react to form the hole. Since the external field is not present during the reading process, hole detection is *nondestructive*. In effect, gated mechanisms add a much-needed "threshold" to the recording process, a property which is commonly found in most successful storage schemes. The external field may be a second photon of a different wavelength λ_2 ("photon-gating" [7.61]), or the gating could perhaps be achieved by any other external field, such as electric field, magnetic field, stress field, and the like.

The limitations on single-photon materials presented in Figs. 7.11, 13 sparked off a search for photon-gated mechanisms in inorganic as well as organic materials. Figure 7.15 shows two possible general schemes for photon-gating. For the three-level case (part (a) of the figure), the frequency-selective wavelength λ_1 excites those centers in the inhomogeneously broadened line that are in resonance with intermediate state 2, which should have a fairly long lifetime in order to build up a large population during the λ_1 irradiation. If the gating wavelength λ_2 is not present, the selected centers return to ground state 1 without photoreaction. However, if λ_2 is present while the centers are in state 2, transitions occur to higher reactive level(s) 3 and the centers then have a nonzero probability of entering a different ground state or permanent reservoir. For example, the reactive levels might be energy levels close to the ionization threshold for the center in the host matrix, and the resulting photoreaction could be liberation of an electron which leaves the center and becomes trapped some distance away, leaving behind an oxidized form of the center.

Part (b) of Fig. 7.15 shows another general scheme that can lead to photon-gating: a four-level system. In this case, the intermediate state is level 3, which may have a different spin quantum number to that of levels 1 and 2. This level diagram is appropriate for most organic molecules, where levels 1 and 2 may be singlet states, and

275

level 3 and the reactive level(s) 4 are triplet states. Excitation with λ_1 from level 1 to level 2 provides the frequency selection of a particular group of centers in the inhomogeneously broadened line. Level 2 is coupled to level 3 with a certain probability due to intersystem crossing. Level 3 usually has a long lifetime for decay back to level 1, hence a large population can build up in level 3. If λ_2 is not present, centers in level 3 eventually decay back to level 1 with no hole formation. However, if the gating light at λ_2 is present while the centers are in level 3, transitions occur to the reactive levels, and the probability for hole formation becomes nonzero. As in the case for Fig. 7.15 (a), the actual reaction could be photoionization, bond scission, or any other type of reaction that causes the entire system to enter a new ground state or permanent reservoir.

To date, several inorganic and organic materials have been discovered that show photon-gating via processes similar to those depicted in Fig. 7.15, and these are summarized in Table 7.3. The first observation of photon-gating resulted from experiments on Sm^{2+} ions in BaClF crystals [7.49]. The mechanism for hole-burning in this material has been described in detail in Chap. 4, and is of the general class shown in Fig. 7.15 (a). The first wavelength, λ_1, is near 690 nm and excites the system from the $^7F_0$ ground state (level 1) to the $^5D_0$ level (level 2). Extended irradiation at 690 nm produces essentially no hole production, but brief periods of simultaneous irradiation with $\lambda_2 = 514$nm produces spectral holes at λ_1. The second photon excites the ion from $^5D_0$ to the conduction band or to an autoionizing level above the conduction band edge and the liberated electron is subsequently trapped in the host matrix leaving behind a Sm^{3+} ion. The gating ratio, or the ratio of gated hole depth to

Table 7.3: Materials showing photon-gated persistent spectral hole-burning

Material System	λ_1	λ_2	Mechanism[a]	Reference
Sm^{2+} in BaClF	690 nm	514 nm	two-step PI	[7.49]
Carbazole in boric acid glass	335 nm	400 nm	two-step PI	[7.62]
Co^{2+} in $LiGa_5O_8$	660 nm	673 nm	two-step PI	[7.63]
Anthracene-tetracene photoadducts in PMMA	326 nm	442 nm	two-step PD	[7.64]
Zn-,Mg-tetrabenzoporphyrin derivatives with halomethanes in PMMA	630 nm	488 nm	two-step DA ET	[7.65-67]
Cr^{+3} in $SrTiO_3$	790 nm	790-1060 nm	two-step PI	[7.68]

[a]PI - photoionization, PD - photodissociation, DA ET - donor-acceptor electron transfer

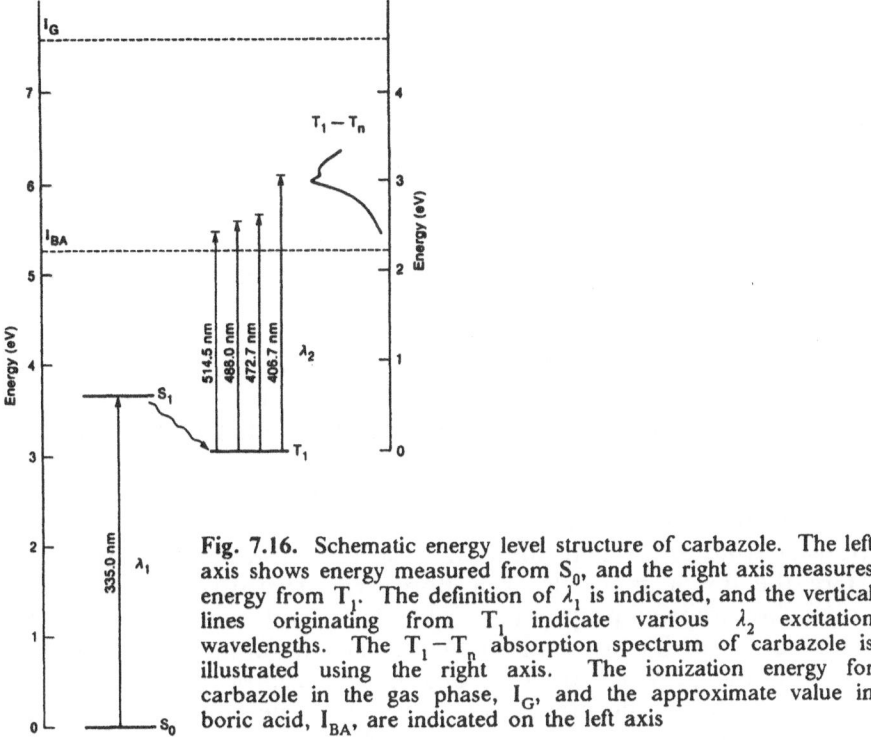

Fig. 7.16. Schematic energy level structure of carbazole. The left axis shows energy measured from S_0, and the right axis measures energy from T_1. The definition of λ_1 is indicated, and the vertical lines originating from T_1 indicate various λ_2 excitation wavelengths. The $T_1 - T_n$ absorption spectrum of carbazole is illustrated using the right axis. The ionization energy for carbazole in the gas phase, I_G, and the approximate value in boric acid, I_{BA}, are indicated on the left axis

single-color hole depth for equal λ_1 burning conditions, has been observed to be greater than 10^4 in preliminary experiments [7.49]. One drawback of this material, however, is the extremely small oscillator strength of the λ_1 transition, which makes high SNR experiments in thin films difficult.

Two-color photon-gated PSHB has also been observed in an organic material composed of carbazole molecules in boric acid glass [7.62], which is a four-level system like Fig. 7.15 (b). The most probable mechanism for this material is depicted in Fig. 7.16. Upon excitation in the singlet-singlet origin with $\lambda_1 = 335$nm, the molecule undergoes intersystem crossing with a high yield to build up a large metastable population in the lowest triplet state, T_1. In the presence of $\lambda_2 = 360 - 514$ nm, holes are formed at the singlet excitation wavelength, λ_1, due to photoionization of the molecule in the triplet manifold and trapping of the ejected electron in the boric acid glass matrix. The observed gating ratio for this material is near 400. The boric acid glass matrix plays an important role: in comparison to less polar hosts like PMMA, boric acid glass lowers the ionization potential of the carbazole guest from the gas phase value (I_G) by 2.3 eV to the more easily attainable value I_{BA}. The absorption cross section σ_1 for carbazole in boric acid is high, and thin film samples with high optical density are easily prepared. The observation of gating in an organic system suggests the use of organic synthetic techniques for generating new materials that exhibit photon-gating.

The third entry in Table 7.3 provides an example of photon-gating for a transition metal ion, Co^{2+}, in an inverse spinel crystal, $LiGa_5O_8$ [7.63] (for details see Chap. 4). This inorganic material utilizes 660 nm for λ_1 and longer wavelengths for λ_2 leading to a gating ratio of $\simeq 20$. The mechanism is similar to that for the Sm^{2+}:BaClF material: two-step photoionization and trapping of the ejected electron in the host crystal. On the other hand, the organic material composed of photoadducts of anthracene and tetracene (A-T) in PMMA [7.64] provides a radically new type of photoreaction leading to photon-gating: two-step photodissociation of the adduct. The gating ratio for this material is small, however, due to one-color photoreactions caused by the absorption of two photons of $\lambda_1 = 326$ nm.

The next to last entry in Table 7.3 provides a particularly important mechanism for photon-gating: donor-acceptor electron transfer from an excited donor to a specific nearby acceptor [7.65]. Here the donor molecule is a porphyrin derivative, meso-tetra(p-tolyl)-Zn-tetrabenzoporphyrin (TZT), shown in the inset of Fig. 7.17. This donor (or its Mg analog) shows photon gating in PMMA thin films in the presence of a halomethane acceptor in high concentration, such as chloroform ($CHCl_3$), methylene chloride, or methylene bromide [7.66]. The frequency-selecting photon $\lambda_1 \simeq 630$ nm excites the lowest singlet-singlet transition, and triplets are produced

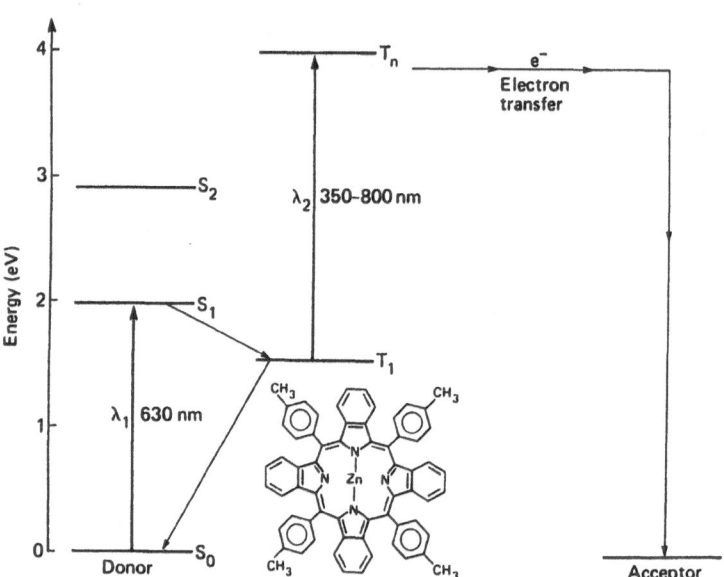

Fig. 7.17. Level diagram for photon-gated PSHB via donor-acceptor electron transfer. The structure of the donor chromophore TZT is shown in the inset

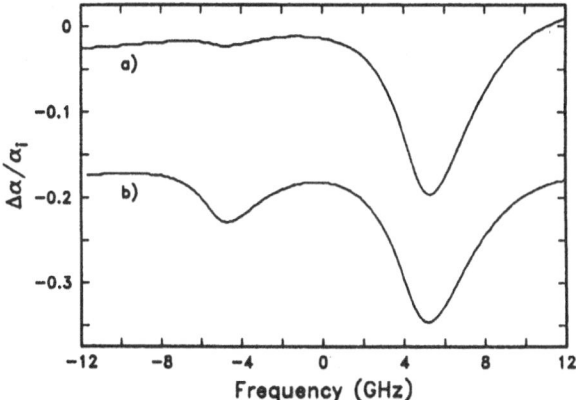

Fig. 7.18. Example of photon-gating for the TZT/CHCl$_3$/PMMA system. (a) Burn time 6 s. The one-color hole is on the left, and the two-color hole on the right was produced with 20 mW at 514 nm. (b) Same as trace (a), except that the irradiation time for the one-color hole was increased to 120 s

efficiently by virtue of the high triplet yield. The triplet lifetime of 40 ms facilitates the buildup of a large population in T$_1$. When the gating light at $\lambda_2 = 350 - 800$ nm is present, the electron is excited to an upper triplet T$_n$ from which it tunnels to a nearby halomethane acceptor.

Figure 7.18 shows an example of the gating enhancement observed for the TZT/CHCl$_3$/PMMA material. In part (a), the shallow one-color hole on the left and the two-color gated hole on the right have both been burned for 6 s. In part (b) of the figure, the burn time for the one-color hole has been increased to 120 s. For this acceptor concentration, the presence of the gating wavelength increases the hole formation rate by a factor in the range 30 – 100.

This first example of photon-gating via a donor-acceptor electron transfer mechanism has the useful property of allowing the gating effect to be controlled by varying the acceptor concentration and/or acceptor electron affinity (by choosing a different acceptor). However, due partly to the PMMA host, the holes are somewhat too wide for applications unless broadly tunable diode lasers or specialized dye lasers can be found and utilized. A further drawback of this specific material is the fact that reversal of the electron transfer is somewhat difficult. This is not a fundamental problem because one can envision directly coupled donor-acceptor molecules in which the electron can be reversibly shuffled back and forth between the two halves of the molecule in a fashion analogous to the proton tautomerization reaction for H$_2$P.

Aside from these specific disadvantages, the TZT/CHCl$_3$/PMMA material and its relatives have a further useful property: the efficiency of gated hole production is high enough that fairly deep ($\simeq$ 1% in transmission) holes can be burned with a *single* 8 ns pulse

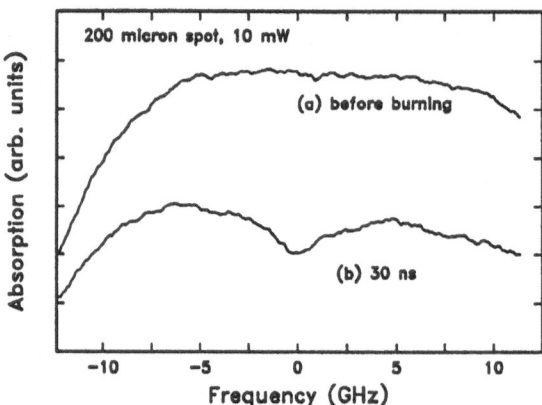

Fig. 7.19. Fast burning in small spots for the TZT/CHCl₃/PMMA material. (a) baseline before burning. The curvature in this trace is due to weak Fabry-Perot resonances in the optical system. The zero of absorption is off-scale at the bottom. (b) After burning with a 10 mW beam at λ_1 for 30 ns, followed immediately with a 200 ms gating beam at 488 nm of 17 mW

at λ_1 followed immediately by a 200 ms gating pulse at λ_2 [7.67]. To the authors' knowledge, this is the fastest spectral hole-burning reported to date for any material, and is a result of several factors: the large absorption cross section at λ_1, the large triplet yield, the reasonably long triplet lifetime, and the large intrinsic probability for electron transfer from the high triplet levels to the nearby acceptor.

Figure 7.19 illustrates a further advantage of this system: with the TZT/CHCL₃/PMMA material, fast burning in small laser spots can be detected. This property has not been observed in any single-photon (or photon-gated) material to date partly because of the fundamental limitations on single-photon processes described in Sect. 7.3.2. The figure shows a clearly detectable spectral hole produced with a 30 ns pulse at λ_1 followed immediately by a 200 ms gating pulse at λ_2. The λ_1 spot diameter was 200 μm (limited by the mechanical stability of the cryostat), and the gating beam was unfocused. The 200 ms gating light pulse acts as a "developer" for centers that were placed in the lowest triplet state by the site-selecting λ_1 pulse. If many bits were to be written in the frequency domain in this material, one would simply inject short pulses at all the λ_1 wavelengths desired and then irradiate with the long λ_2 pulse to make the spectral holes permanent. This result directly illustrates the superiority of photon-gated over single-photon mechanisms for FDOS applications. The one remaining performance requirement of fast detection of the written holes in 30 ns/bit is within the realm of possibility considering the depth of the holes and the relative insensitivity of the material to probing light at λ_1, although fast detection would be easier if the gating ratio were higher.

These first examples of gated spectral hole-burning have opened up a novel class of mechanisms for PSHB, and considering the limitations on single-photon materials for frequency domain optical storage applications, the search for gated mechanisms should be an important area for future study.

We now turn to a discussion of the *optimal* materials parameters for practical gated recording media. This will provide direction for future research on gated mechanisms. In order to identify the required characteristics of a practical photon-gated storage medium, some assumptions must be made about the performance characteristics of the frequency domain optical storage system. For the case of frequency domain readout, the relevant characteristics are identical to those used in Sect. 7.3.2 above: 30 ns reading and writing per bit, shot-noise-limited transmission detection with at least 26 dB wideband signal-to-noise ratio, and 10μm diameter laser spots. The reader is referred to [7.69] for complete details.

A useful quantity for modeling purposes is the effective hole-burning yield for gated systems, η_e, which is defined as the relative absorption change $\eta_e = (\Delta\alpha/\alpha)$, that is produced by the best-case two-color hole-burning during the writing time of 30 ns. Due to the complex nature of gated PSHB processes the hole-burning yield depends critically on the specific microscopic properties of the gated PSHB mechanism as well as on the writing conditions. This parameter then effectively replaces the microscopic η of Fig. 7.10 for single-photon mechanisms.

The analysis seeks appropriate values for the absorption cross section σ_1 at λ_1, the density of centers within a homogeneous width of the laser frequency, N_ω, sample length L, and yield η_e which result in SNR $\geq$ 26 dB in a 16 MHz bandwidth. The thickness L of the storage material and the concentration N_ω of the active centers are not intrinsic material properties and can be controlled when fabricating the storage medium. Thus, it is meaningful to classify materials by a concentration-thickness product $N_\omega L$. However, well-defined spatial resolution is required when accessing the stored information with the laser beam. Therefore, the material thickness L should not exceed the depth of field associated with the focused laser beam of 10μm diameter, i.e. L $\leq$ 100μm.

Identifying materials by suitable combinations of σ_1 and $N_\omega L$ using the hole-burning yield η_e as a parameter is particularly helpful for establishing guidelines for the search for promising gated PSHB materials, and this is done in Fig. 7.20. Materials within the shown boundaries for particular values of η_e permit data readout with SNR equal to or greater than the required value with a maximum of 10 mW reading power. For high cross sections σ_1 and large values $N_\omega L$ the optical absorption becomes so strong as to prevent sufficient light from reaching the detector. For high σ_1 and low values $N_\omega L$ the achievable signal-to-noise is limited by power broadening of the detected holes. For $\sigma_1 \leq 10^{-15}$ cm^2 the available laser power of 10 mW is lower than the saturation power; thus the available

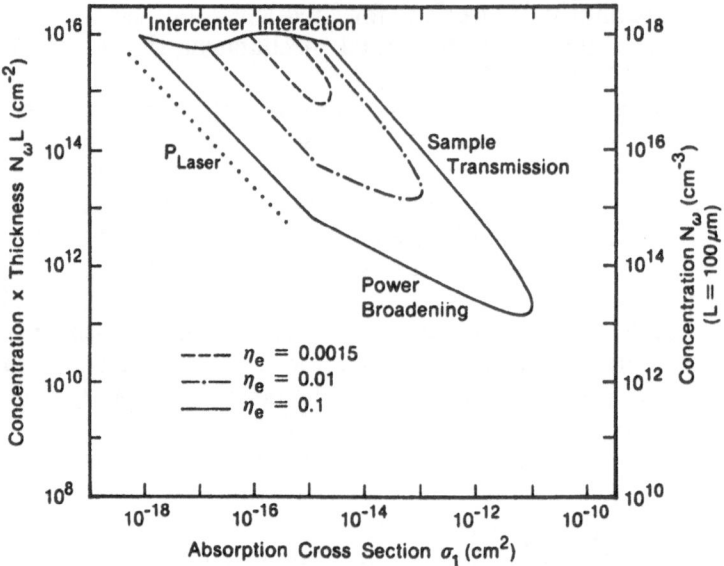

Fig. 7.20. Materials constraints for gated PSHB materials in order to achieve practical S/N ratios. The various regions and symbols are defined in the text. To illustrate the effect of increased laser power, the dotted line shows the $\eta_e = 0.1$ boundary for 100 mW reading power

reading power defines the S/N limit. At the top of the figure, undesirable interactions between the optically active centers, e.g. energy migration and cooperative excited-state quenching, can restrict the usable concentration of centers; this limit is highly material dependent.

Spectral broadening of the produced hole, either by saturation of the transition or by excessive hole-burning, imposes an upper limit on yield η_e. For this analysis, it seems justified to restrict the hole-burning yield to $\eta_e \leq 0.1$. The $\sigma_1 - N_\omega L$ parameter space shrinks rapidly as η_e decreases. In order to successfully implement a FDOS system based on gated mechanisms, it is critical to find a material that permits gated PSHB with very high efficiency in the short writing times required for fast data transfer rates.

These considerations apply to the photophysical characteristics of the gated material, whatever the details of the gating process. To date, gated PSHB has been observed only in the form of photon-gating. Photon-gated PSHB is attractive for optical storage applications since there are no fundamental technical barriers for implementing such a process in the design of a practical device. For the case of photon-gating, it is possible to state several general requirements on the actual microscopic writing process, in order to optimize the value of η_e.

Figure 7.21 illustrates some general requirements on three-level and four-level photon-gated PSHB mechanisms in order to achieve

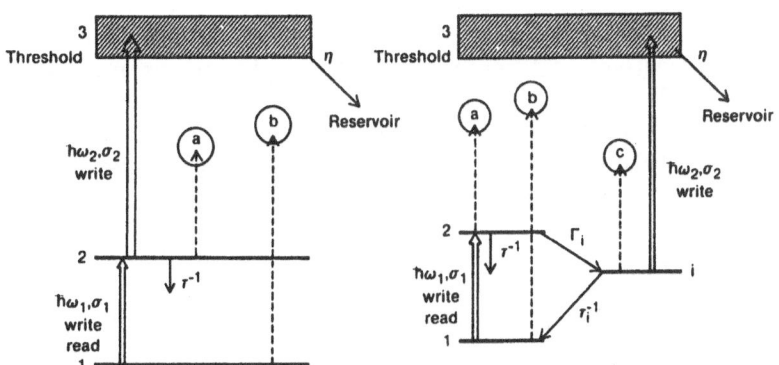

Fig. 7.21. (a) Optimal level structure of a three-level photon-gated PSHB mechanism. The absorption should be small in the circular regions. **(b)** Optimal level structure of a four-level photon-gated PSHB mechanism

high overall efficiency and high gating ratio. In the case of a three-level system (Fig. 7.21 (a)), λ_1 should be larger than λ_2, and the absorption in the circular regions should be small to reduce one-color burning. The lifetime τ should be long enough to allow large populations to build up in level 2. This is the general level scheme for Sm^{2+} in BaClF, in which $N_\omega L \simeq 10^{14}$ cm^{-2} and $\sigma_1 \simeq 10^{-18}$ cm^2. In cases where τ is much larger than the data access time the storage material does not have to be exposed to both photon energies simultaneously. There is no fundamental need for a frequency-selective narrow-band transition from level 2 to level 3; however, in certain instances, narrow-band levels 3 involving transitions with high peak cross sections may be preferable to a continuum absorption such as a conduction band.

In principle, systems with $\omega_1 = \omega_2$ can exhibit gated PSHB in the sense that the hole-burning yield is nonlinear with laser intensity. However, the requirement of non-destructive reading makes it desirable to use systems with $\omega_1 \neq \omega_2$ permitting complete decoupling of reading and writing processes.

Figure 7.21 (b) illustrates optimized photon-gated PSHB involving a four-level system. For the case of the TZT/CHCl$_3$/PMMA material, $N_\omega L \simeq 5 \times 10^{12}$ cm^{-2}, $\sigma_1 \simeq 5 \times 10^{-12}$ cm^2, and $\eta_e \simeq 0.01$ [7.66]. For efficient gated PSHB the ($2 \rightarrow i$) rate Γ_i should be as large as possible consistent with the requirement that the lifetime of level 2 not be too short. Further, a long intermediate state lifetime τ_i would be advantageous in achieving a large population in level i. It is evident that absorptions 2 → (a), 1 → (b), and i → (c) involving photons of frequency ω_1, ω_2, and ω_1, respectively, should not be large. Of course, the microscopic yield η should be as large as possible for efficient photon gating.

7.3.4 Limitations on Storage Density

The high storage density made possible by PSHB is based on the capability of selectively addressing subclasses of ions or molecules contained in each irradiated spot of recording medium. The stress-induced variations in the local environment of the storage centers are random in nature and consequently centers with a particular resonance frequency are randomly distributed in the host material. A tightly focused laser spot may interact with such a small ensemble of centers that statistical number fluctuations on the inhomogeneously broadened line become noticeable. This fundamental source of noise imposes a limit on the achievable storage density for a given concentration of storage centers[1].

The essential requirement is that the number-fluctuation noise is sufficiently smaller than the depth of the written holes so that the required SNR can be obtained for data readout. The number-fluctuation noise is given by $\sqrt{N_\omega V}$, where V is the focal volume in the recording medium, i.e., $V = \pi w_0^2 L$ with where L is active sample thickness and w_0 the radius of the focused beam waist. The sample thickness is approximately limited to the Rayleigh range of the waist, i.e., $L \simeq \pi w_0^2/\lambda$. The achievable areal storage density is approximately given by $D_s \simeq f_\omega/(2w_0)^2$, where f_ω is the storage capacity of the frequency domain. Since $f_\omega \simeq \Delta\omega_{inh}/\Delta\omega_{hom}$, the total concentration of storage centers is $N_{tot} \simeq f_\omega N_\omega$. The relative hole depth $(\Delta N_\omega)/N_\omega$ is given, as before, by the effective hole-burning yield η_e (see Sect. 7.3.3). For each center concentration, the maximum storage density is estimated by applying the condition that the relative hole depth be larger than the relative statistical fluctuations of the number of centers in the volume V by a factor equal to the required SNR value. This is equivalent to requiring that the photon shot noise is larger than the statistical fluctuation noise in the readout signal.

Based on these fundamental considerations, it can be shown readily that for a given center concentration N_{tot}, a maximum storage density of

$$D_s \leq \frac{\pi\eta_e}{4SNR_0} \left(\frac{N_{tot}f_\omega}{\lambda} \right)^{1/2} \tag{7.2}$$

can be obtained, where SNR_0 is the required value of SNR (voltage ratio). Figure 7.22 shows the predictions of (7.2) with η_e as a parameter, $SNR_0 = 20$, and $f_\omega = 1000$. For this value of f_ω, the storage density is furthermore limited to $\leq 10^{11}$ bits/cm^2 due to the $\simeq 1\mu$m diffraction limit for the laser beam diameter. The dotted line gives the maximum storage density for $f_\omega = 10^4$ and $\eta_e = 0.1$. These results indicate that it is advantageous to maximize the storage capacity of

---

[1]Statistical fine structure on inhomogeneously broadened lines has recently been observed: see W.E. Moerner, T.P. Carter: Phys. Rev. Lett. **59**, 2705 (1987).

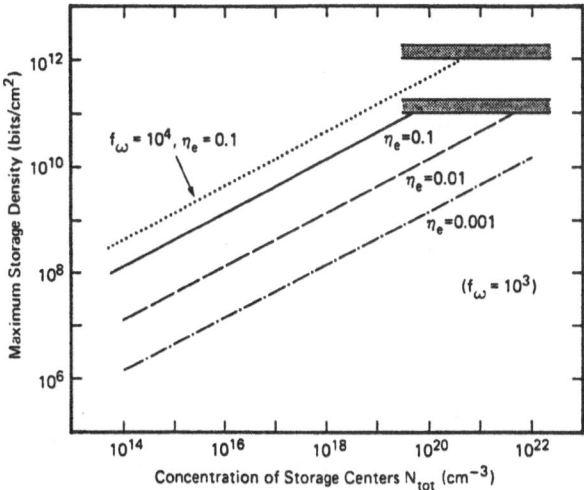

Fig. 7.22. Maximum storage density in FDOS achievable versus total concentration of centers, with effective yield η_e and frequency-domain storage factor f_ω as parameters

the frequency domain and the hole-burning yield while keeping the spatial resolution of the focused laser beam at moderate levels. For simplicity, contributions from other noise sources have been neglected for this rather fundamental discussion of statistical noise. Thus, these considerations are applicable to any FDOS configuration independent of the particular readout scheme.

For example, assuming a laser spot size of $10\mu m$ diameter and $f_\omega = 1000$ (i.e., a storage density of $\simeq 10^9$ bits/cm^2), minimum center concentrations of $N_{tot} = 5 \times 10^{19}$ cm^{-3}, 5×10^{17} cm^{-3}, and 5×10^{15} cm^{-3} are required for a hole-burning yield of $\eta_e = 0.001$, 0.01 and 0.1, respectively. Comparison of these values with Fig. 7.20 shows that statistical number fluctuations are not the dominant noise source for this storage density. However, Fig. 7.22 indicates that very high storage densities in excess of 10^{10} bits/cm^2 can only be achieved with PSHB materials with very high yield and center concentration. In order to achieve a given storage density, a suitable recording material has to satisfy both Eqns. (7.1) and (7.2).

This section has presented a variety of issues relating to materials for frequency domain optical storage applications. Areas still requiring future study involve reversibility and the source of the large hole widths for the electron-transfer and photoionization systems. The limitations of single-photon materials underscored the importance of finding suitable gated mechanisms, and recent discoveries of high efficiency photon-gating as well as demonstrations of fast burning in small laser spots suggest that finding a suitable material for FDOS is within the realm of possibility in the near future.

7.4 Alternative Data-Storage Configurations

7.4.1 Time Domain Storage

The preceding sections have dealt primarily with a PSHB optical storage system in which reading and writing are performed in the frequency domain; i. e. a narrowband laser beam is scanned across the inhomogeneous line and spectral holes are written at those frequencies at which "1"s (or "0"s) are desired, and the written data are read out by again scanning a single-frequency laser and detecting the changes in the sample transmission. It has been implicitly assumed that in all cases the writing and reading times for a single bit are longer than the homogeneous dephasing time of the transition, T_2, so that a rate equation analysis can be used to describe the kinetics of the laser-material interaction.

An alternative scheme, time-domain reading and writing [7.70], occurs if the sample is excited by properly chosen pulses with widths τ much less than T_2. Since T_2 is of the order of $1-10$ ns for $100-1000$ MHz homogeneous widths, this time-domain or coherent transient regime occurs when excitation pulses with durations in the picosecond range are utilized. A variety of interesting and useful effects occur in this regime, such as free-induction decay, optical nutation, and photon echoes, to name a few, and the reader is referred to any of several treatises on coherent transient effects for background material [7.71 − 73].

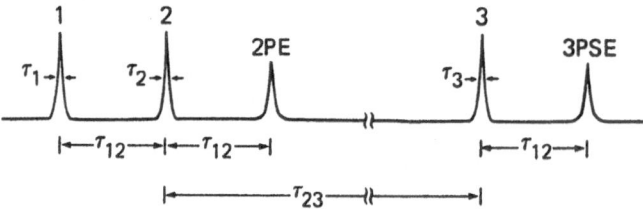

Fig. 7.23. Basic scheme of two-pulse echo (2PE) and three-pulse stimulated echo (3PSE) generation. Pulses 1, 2, and 3 have widths as shown. τ_{12} is the time interval between pulses 1 and 2, and τ_{23} is the time interval between pulses 2 and 3

The particular coherent transient effect of interest here is the three-pulse stimulated photon echo, illustrated in Fig. 7.23. Three pulses, labeled 1, 2, and 3, are incident on the sample, which may generally be regarded as an ensemble of two-level systems. The pulse widths τ_p satisfy $\tau_p << T_2$ (p = 1,2,3), and further, the spacing between pulses 1 and 2 (τ_{12}) is also less than T_2. Pulse 1 prepares the ensemble by coherently exciting the two-level systems into a superposition of ground and excited states. The ensemble starts out in phase due to the effect of pulse 1, and then begins to dephase because centers with different detunings from the laser frequency evolve in time with different rates. For the proper choice of the pulse area, pulse 2 will produce a reversal of the phase of each of the various excited centers so that later all the centers will add up in phase and emit a burst of light, the photon echo. This first echo is called the two-pulse photon echo (2PE), and it appears at a time after the second pulse equal to τ_{12}. Note that since T_2 is usually less than the excited state lifetime T_1, all this necessarily occurs before decay of the excited population.

Now, if a third pulse is applied to the sample at a time after the second pulse designated τ_{23}, then under certain conditions the sample will emit another photon echo, called the three-pulse stimulated echo (3PSE). This echo occurs at a time delay after pulse 3 equal to τ_{12}, in a fashion similar to the "production" of the 2PE by pulse 2. The 3PSE is particularly interesting because the only requirement on τ_{23} is that the ground- and/or excited-state distributions have not returned completely to the original equilibrium distribution, with all centers in the ground state. For example, for simple two-level systems a 3PSE will always occur if $\tau_{23} < T_1$, and in fact if additional "3" pulses are applied within the excited state lifetime, more 3PSE's will occur, with their amplitudes falling off as $\exp(-t/T_1)$. It is this "memory effect" in the spin analog of the optical 3PSE that formed the foundation for the "spin-echo serial storage memory" of *Garwin* et al., proposed in the 1950s [7.74].

Without persistent spectral hole-burning or a long-lived intermediate state, the 3PSE effect would stop after several excited state lifetimes have elapsed. However, if for example bottlenecks exist in the level structure that cause population to enter a long-lived intermediate state, 3PSE's will occur for $\tau_{23} > T_1$, as long as τ_{23} is less than the intermediate state lifetime. Photon echoes stimulated from an accumulated grating in the ground state population caused by a triplet intermediate level were observed by *Hesselink* et al. for pentacene in naphthalene single crystals [7.75]. For theoretical and numerical treatments of 3PSE formation due to accumulation in a bottleneck level, the reader is referred to [7.76,77].

On the other hand, if *persistent* spectral hole-burning occurs during optical excitation so that some of the centers in the excited state "photoreact" and do not refill the original ground state, 3PSE's will continue to appear after each new "3" pulse (unless the material is a single-photon material with destructive reading). It is this form of 3PSE that offers the most promise for optical storage applications. An experimental observation of 3PSE effects due to photochemical accumulation (spectral hole-burning) was reported in 1983 for the case of porphyrazine in poly(styrene) [7.78,79] and later for octaethylporphyrin in poly(styrene) [7.80]. In these cases, in order to accumulate sufficient contrast in the modulated ground-state distribution, many pairs of pulses 1 and 2 had to be applied to the sample (10^{10} such pairs) before stimulation with pulse 3 would produce an easily detectable 3PSE, for reasons to be described below.

The first clear description of how to implement the 3PSE for time-domain frequency-selective optical data storage (TDOS) was presented by *Mossberg* in 1982 [7.70]. The general scheme involves modifying the envelope of pulse 2 in Fig. 7.23 so that it contains information, such as a sequence of narrow pulses representing digital data. Mossberg described the general conditions under which pulse 3 becomes a faithful replica (or a time-reversed replica) of all the temporal structure in pulse 2 for an inhomogeneously broadened absorption line. Following reference [7.81], the three laser pulses are taken to have the form

$$E_p(\mathbf{r}, t) = \varepsilon_p(t - \eta_p) \cos[\omega_0(t - \eta_p)], \qquad (7.3)$$

where $p = 1,2,3$, $\varepsilon_p(t)$ is a slowly varying envelope function, $\eta_p \equiv (\mathbf{k}_p \cdot \mathbf{r})/c + t_p$ is the phase factor, $\mathbf{k}_p$ is the unit wave vector of pulse p, and t_p is the time at which pulse p passes the arbitrary location $\mathbf{r} = 0$. Under the conditions that (i), the lengths of pulses 1-3 are $<< T_2$, (ii) the finest temporal structure in any pulse is longer than T_2^*, where $T_2^* = (\Delta \nu_{inh})^{-1}$ with $\Delta \nu_{inh}$ the inhomogeneous width, and (iii) the three pulses are weak in the sense that they separately excite only a small fraction of the initial ground-state population to the excited state, then the resulting coherent echo (3PSE) signal E_c may be written using the Fourier transforms of each of the three excitation pulses $E_p(\omega)$ as

$$E_c(t) \sim \int_{-\infty}^{+\infty} E_1(\omega)E_2^*(\omega)E_3^*(\omega)e^{-i\omega t}d\omega, \qquad (7.4)$$

where an infinite inhomogeneous width has been assumed. This relation states the simple result that the final output pulse is basically the inverse Fourier transform of the product of the Fourier transforms of the three input pulses. In fact, the group of assumptions leading to (7.4) are called the Fourier transform assumptions, and in this regime a number of interesting optical effects can occur. The resulting output pulse is coherent emission, which means that it must satisfy a phase-matching condition dependent upon the phases of the input pulses 1, 2, and 3 of the form

$$\eta_c \equiv \eta_2 + \eta_3 - \eta_1. \qquad (7.5)$$

This means that the stimulated echo phenomenon can lead to space-time holographic images of objects, an effect that has been described in an alternative way in Chap. 2 of this book, with several examples using photochemically accumulated 3PSE's (See also [7.82–85]).

Equation (7.4) has several interesting special cases. For example, if pulses 1 and 3 are sufficiently short (effectively δ-functions) so that

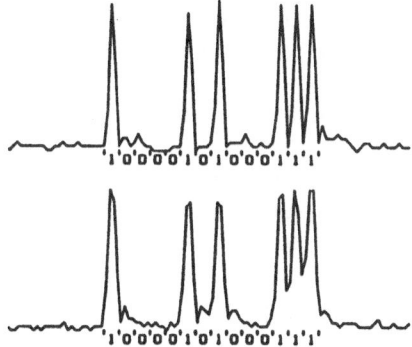

Fig. 7.24. Example of time-domain storage. The upper trace shows the input pulse 2, and the lower trace shows the resultant output 3PSE. A decoding of the time structure into digital data is shown under each trace. (after [7.86])

their Fourier transforms are constant over the inhomogeneous line-width, then E_c is a faithful time-domain representation of the temporal structure of pulse 2. This is one configuration that could be used for time-domain optical storage and readout, which has been illustrated in Fig. 2.27 for the case of photochemical accumulation by many pulse pairs 1 and 2. This experiment has also been performed in a model system composed of ^{174}Yb atoms, where the pulse-shape information is stored in the sublevel coherence of the $^3P_1$ state [7.86] (see Fig. 7.24). The upper trace shows the temporal input information (pulse 2), and the lower trace shows the resulting (single event) 3PSE pulse. For this model system, the 3PSE effect lasts only as long as the excited state lifetime (850 ns). If the roles of pulses 1 and 2 are reversed with the data stream on pulse 1 and a δ-function for pulse 2, the data read out by pulse 3 will be time-reversed [7.87]. Using the phase-matching condition, phase conjugation of spatially aberrated beams has also been observed in model systems without photochemistry [7.88]. Finally, if only one pulse has effective δ-function character, the output can be either a convolution or a cross-correlation [7.81] of the other two pulses.

An alternative way of viewing the storage of information using 3PSE's is as follows: consider the pair of pulses 1 and 2 in Fig. 7.23. The Fourier transform of this pulse pair contains a sinusoidal modulation in frequency space at the frequency $(\tau_{12})^{-1}$. Now if the sample can be regarded as a linear system that records the Fourier transform of any incident light, then this sinusoidally modulated spectrum can be "burned" into the absorption line. This can only occur if the modulation frequency is bounded above by the inhomogeneous linewidth and below by the homogeneous width. Multiple pulse pairs 1 and 2 can be utilized to increase the contrast of the stored information, if the one-cycle quantum efficiency of spectral hole-burning is small. In fact, the sinusoidal modulation of the inhomogeneous absorption spectrum due to hole-burning by pairs of picosecond pulses was observed directly in the frequency domain [7.89] for cryptocyanine in alcohol before the first experiment showing true time-domain readout by a δ-function pulse 3 (see Chap. 2). In this picture, pulse 3 has the effect of stimulating the δ-function response of the system, which in this case is simply a time-delayed version of pulse 2. To get storage of many bits, pulse 2 simply becomes a serial sequence of pulses to encode digital information.

The time-domain readout method (TDOS) has several advantages and disadvantages relative to the frequency-domain configurations (FDOS) described in the first part of this chapter. On a very general level, both schemes require persistent spectral hole-burning in an inhomogeneously broadened line, so the issues of single-photon versus photon-gated materials and nondestructive reading are relevant for both approaches. Unlike the case for FDOS (see the preceding sections and [7.58,69]), no careful analysis of the optimal materials requirements has been performed for the TDOS configuration. However, several general comments can be made.

One critical materials parameter is the dephasing time T_2, which effectively determines the time scale of the TDOS scheme. For many organic PSHB materials, T_2 is of the order of $1 - 10$ ns, and in fact, if the hole widths do not reflect the homogeneous linewidth (See Chaps. $3 - 5$) the pulse width requirement becomes $\tau_p << 2/(\pi \Delta \nu_{hole})$. For weak, partially forbidden transitions with longer excited-state lifetimes such as in the rare earth or transition metal ion materials, T_2 values can be larger and larger pulse widths can be used.

FDOS requires a single-frequency rapidly-tunable laser source for reading and writing, whereas TDOS requires a (non-tunable) laser source with possibly subnanosecond or subpicosecond pulses, depending upon the T_2 of the storage material. It may be difficult to find lasers that tune over a large enough wavelength range for the former case. Mode-locked diode lasers could be used for a TDOS system.

The differing time scales of TDOS and FDOS present another problem. Modulation of the input beams for FDOS to write the information simply requires amplitude modulation of the tunable laser beam during the laser scan at $10 - 100$ MHz rates, which is fairly easy to accomplish with standard acousto-optic or electro-optic modulators. The readout data rate of FDOS is in the $10 - 100$ MHz range, which is comparable to the data rates for conventional optical or magnetic storage. On the other hand, for writing, TDOS requires assembling many fast pulses in a specific sequence, which may require sophisticated electro-optical multiplexing devices [7.90]. For readout, all the data (perhaps $100 - 1000$ bits) arrive at the detector in a time less than T_2, which may yield an output data rate of up to 1 THz for nanosecond dephasing times. This data rate is out of the range of conventional electronics and would require either an all-optical computing system or sophisticated demultiplexing of the output pulse train [7.90].

Another consideration is the signal-to-noise achievable for fast burning in small spots. The analysis of single-photon and photon-gated PSHB materials in Sect. 7.3 and the experimental results reported show that 1% deep spectral features can be produced in 30 ns burning times if the site-selecting transition is almost saturated during the burning. However, experimental studies to date of the 3PSE effect in PSHB materials [7.78 – 80] have always required a very large number of pulse pairs 1 and 2 (10^{10} in fact) in order to achieve sufficient readout signal-to-noise. This is partly due to the $\sim 10^{-3}$ hole-burning efficiency, and partly due to the Fourier transform, weak excitation requirement on the input pulses necessary for (7.4) to be valid. (Actually, with short pulses, only the data pulse need be weak, because the preparation and readout pulses can induce "shape-locking" of the data pulse even if they violate the weak-excitation requirement [7.91].) One intriguing way out of the weak excitation requirement involves linearly chirped excitation pulses [7.86], such that the product $E_1(\omega)E_3^*(\omega)$ [see (7.4)] is frequency-independent. The validity of this concept has been experimentally demonstrated recently for the model system composed of Yb atoms [7.92] as shown in Fig. 7.25. Trace (a) of the Figure shows successively the chirped pulse 1, the input data on pulse 2, and the chirped readout pulse 3. Trace (b) shows the resulting output pulse

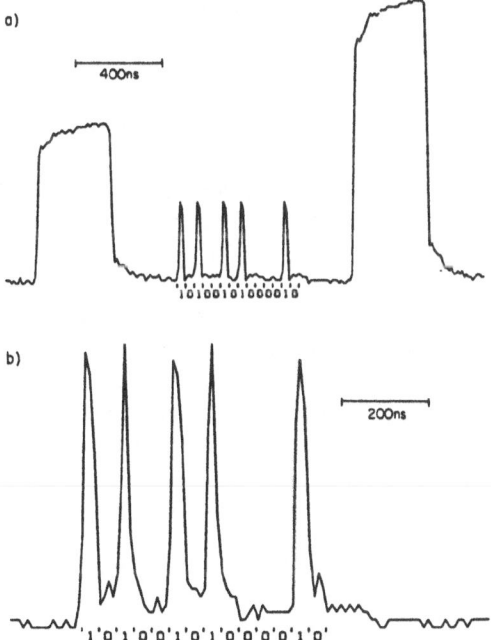

Fig. 7.25. Illustration of time-domain storage using chirped pulses 1 and 3. Trace (a) shows the input pulses 1 – 3, with temporal information on pulse 2. Trace (b) shows the resulting output 3PSE. (after [7.86])

on a single-event basis. The advantage of this scheme is that the pulse areas of pulses 1 and 3 can be increased greatly, providing more overall excitation of each of the frequency packets in the inhomogeneously broadened line, thus more net photochemistry, and hence higher overall contrast in the readout signal. Tests of this concept in real PSHB materials would provide important confirmation of the importance of chirped excitation pulses in TDOS.

The idea of using 3PSE effects in PSHB materials to achieve time-domain readout provides a possible alternative to the more thoroughly analyzed frequency-domain readout configuration. Future efforts should consider the requirements placed on the material properties in order to determine which combinations of cross section, hole widths, and quantum efficiencies are optimal for time-domain readout. Implications of the very high writing and reading data rates for the system configuration should be considered, as well as spatially parallel (holographic) configurations. To prevent destructive reading, it is likely that photon-gated processes will be important in providing a threshold in the writing process in a manner similar to the case for FDOS.

7.4.2 Electric-Field Readout

In principle, any external field that perturbs the absorbing centers composing a spectral hole can be used to develop an alternative hole detection scheme. External fields usually cause the spectral holes to shift, split, or broaden in a reversible manner; thus a time-varying external field gives rise to a time-varying sample transmission that may be detected. Since hole widths are much narrower than the entire inhomogeneously broadened line, effects of external fields on spectral holes can be observed with very high sensitivity. For example, magnetic and electric-field perturbations acting through the Zeeman and Stark effects, respectively, have been suggested as methods of addressing the information stored in the frequency domain [7.93]. Shifting and/or splitting of spectral holes due to the uniaxial stress caused by ultrasonic modulation has also been suggested as a means of detecting spectral holes at high sensitivity with zero background [7.22–24]. A recent comprehensive review may be consulted for a detailed description of experimental data and model calculations of the effects of various external fields on persistent spectral holes [7.94].

The hole readout scheme utilizing external fields that has received the most attention to date is electric-field readout via the Stark effect. The first measurement of the Stark effect in a PSHB material was reported in 1977 by *Marchetti* et al. [7.95] for the anionic dye resorufin in a poly(methyl methacrylate) film. Since that time, Stark effect measurements have been reported for a variety of materials, including inorganic crystalline systems [7.96] (see also Chapter 4), chlorin in polymers [7.97,98], and perylene in poly(vinyl butyral) [7.99], to name a few of the early examples. The essential observation for the case of amorphous hosts is that a spectral hole burned in zero field will "smear out" or broaden roughly symmetrically as an electric field is applied, because the many centers comprising the spectral hole experience different level shifts due to the distribution of local environments and center orientations with respect to the applied field. The rate at which a hole broadens with field depends upon the difference in the dipole moments and the difference in polarizabilities between the ground and excited states, averaged over the many molecular orientations in the sample. Even for the case of perylene in which the molecule has no electric dipole moment in the ground or excited states, a linear Stark effect is nevertheless observed due to matrix-induced electronic level shifts [7.99]. For crystalline hosts, spectral holes usually shift and split more than they broaden, and very precise stress-splitting coefficients may be determined in this fashion (see Chap. 4). Application of *inhomogeneous* electric fields may provide a means of producing different splitting rates for the different centers in a crystalline sample in a manner similar to the amorphous hosts.

Suppose that several holes are written in a material in the frequency domain with zero electric field in order to represent digital data. As the applied electric field is increased, the holes will each symmetrically broaden or "smear", effectively removing the frequency-dependent structure in the inhomogeneously broadened line at sufficiently high field. With the field applied, a new sequence of spectral holes may be written in the frequency domain, and this process may be repeated until the electrical breakdown field of the sample is reached. The net effect is a limited type of multiplexing of spectral holes in both the electric-field and spectral domains. This scheme for optical data storage has been studied experimentally and theoretically for several molecules in polymer hosts [7.100 - 102]. Figure 7.26 shows an example of multiplexed holes in the frequency and electric-field dimensions for the case of chlorin molecules in poly(vinyl butyral) films [7.100].

One interesting feature of electric-field readout is that a tunable laser is not absolutely essential. Figure 7.27 shows a recent demonstration of electric-field readout for the case of the nonphotochemical hole-burning material composed of 9-aminoacridine in poly(vinyl butyral) films [7.101]. Here a fixed frequency He-Cd laser was used to burn and detect the spectral holes in the sample by varying the applied electric field. Part (a) shows the fluorescence signal as a function of electric field before burning. Then the laser power was increased and the electric field was held fixed at those voltages at which holes were to be recorded. Reading of the stored information (part (b) of the Figure) only required attenuation of the

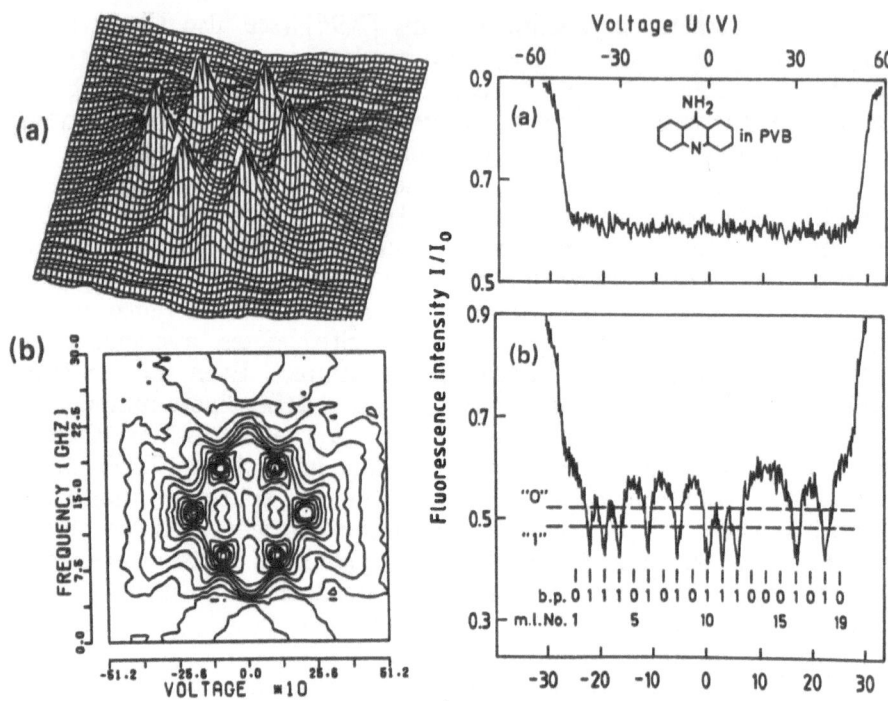

Fig. 7.26. Illustration of multiplexing of spectral holes in both the frequency and electric-field domains for chlorin in poly(vinyl butyral) films. **(a)** Transmittance profile of the six holes. **(b)** Contour plot of the change in optical density. (after [7.100])

Fig. 7.27. Electric-field readout of spectral holes for 9-aminoacridine in poly(vinyl butyral) films using a fixed frequency HeCd laser. **(a)** Background fluorescence signal before burning. **(b)** Fluorescence signal after burning spectral holes at various values of the applied field. (after [7.101])

laser beam and ramping of the electric field while monitoring the emitted fluorescence. A sequence of 19 bits of information has been recorded into the inhomogeneously broadened line in one laser spot in this case.

The electric-field readout scheme offers several advantages and a few disadvantages. The laser need not be tunable if the number of bits that can be recorded in the electric-field dimension is sufficient for practical applications. If more than perhaps 25 – 50 bits are needed in each spot, laser tunability will probably be essential, however. Changing the electric field impressed upon the sample can be quite fast, limited by the capacitance of the sample. By applying an electrode pattern to the surface of the recording material, different electric fields can be applied to different spatial locations, and various parallel writing schemes can be envisioned [7.101,103]. It is important to realize that the electric-field and frequency dimensions are not independent – the same absorbing centers are being used for both. For each specific material, the trade-off between signal-to-noise ratio for holes written in the electric-field domain and in the frequency domain due to the limited number of absorbing centers will have to be carefully evaluated. Finally, use of the electric field for readout does not solve the problem of destructive reading with single-photon materials. The SNR considerations for electric-field readout are quite analogous to those for frequency-domain readout described in Sect. 7.3.2, thus gated PSHB materials (with cross sections and concentrations similar to those described by Fig. 7.20) will probably be required for practical applications.

7.4.3 Holographic Readout

Holographic readout is another intriguing method of hole detection that would affect the organization of a frequency domain optical storage system. In this section the emphasis is on frequency-domain holographic readout, as opposed to the time-domain holographic techniques described in Chap. 2 and in Sect. 7.4.1. A frequency-domain holographic readout scheme was proposed in 1978 [7.104] in which the reading light would sense changes in dispersion at frequencies far away from the actual spectral holes in order to achieve nondestructive reading; however, this idea has not been experimentally implemented. Strong holographic signals have been detected in resonance with the spectral hole by *Wild* et al. [7.105] using the perturbations in both the real and imaginary parts of the index of refraction.

Figure 7.28 shows the apparatus used for the first experimental demonstration of holographic readout of spectral holes [7.105]. In part (a), holes are burned at desired laser frequencies using two coherent beams overlapped in the sample. The interference pattern thus formed burns a spatial grating (hologram) in the population of the absorbing centers. In general, the result of such a population grating is a grating in both the optical absorption and in the index of refraction, both of which are dependent on frequency. For readout (Fig. 7.28 (b)), only one tunable probing beam is used (with attenuated power). The diffracted signal is then detected with essentially zero background as a function of the frequency of the probing beam.

To see the way in which the diffracted signal is generated, the burning beams form an interference pattern given by (following [7.105])

$$I = I_0 [1 + V \cos(2\pi x/\Lambda)], \tag{7.6}$$

where $\Lambda = \lambda/(2 \sin \theta)$, V is the fringe visibility, I_0 is the average intensity, and Λ is the grating period determined by the optical

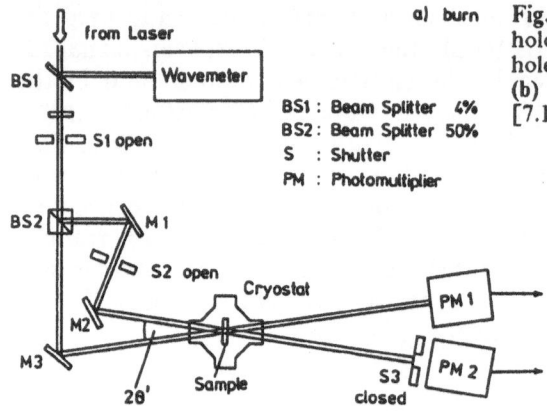

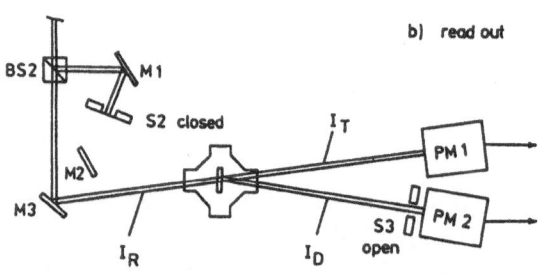

a) burn

BS1 : Beam Splitter 4%
BS2 : Beam Splitter 50%
S   : Shutter
PM : Photomultiplier

b) read out

Fig. 7.28. Optical apparatus for holographic detection of spectral holes. (a) Burning configuration, (b) Reading configuration. (after [7.102])

wavelength λ and the angle 2θ between the interfering beams. This intensity distribution burns a spatially dependent hole pattern that has a frequency-dependent absorption coefficient and a frequency-dependent index of refraction; both contain a background term plus a spatially varying term proportional to $\cos(2\pi x/\Lambda)$. As was first derived by *Kogelnik* [7.106], the resulting thick hologram diffraction efficiency, $\eta(\omega)$, for $\eta(\omega) << 1$ is

$$\eta(\omega) = \frac{C(\gamma/2)^2}{[(\omega_0 - \omega)^2 + (\gamma/2)^2]^2} + \frac{C(\gamma/2)(\omega_0 - \omega)^2}{[(\omega_0 - \omega)^2 + (\gamma/2)^2]^2}, \qquad (7.7)$$

where γ is the hole width, ω_0 is the center frequency of the hole, and C is a constant depending upon the background absorption, the hole depth, and the geometrical properties of the hologram. The first term in this expression is proportional to the square of the absorption profile of the hole, and the second term is proportional to the square of the dispersion profile of the hole. When the two terms in (7.7) are combined, the result is

$$\eta(\omega) = C(\gamma/2)^2/[(\omega_0 - \omega)^2 + (\gamma/2)^2], \qquad (7.8)$$

which is just a Lorentzian-shaped resonance with the same spectral

296

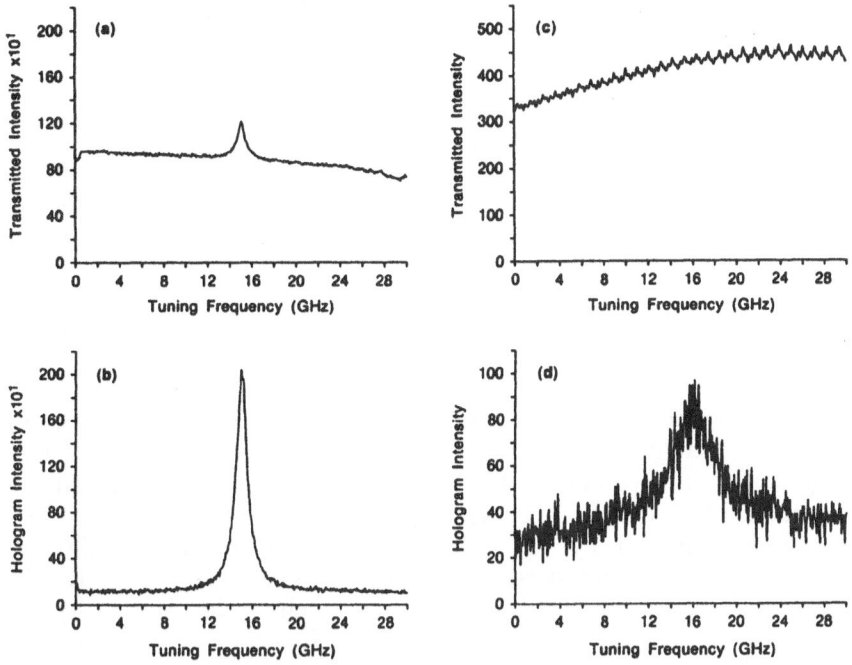

Fig. 7.29. Comparison of transmission and holographic detection. (**a**) Transmission detection of a fairly deep hole, (**b**) Holographic detection of the same hole, (**c**) transmission signal for a very shallow hole, (**d**) holographic detection for the same hole as (**c**). (after [7.105])

linewidth as the absorption profile of the spectral hole. More detailed expressions for the diffraction efficiency are given in [7.102].

Figure 7.29 compares the detection of a spectral hole using both transmission detection and holographic detection for the case of chlorin in poly(vinyl butyral) [7.105]. Parts (a) and (b) show the same hole detected using transmission and holographic methods, respectively; the holographic signal appears near zero background with higher signal value. Parts (c) and (d) show transmission and holographic detection, respectively, for a much weaker hole; the signal-to-noise is superior in the holographic case, although the low-frequency noise in the transmission case is not fundamental (see below) and could be partially removed using optical normalization. Holographic detection has also been used to achieve useful SNR values for a careful study of Stark effects on spectral holes [7.102].

An important question is whether or not holographic readout is fundamentally superior to transmission detection, if shot-noise-limited performance is assumed for both. To be general, we assume that during burning a hole is produced of relative depth $\Delta\alpha/\alpha_0 = H$. Under fast-burning conditions, H is on the order of η, so that by using the variable H here we allow for the possibility that burning many holes in parallel over a longer time period than 30 ns could produce a deeper hole. Then the constant C in (7.8) can be computed to be

$$C = \exp(-\alpha_o L / \cos \theta)(H\alpha_o L/4 \cos \theta)^2. \qquad (7.9)$$

As in Sect. 7.3.2, a readout beam of photon flux F_R is assumed, and the peak signal at the detector can be computed using the diffraction efficiency from (7.8). Assuming that the shot-noise of the detected signal dominates all other noise sources, the SNR for holographic readout of one hole then becomes

$$\text{SNR} = (\eta_Q F_R A \tau_R/2)^{1/2} \left(\frac{H\alpha_o L}{\cos \theta} \right) \exp(-\alpha_o L / 2 \cos \theta), \qquad (7.10)$$

where the symbols have the same meaning as in Sect. 7.3.2. This expression is identical to the SNR for transmission readout with the sample length L replaced by the effective sample length at the angle of incidence, $(L/\cos \theta)$, and with the relative hole depth for fast burning, η, replaced by H. Thus, under fast-burning conditions where parallelism is not used, holographic readout has no SNR advantage over quantum-limited transmission techniques. If parallelism is used, the entire configuration of the optical storage system changes dramatically.

A more subtle disadvantage results from the fact that the holographic signal depends upon *both* the absorption and the dispersion of the hole. The dispersion due to a Lorentzian hole at ω_0 falls off with frequency as $(\omega - \omega_0)^{-1}$, whereas the absorption due to the same hole falls off as $(\omega - \omega_0)^{-2}$. As more and more holes are written into the absorption line, the cumulative effects of the slowly decaying dispersion tails cause a nonzero diffraction background to build up in between the written holes, which degrades the SNR and limits the number of bits that can be stored in the frequency dimension. The detailed implications of this effect for the readout SNR should be addressed in future studies.

Of course, parallel writing schemes are definitely possible and would require the use of page composers and array detectors. Furthermore, motion of the recording medium would probably be necessary, because the information-bearing beams in such a storage device could not be deflected easily by simple angular beam deflection to many different spatial locations without loss of resolution. The detailed implications of storage of high-spatial-frequency holograms on the system configuration and on the properties of the recording materials in a frequency domain optical storage system have not been determined to date. As with the other readout techniques, gated PSHB materials will probably be required for practical applications.

7.5 Other Applications of Persistent Spectral Hole-Burning

7.5.1 General Remarks

In a general sense, the phenomenon of PSHB makes it possible to burn away portions of an inhomogeneously broadened absorption band to form an absorption versus optical frequency lineshape of almost arbitrary profile. Since absorption is always accompanied by dispersion, the index of refraction as a function of optical frequency can also be controlled to some extent. Additional flexibility can be achieved when an external electric or magnetic field can be used to modulate or shift the spectral holes. The ability to "machine" complicated absorption lineshapes and to control them with external fields leads to several interesting device applications where PSHB optical elements could be used for time-domain laser-pulse shaping and frequency-multiplexed optical spatial filters. Since such PSHB optical elements would be exposed to high fluences in any real application, strongly gated PSHB materials would be required to prevent deleterious additional hole-burning.

7.5.2 Laser Pulse Shaping Based on Fourier Synthesis

Given a PSHB recording material that exhibits a strong absorption line into which deep holes can be burned, it should be possible to produce a novel passive optical element that performs generalized Fourier synthesis of laser pulses with arbitrary temporal profiles. This optical element would consist simply of a sample of the PSHB material prepared to contain multiple spectral holes at carefully selected optical frequency locations and with carefully chosen depths. These holes introduce arbitrary and controllable changes in the relative phases and amplitudes of the frequency components of a multifrequency laser beam passed through the sample and thus make possible a generalized Fourier synthesis of laser pulses with arbitrary temporal profiles. Optical elements of this type have been proposed in the patent literature [7.107,108], but have not yet been experimentally implemented.

The incident multifrequency laser beam must have frequency components that are stationary in frequency and maintain constant relative amplitude and phase. Such a laser beam can be obtained by directly utilizing the output of a mode-locked laser or by passing a single-frequency laser through an amplitude or phase modulator driven by a periodic waveform. The inhomogeneous absorption band in the PSHB sample must completely overlap the spectrum of the laser and provide sufficient absorption to significantly attenuate those frequency components that are not coincident with spectral holes.

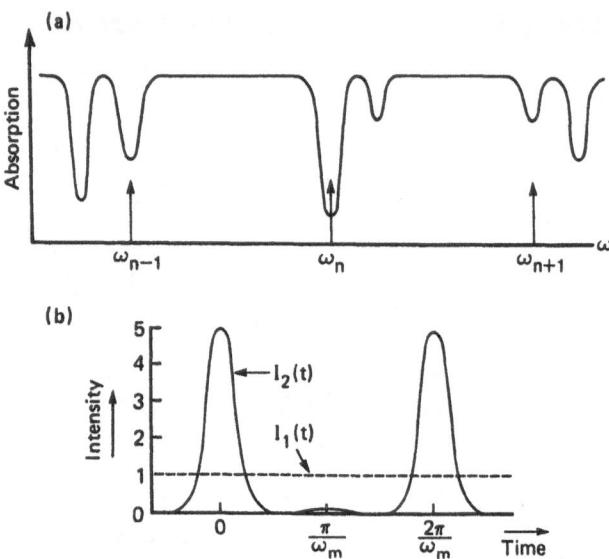

Fig. 7.30. Illustration of laser pulse shaping based on Fourier synthesis. **(a)** Each laser frequency component ω_n has two holes associated with it: one to adjust the attenuation, and one to adjust the phase. **(b)** Temporal structure of a particular frequency-modulated laser beam before $(I_1(t))$ and after $(I_2(t))$ the spectral filter

Figure 7.30 (a) shows the relative positions of the spectral holes and of the laser frequency components. Two holes would be associated with each frequency component ω_n. One hole would have center frequency exactly equal to ω_n, and the hole depth would control the amplitude of the frequency component at ω_n. The other hole would be offset several hole-widths from ω_n (see the shallower hole adjacent to ω_n in the figure). The depth of this hole would control the magnitude of the phase shift at ω_n and the sign of the phase shift could be controlled by the direction of the offset. In order to be able to achieve phase shifts of the order of π, the required hole depths correspond to changes in the optical density on the order of 3.0.

Figure 7.30 (b) shows an example of the type of pulse shaping that could be achieved. In this case, the multifrequency laser beam is assumed to be provided by passing a single frequency laser beam through a sinusoidally driven phase modulator with a modulation index of 1.5. As is true for pure phase modulation at any modulation index, the slowly varying envelope intensity, $I_1(t)$, of the beam emerging from the modulator is exactly constant, as shown by the dashed curve in the figure. The PSHB sample is assumed to contain a pattern of holes such that the amplitudes of the laser frequency components are unaffected, but the phases of several of the components are shifted by π. The slowly varying envelope intensity, $I_2(t)$, describing the beam emerging from the sample is no longer constant in time, but takes the form of a train of clean and sharp pulses of duration T/5, period T, and peak intensity $5I_1$, where T is the period of the sinusoidal phase modulation. Thus in this example, a constant intensity laser beam has been converted to a sequence of pulses using the pattern of holes burned into the PSHB material.

7.5.3 Laser Pulse Shaping Based on Voltage Modulation

Recently, *Schätz* et al. [7.109], have demonstrated intensity modulation of laser light using electric-field-induced changes of persistent spectral holes. As described in Sect. 7.4.2, this technique makes use of electric-field-induced hole filling and broadening to produce a variable absorption in the PSHB sample. The dependence of the transmission of the sample on the applied voltage can be adjusted to match specific applications.

The PSHB sample in the experiments consisted of a 30 μm thick poly(vinylbutyral) (PVB) film doped with perylene molecules at a concentration of 3×10^{-3} mole/l. The film was contained between two electrodes made of transparent indium tin oxide and indium, respectively. An additional metal reflecting layer was used to multipass the laser beam, resulting in a total interaction length of about 400 μm. The spectral holes were burned with He-Cd laser light (441.6 nm) which is absorbed in the $0-0'$ band of the $S_1 \leftarrow S_0$ transition of perylene. The voltage modulated transmission of the sample was probed using an attenuated He-Cd laser beam. All measurements were performed at 1.3 K.

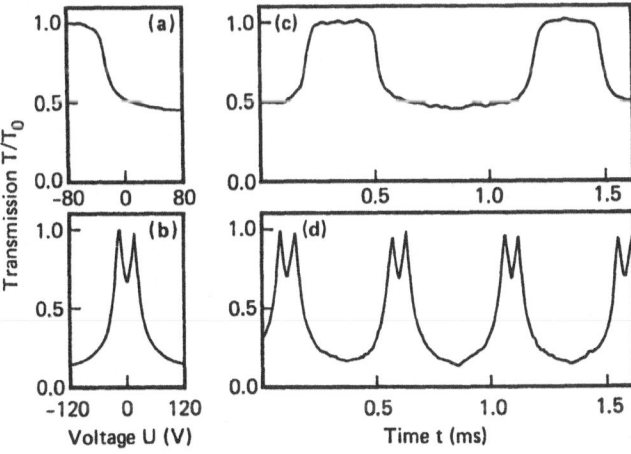

Fig. 7.31. Example of laser pulse shaping using voltage modulation of spectral holes. (a), (b) transmission vs voltage characteristic curves that have been burned into the sample, (c), (d) Resulting time-dependent sample transmission if a sinusoidal (c) or (d) triangular time-dependent voltage is applied to samples with transmission characteristics like (a) and (b), respectively. (after [7.109])

Figures 7.31 (a) and (b) show examples of some of transmission vs voltage characteristic curves that could be obtained when the hole-burning was performed in the presence of a time varying voltage. In the case of Fig. 7.31 (a), the voltage was periodically linearly ramped in a triangular manner between -88 V and -40 V, while for Fig. 7.31 (b), the dc voltage was alternately switched between $+15$ V and -15 V. Figures 7.31 (c) and (d) show the transmitted laser intensity vs time that was observed for an applied sinusoidal voltage and periodic triangular voltage, respectively.

7.5.4 Frequency Multiplexed Optical Spatial Filters

Matched optical spatial filters have a variety of important applications in coherent and incoherent optical pattern recognition systems. The use of spatial filters that are wavelength or optical frequency multiplexed introduces additional advantages, in that several objects can be simultaneously recognized and the requirements for filter positioning devices are reduced [7.110].

It has been proposed [7.110,111], but not experimentally demonstrated, that such a multiplexed filter could be produced using a sample of material that exhibits PSHB. The filter would be produced by illuminating a series of master transparencies with the coherent output of a narrowband tunable laser source and imaging the transmitted light onto a thin film sample of PSHB material. Each master transparency would have a unique spatially varying transmissivity that corresponds to one of the spatial filtering functions desired to be contained in the multiplexed filter. The laser would be tuned to a different optical frequency within the inhomogeneous absorption line of the PSHB sample for each master transparency. Thus the image of each master transparency would be recorded in terms of a spatial variation of the depth of the spectral hole burned at the particular laser frequency corresponding to that transparency. Once the PSHB sample is then exposed in this manner for the complete set of master transparencies, the frequency multiplexed optical spatial filter is produced.

The burning of a spectral hole at a particular optical frequency enhances the transmissivity of the sample for a narrow band of frequencies centered about that frequency. Thus the multiplexed filter will provide an optical intensity spatial filtering function which corresponds to that of any one of the master transparencies when illuminated with light at the wavelength required to select the desired master transparency. Provided that the spacing between the optical frequencies of which the master transparencies are recorded is greater than the spectral hole width, there will be no significant crosstalk between the multiplexed filters. Since the frequency width of spectral holes is typically a factor of 1000 less than the width of the zero phonon line, up to 1000 filters can be multiplexed in a single sample.

302

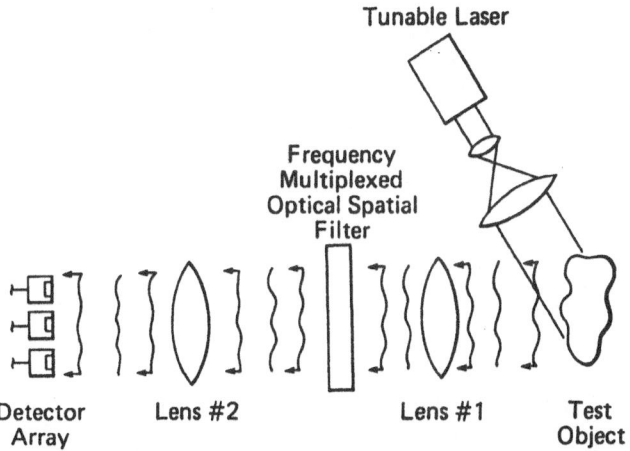

Tunable Laser

Frequency Multiplexed Optical Spatial Filter

Detector Array

Lens #2

Lens #1

Test Object

Fig. 7.32. Scheme of frequency multiplexed optical spatial filtering using PSHB

Figure 7.32 shows how such a multiplexed filter could be used for spatial filtering of the light scattered by a test object. In this case a rapidly tunable laser, such as a current-tuned GaAlAs diode laser, is utilized to illuminate the test object. The light scattered by the object is collected by a lens and directed through the multiplexed spatial filter. Since the frequency of the light scattered by the object is the same as the laser frequency, the desired spatial filtering function would be selected by tuning the laser. Specific potential applications of this concept include optical character recognition and optical image processing.

7.6 Summary and Future Prospects

This chapter has presented a review of possible technological applications of PSHB. Frequency domain optical data storage based on PSHB has been discussed in detail and other proposed applications have been described briefly. The engineering and materials requirements for a practical FDOS system have been analyzed for the case of frequency domain readout. An experimental investigation of the principal engineering issues for FDOS was described. In addition, a general analysis of single-photon mechanisms has underscored the importance of finding gated, two-step mechanisms for PSHB. Recent discoveries of gated PSHB in inorganic and organic systems show that these mechanisms exist. Other configurations for data storage including time domain and electric field have been considered, and these schemes as well as other applications of PSHB work best with photon-gated materials. Additional applications of PSHB for laser pulse shaping and multiplexed optical spatial filters were summarized. The practical realization of these technological applications of PSHB

303

presents a challenge for interdisciplinary research: identification and characterization of gated hole-burning mechanisms in materials that will satisfy all the requirements for frequency domain optical storage as well as other useful applications of persistent spectral hole-burning.

Acknowledgements

The authors thank D. Haarer for permission to reprint Fig. 7.1 from [7.6], T. Mossberg for permission to reprint Figs. 7.24 and 7.25 from [7.86], U. Bogner for permission to reprint Figs. 7.27 and 7.31 from [7.101] and [7.109], respectively, and U. Wild for permission to reprint Fig. 7.26 from [7.100], Fig. 7.28 from [7.102], and Fig. 7.29 from [7.105]. This research has been supported in part by the U. S. Office of Naval Research.

References

7.1   See for example A. E. Bell: Laser Focus/Electro-Optics **19**, 61 (August 1983), A. E. Bell: Laser Focus/Electro-Optics **19**, 125 (September 1983), A. E. Bell: Proc. Soc. Infor. Disp. **24**, 17 (1983), and references therein

7.2   A. Szabo: "Frequency selective optical memory," U. S. Patent No. 3,896,420 (1975)

7.3   G. Castro, D. Haarer, R. M. Macfarlane, H. P. Trommsdorff: "Frequency selective optical data storage system," U. S. Patent No. 4,101,976, (1978)

7.4   D. Haarer: Proc. Soc. Photo-Opt. Inst. Engr. **177**, 97 (1979)

7.5   G. C. Bjorklund, W. Lenth, C. Ortiz: Proc. Soc. Photo-Opt. Instr. Eng. **298**, 107 (1981)

7.6   A. R. Gutierrez, J. Friedrich, D. Haarer, H. Wolfrum: IBM J. Res. Devel. **26**, 198 (1982), and references therein

7.7   W. E. Moerner: Proc. Soc. Photo-Opt. Instr. Engr. **541**, 60 (1985)

7.8   W. Lenth, R. M. Macfarlane, W. E. Moerner, F. M. Schellenberg, R. M. Shelby, G. C. Bjorklund: Proc. Soc. Photo-opt. Instrum. Engr. **695**, 216 (1986)

7.9   W. E. Moerner: J. Molec. Elec., **1**, 55 (1985)

7.10 Oxford Instruments Ltd., Oxford, England

7.11 A. Winnacker, R. M. Shelby, R. M. Macfarlane: J. de Phys. Colloq. C7, Suppl. 10, **46**, C7-543 (1985)

7.12 F. M. Schellenberg, W. Lenth, G. C. Bjorklund: Appl. Opt. **25**, 3207 (1986)

7.13 S. Kobayashi, Y. Yamamoto, M. Ito, T. Kimura: IEEE J. Quant. Elec. **QE-18**, 582 (1982)

7.14 W. T. Tsang, N. A. Olsson, R. A. Logan: Appl. Phys. Lett. **42**, 650 (1983)

7.15 K. J. Ebeling, L. A. Coldren: Appl. Phys. Lett. **44**, 735 (1984)

7.16 G. C. Bjorklund: Opt. Lett. **5**, 15 (1980)

7.17 G. C. Bjorklund, M. D. Levenson, W. Lenth, C. Ortiz: Appl. Phys. **B32**, 145 (1983), and references therein

7.18 M. D. Levenson, W. E. Moerner, D. E. Horne: Opt. Lett. **8**, 108 (1983)

7.19 E. A. Whittaker, H.R. Wendt, H. E. Hunziker, G. C. Bjorklund: Appl. Phys. B **35**, 105 (1984)

7.20 M. Gehrtz, W. E. Moerner, G. C. Bjorklund: IBM Research Report #RJ 4678, April 30, 1985

7.21   M. Gehrtz, G. C. Bjorklund, E. A. Whittaker: J. Opt. Soc. Am. B 2, 1510 (1985), and references therein
7.22   A. L. Huston, W. E. Moerner: J. Opt. Soc. Am. B: Opt. Phys. 1, 349 (1984)
7.23   A. L. Huston, W. E. Moerner: U. S. Patent 4,614,116, "Phase Sensitive Ultrasonic Modulation Method for the Detection of Strain-Sensitive Spectral Features", September 30, 1986
7.24   W. E. Moerner, A. L. Huston: Appl. Phys. Lett. 48, 1181 (1986)
7.25   M. Romagnoli, M. D. Levenson, G. C. Bjorklund: Opt. Lett. 8, 635 (1983)
7.26   M. Romagnoli, M. D. Levenson, G. C. Bjorklund: J. Opt. Soc. Am. B: Opt. Phys. 1, 571 (1984)
7.27   P. Pokrowsky, W. E. Moerner, F. Chu, G. C. Bjorklund: Opt. Lett. 8, 280 (1983)
7.28   P. Pokrowsky, W. Zapka, F. Chu, G. C. Bjorklund: Opt. Commun. 44, 175 (1983)
7.29   W. Lenth: Opt. Lett. 8, 575 (1983)
7.30   W. Lenth: IEEE J. Quant. Elec. QE-20, (1984)
7.31   M. Gehrtz, W. Lenth, A. T. Young, H. S. Johnston: Opt. Lett. 11, 132 (1986)
7.32   W. Lenth and M. Gehrtz: Appl. Phys. Lett. 47, 1263 (1985)
7.33   D. J. Bernays: Proc. Soc. Photo-Opt. Instrum. Engr., 498, 175 (1984)
7.34   W. E. Moerner, F. M. Schellenberg, G. C. Bjorklund: Appl. Phys. B28, 263 (1982)
7.35   W. E. Moerner, P. Pokrowsky, F. M. Schellenberg, G. C. Bjorklund: Phys. Rev. B. 32, 1270 (1985)
7.36   S. Yamaguchi, M. Suzuki: Appl. Phys. Lett. 41, 597 (1982)
7.37   W. E. Moerner: *Proceedings of the International Conference: Lasers '83*, R. C. Powell, editor, (STS Press, McLean, VA 1983), p. 489
7.38   R. M. Macfarlane, R. T. Harley, R. M. Shelby: Rad. Effects 72, 1 (1983), and references therein. (See also Chap. 4 of this volume)
7.39   H. P. H. Thijssen, R. E. van den Berg, S. Völker: Chem. Phys. Lett. 103, 23 (1983)
7.40   A. Guiterrez, G. Castro, G. Schulte, D. Haarer: "Dynamical Linewidth Effects of Hole Burning of Free Base Phthalocyanine in Polymers: Spectral Diffusion and Exchange Narrowing", in *Organic Molecular Aggregates*, Vol 49 , P. Reineker, H. Haken, and H. C. Wolf, eds. (Springer, Berlin, Heidelberg 1983) pp. 206-214
7.41   H. W. H. Lee, A. L. Huston, M. Gehrtz, W. E. Moerner: Chem. Phys. Lett. 114, 491, (1985)
7.42   W. E. Moerner, F. M. Schellenberg, G. C. Bjorklund, P. Kaipa, F. Luty: Phys. Rev. B. 32, 1270 (1985)
7.43   D. Botez: IEEE Spectrum, June 1985, pp. 43-53
7.44   J.-C. Baumert, P. Günter, H. Melchior: Opt. Commun. 48, 215 (1983)
7.45   D. M. Burland, D. Haarer: IBM J. Res. Devel. 23, 534 (1979), and references therein
7.46   L. A. Rebane, A. A. Gorokhovskii, J. V. Kikas: Appl. Phys. B29, 235-250 (1982), and references therein
7.47   S. Völker, J. H. van der Waals: Molec. Phys. 32, 1703 (1976)
7.48   S. Völker, R. M. Macfarlane: IBM J. Res. Devel. 23, 547 (1979)
7.49   A. Winnacker, R. M. Shelby, R. M. Macfarlane: Opt. Lett. 10, 350 (1985)
7.50   W. Breinl, J. Friedrich, D. Haarer: Chem. Phys. Lett. 106, 487 (1984)
7.51   W. Breinl, J. Friedrich, D. Haarer: J. Chem. Phys. 81, 3915 (1984)
7.52   C. Ortiz, R. M. Macfarlane, R. M. Shelby, W. Lenth, G. C. Bjorklund: Appl. Phys. 25, 87 (1981)
7.53   C. Ortiz, C. Alfonso, P. Pokrowsky, G. C. Bjorklund: Appl. Phys. Lett. 43, 1102 (1983)
7.54   For a clear definition, see W. E. Moerner, M. Gehrtz, A. L. Huston: J. Phys. Chem. 88, 6459 (1984)

7.55   L. Kador, G. Schulte, D. Haarer: J. Phys. Chem. **90**, 1264 (1986)
7.56   M. Romagnoli, W. E. Moerner, F. M. Schellenberg, M. D. Levenson, G. C. Bjorklund: J. Opt. Soc. Am. B: Optical Physics **1**, 341 (1984)
7.57   The excited state lifetime cannot be much shorter, or the hole width will become greater than 500 MHz. See reference [7.58]
7.58   W. E. Moerner, M. D. Levenson: J. Opt. Soc. Amer. B: Optical Physics **2**, 915 (1985)
7.59   W. E. Moerner, A. R. Chraplyvy, A. J. Sievers, R. H. Silsbee: Phys. Rev. B **28**, 7244 (1983), and references therein
7.60   F. M. Schellenberg: IBM Research Report #RJ4687, May 3, 1985
7.61   D. M. Burland, F. Carmona, G. Castro, D. Haarer, R. M. Macfarlane: IBM Tech. Discl. Bull. **21**, 3770 (1979)
7.62   H. W. H. Lee, M. Gehrtz, E. Marinero, W. E. Moerner: Chem. Phys. Lett. **118**, 611 (1985)
7.63   R. M. Macfarlane, J. C. Vial: Phys. Rev. B **34**, 1 (1986)
7.64   M. Iannone, G. W. Scott, D. Brinza, D. R. Coulter: J. Chem. Phys. **85**, 4863 (1986)
7.65   T. P. Carter, C. Bräuchle, V. Y. Lee, M. Manavi, W. E. Moerner: Opt. Lett. **12**, 370 (1987)
7.66   T. P. Carter, C. Bräuchle, V. Y. Lee, W. E. Moerner: J. Phys. Chem. **91**, 3998 (1987)
7.67   W. E. Moerner, T. P. Carter, C. Bräuchle: Appl. Phys. Lett. **49**, 430 (1987)
7.68   A. J. Silversmith, W. Lenth, and R. M. Macfarlane: to be published
7.69   W. Lenth, W. E. Moerner: Opt. Commun. **58**, 249 (1986)
7.70   T. W. Mossberg: Opt. Lett. **7**, 77 (1982)
7.71   L. Allen, J. H. Eberly: *Optical Resonance and Two-Level Atoms*, (Wiley, New York 1975)
7.72   R. G. Brewer: "Coherent Optical Spectroscopy," in Proc. Int. Summer School "Enrico Fermi" LXIV: *Nonlinear Spectroscopy*, Varenna, Italy, ed. by N. Bloembergen, (North-Holland, Amsterdam 1977), pp. 87-137
7.73   R. L. Shoemaker: "Coherent Transient Infrared Spectroscopy," in *Laser and Coherence Spectroscopy*, ed. by J. I. Steinfeld, (Plenum, New York 1978), pp. 197-371
7.74   A. G. Anderson, R. L. Garwin, E. L. Hahn, J. W. Horton, G. L. Tucker, R. M. Walker: J. Appl. Phys. **26**, 1324 (1955)
7.75   W. H. Hesselink, D. A. Wiersma: Phys. Rev. Lett. **43**, 1991 (1979)
7.76   W. H. Hesselink, D. A. Wiersma: J. Chem. Phys. **75**, 4192 (1981)
7.77   H. de Vries, D. A. Wiersma: J. Chem. Phys. **80**, 657 (1984)
7.78   A. Rebane, R. Kaarli, P. Saari, A. Anijalg, K. Timpmann: Opt. Commun. **47**, 173 (1983)
7.79   A. K. Rebane, R. K. Kaarli, P. M. Saari: JETP Lett. **38**, 383 (1983)
7.80   A. Rebane, R. Kaarli: Chem. Phys. Lett. **101**, 317 (1983)
7.81   Y. S. Bai, W. R. Babbitt, N. W. Carlson, T. W. Mossberg: Appl. Phys. Lett. **45**, 714 (1984)
7.82   P. M. Saari, R. K. Kaarli, A. K. Rebane: Kvantovaya Elektron. (Moscow) **12**, 672 (1985)
7.83   K. K. Rebane: Cryst. Latt. Def. Amorph. Mater. **12**, 427 (1985)
7.84   P. Saari, R. Kaarli, A. Rebane: J. Opt. Soc. Am. B **3**, 527 (1986)
7.85   K. K. Rebane: Sov. Phys. Usp. **29**, 290 (1986)
7.86   W. R. Babbitt, Y. S. Bai, T. W. Mossberg: Proc. Soc. Photo-Opt. Instrum. Engr. **639**, 240 (1986)
7.87   N. W. Carlson, L. J. Rothberg, A. G. Yodh, W. R. Babbitt, T. W. Mossberg: Opt. Lett. **8**, 483 (1983)
7.88   N. W. Carlson, W. R. Babbitt, T. W. Mossberg: Opt. Lett. **8**, 623 (1983)
7.89   A. K. Rebane, R. K. Kaarli, P. M. Saari: Opt. Spektrosk. **55**, 405 (1983)
7.90   P. W. Smith: Phil. Trans. R. Soc. Lond. A **313**, 349 (1984)
7.91   N. W. Carlson, W. R. Babbitt, Y. S. Bai, T. W. Mossberg: Opt. Lett. **9**, 232 (1984)

7.92   Y. S. Bai, W. R. Babbitt, T. W. Mossberg: Opt. Lett. 11, 724 (1986)
7.93   G. Castro, R. H. Dicke, D. Haarer: IBM Tech. Discl. Bull. 21, 3333 (1979)
7.94   M. Maier: Appl. Phys. B 41, 73 (1986)
7.95   A. P. Marchetti, M. Scozzafava, R. H. Young: Chem. Phys. Lett. 51, 424 (1977)
7.96   R. M. Macfarlane, R. M. Shelby: Phys. Rev. Lett. 42, 788 (1979)
7.97   V. D. Samoloilenko, N. V. Razumova, R. I. Personov: Opt. Spectrosc. (USSR) 52, 346 (1982)
7.98   F. A. Burkhalter, G. W. Suter, U. P. Wild, V. D. Samoilenko, N. V. Rasumova, R. I. Personov: Chem. Phys. Lett. 94, 483 (1983)
7.99   U. Bogner, P. Schätz, R. Seel, M. Maier: Chem. Phys. Lett. 102, 267 (1983)
7.100 U. P. Wild, S. E. Bucher, F. A. Burkhalter: Appl. Opt. 24, 1526 (1985)
7.101 U. Bogner, K. Beck, M. Maier: Appl. Phys. Lett. 46, 534 (1985)
7.102 A. J. Meixner, A. Renn, S. E. Bucher, U. P. Wild: J. Phys. Chem. 90, 6777 (1986)
7.103 W. E. Moerner: "Use of Homogeneous Electric Fields to Access the Longitudinal Spatial Dimension and to Provide Transverse Random Access in Frequency Domain Optical Memories," Research Disclosure, No. 25333, May 1985
7.104 D. Haarer, R. V. Pole, S. Völker: "Non-destructive Readout Scheme for Holographic Storage System," U. S. Patent 4,103,346, July 25, 1978
7.105 A. Renn, A. J. Meixner, U. P. Wild, F. A. Burkhalter: Chem. Phys. 93, 157 (1985)
7.106 H. Kogelnik: Bell Syst. Tech. J. 48, 2909 (1969).
7.107 G. C. Bjorklund: U.S. Patent No. 4,306,771, "Optical Pulse Shaping Device and Method," December 22, 1981
7.108 G. C. Bjorklund, M. D. Levenson: "Laser Pulse Shaping Device Based on Fourier Synthesis Using Optical Anisotropies Produced by Spectral Hole Burning," IBM Tech. Discl. Bull. 23, 2517 (1980)
7.109 P. Schätz, U. Bogner, M. Maier: Appl. Phys. Lett. 49, 1132 (1986)
7.110 S. K. Case: Appl. Opt. 18, 1890 (1979)
7.111 G. C. Bjorklund, G. T. Sincerbox: "Frequency Multiplexed Optical Spatial Filter Based Upon Photochemical Hole Burning," U. S. Patent No. 4,533,211, August 6, 1985

Subject Index

silicate
    Eu^{3+}-doped 138
    Nd^{3+}-doped 135
    Pr^{3+}-doped 135, 137
structure 47

H$_2$-octaethylporphyrin in polystyrene
    (OEP-PS) 49, 50, 64, 70, 287
H$_2$-phthalocyanine (H$_2$-Pc)
    also called free-base phthalocyanine
    hole-burning 7, 8, 90, 96, 252, 280
    in n-octane 35
    in poly(ethylene) 266
    protonated, in sulfuric acid glass 266
H$_2$-tetra-4-t-butyl-phthalocyanine (H$_2$-Pc*)
    in isopropylene-ether glass 47
    in n-nonane 53
    in poly(ethylene) 267
    in tetradecane 44, 47
H$_2$-porphine (H$_2$-P) 8, 90, 91, 97, 98
    in n-decane 46
H$_2$-tetra(t-butyl)-porphirazine (TAP*) 48
    in n-decane 46
    in n-octane 54
    in poly(styrene) 287
Harmonic oscillator 88
High-density recording 262
Hole, persistent spectral
    broadening 5, 40, 183, 196, 265
    detection *See Detection techniques*
    growth rate 6, 37-42
    laser-induced filling 188
    lifetime 6, 100-107
    multiplets 56
    width 5, 40, 183, 196, 265
Hole-burning
    effective yield 281
    efficiency 6, 263, 270
    fast 267, 280
    infrared 203-248
    lifetime 100-107, 267
    mechanisms 8, 86-92, 129-131, 153,
        157
    non-photochemical (NPHB) 90,
        153-202, 207
    persistent spectral (PSHB)
        See also Persistent spectral hole-burning
    photochemical (PHB) 86-92
    photon-gated 129, 141, 146, 274-284
    polarized spectral 236
    reversal 266

spectroscopy 219
    transient 6, 7, 85, 206, 219
    vibrational 209
Hologram diffraction efficiency, thick 296
Holographic detection 63, 295-299
Homogeneous lineshape 18, 112
Homogeneous linewidth
    definition 2, 19, 20, 84, 105
    in glasses 47-52, 97, 165-182
    temperature dependence 23, 50, 51,
        95, 96, 136
    time dependence 108, 109
Hot-forging 218
Hydrogen-bonded crystals 158
Hydrogen bond rearrangement 9
Hyperfine interactions 32

IR (infrared)
    persistent spectral holes (PIRSH)
        203-248
Impurity-impurity interactions 14
Incoherent saturation 205
Index of refraction 295, 299
Inhomogeneous
    broadening 1-3, 19
    distribution function 27, 29, 31, 42
    site distribution 95
    width 3, 26, 266
Inorganic materials 127-150, 203-248
Interference pattern 295
Ions
    cyanide 203
    nitrite 203
    perrhenate 203, 206, 216
    samarium 276

Kinetics
    basic equations 38
    general 6
    high concentration 41
    hole width 40
    in glasses 100-112
    metastable states 41
    optically dense samples 41
    overlap of product and educt 43
    reverse reactions 42
    two-photon reaction 42
Kr matrix 207-216

Laser frequency tuning 255
Laser saturation 206

311

Laser-induced hole filling 6, 188-196
Latency in data access 253
Librational modes 98, 99
Lifetime-limited width 44, 209, 219
$LiGa_5O_8:Co^{2+}$ 146, 278
Limitations
    destructive reading 269
    shot-noise 270
    single photon 269
Linear intensity dependence 148
Linewidths 165, 166, 173, 175, 177, 180
    See also Hole widths, dephasing
Liquid helium cryostat 256
Local disorder 121
Local phonon modes 25, 221
Lorentzian lineshape 87, 95

Matrices
    rare gas 203-215
    Ar 207-216
    Kr 207-216
    N_2 207-216
    Xe 207-216
Matrix-isolated CO 205
Matrix rearrangement 52
Mechanisms
    donor-acceptor electron transfer 278
    hole-burning 8, 86-92, 129-131, 153,
       157, 274-284
    monophotonic 269-274
    nonphotochemical 158-165, 203
    photochemical 86-92, 129-131
    photoionization 139
    photon-gated 129, 274-282
    photophysical

       See Mechanisms: nonphotochemical
    reorientation 47
    single photon 269, 270
Memory function, site 109, 112
Meso-tetra(p-tolyl)-Zn-tetrabenzoporphyrin
    (TZT) 278-280, 283
Metastable hole-burning 41, 49, 58, 254
Methanol dimers 216
Microscopic interactions 6
Mode
    gap 206
    librational 206
    local 206
    resonant 206
    rotational 206

    tunneling 206
Modulation
    diagonal 166, 170, 171, 176, 177, 180
    off-diagonal 166, 170, 171, 173, 180
Mössbauer analogy 17
Molecular aggregates 213-215
Molecular reorientation 208
Molecule-lattice interaction 205
Monophotonic mechanisms 269
Motional narrowing 172, 177, 182
Multiphonon decay 221

NCO^- 243
Nitrite ions 203
NO_2 261
Non-correlation 55-58
Nonphotochemical hole-burning (NPHB)
    chlorophyll in glasses 59
    definition 8, 11
    examples 10, 11, 37
    to study TLS relaxation 153-198
    vibrational modes 216-240
NPHB *See Nonphotochemical hole-burning*

O_2^- 28
OEP-PS
    See H_2-octaethylporphyrin in polystyrene
One-photon mechanism 37, 269-274
Optical filters 19
Optical recording
    conventional 253
    magneto-optic 256
Optical spatial filters, frequency multiplexed
    299, 302, 303
Optically dense samples 41, 62
Oscillator strength 23, 112, 274

Parallel writing 298
Peg (antihole) 225-228
PEL *See Purely electronic line*
Pentacene in benzoic acid 11
Pentacene in p-terphenyl 46
Permittivity 68
Perrhenate ion in alkali halides 11, 216-230
Persistence, definition 4, 34
Persistent elastic polarization 228
Persistent spectral hole-burning (PSHB)
    applications 6, 71-73, 251-304
    basic principles 1-5, 33-37
    characteristic time 34

Topics in Current Physics

Founded by Helmut K. V. Lotsch